Nanomaterials: Evolution and Advancement towards Therapeutic Drug Delivery *(Part I)*

Edited by

Surendra Nimesh
Department of Biotechnology
Central University of Rajasthan
India

Nidhi Gupta
Department of Biotechnology
IIS (Deemed to be University)
India

&

Ramesh Chandra
Department of Chemistry
University of Delhi
India

Nanomaterials: Evolution and Advancement Towards Therapeutic Drug Delivery (Part I)

Editors: Surendra Nimesh, Nidhi Gupta and Ramesh Chandra

ISBN (Online): 978-1-68108-781-8

ISBN (Print): 978-1-68108-782-5

ISBN (Paperback): 978-1-68108-783-2

Bentham Science Publishers Ltd.
Executive Suite Y - 2
PO Box 7917, Saif Zone
Sharjah, U.A.E.
Email: subscriptions@benthamscience.net

CONTENTS

FOREWORD

Engineered nanomaterials (ENMs), due to their interesting physicochemical properties such as smaller size, larger surface area, electrical, optical and magnetic properties are being sought in a wide range of applications including technology, cosmetics, food packaging, medical imaging and drug delivery. Carbon nanotubes (CNTs), quantum dots, mesoporous and amorphous nanosilica, nanosilver, nano titanium and zinc oxides are some of the ENMs currently in commerce. Nevertheless, the attractive physicochemical characteristics of the ENMs also create concerns when exposed to, with respect to human and ecosystem health. This book on "nanomaterials" is very timely, and touches upon the different aspects of application of ENMs in drug delivery. The chapters in this book discuss the use of a spectrum of nanomaterials in drug delivery including nano metal oxides, CNTs and lipid nanoparticles, their various nanoforms, synthesis, characterization, efficacy in terms of drug delivery and the need for toxicity testing. Physicochemical characterization is an important aspect in nanotechnology, especially, in the realm of drug delivery. The synthesis of ENMs can introduce batch to batch variation in terms of size, shape, surface characteristics and chemical composition based on source materials and synthetic routes. Moreover, the stability of ENMs can be affected based on storage conditions. This book has thus given an importance to the aspect of physicochemical testing and discusses the different analytical methods to assess morphology, surface functionalities, behavior in solution, stability, etc. This book on "nanomaterials" also identifies the need for toxicity testing of the ENMs in drug delivery. Toxicity testing is a critical component for the selection of safer ENMs for application in drug delivery and to meet regulatory standards. This book has done a fantastic job in familiarizing the reader with the scope and application of the various ENMs and their nanoforms in drug delivery along with some insight into medical imaging and computational aspect of structure-activity relationships. I congratulate the editor Dr. Surendra Nimesh on doing a fantastic job with this book on the application ENMs in drug delivery, which is one of the promising emerging medical technologies.

Prem Kumarathasan
Analytical Biochemistry and Proteomics Laboratory
Mechanistic Studies Division
Environmental Health Science and Research Bureau
Healthy Environments and Consumer Safety Branch
Health Canada, Ottawa, Ontario, Canada K1A0K9

PREFACE

Chemically synthesized drugs have been one of the major tools to combat several diseases, including bacterial and viral infections. These drug molecules however, face several barriers that include poor cellular uptake and instability in the physiological environment that mask the therapeutic potential. In order to circumvent these issues, there is a need to develop vehicles that could effectively and safely transport the drug molecules to the target sites. Nanotechnology has come up as one of the potent and viable strategies due to which, several candidates have been proposed, such as nanoparticles, liposomes, carbon nanotubes, mesoporous silica nanoparticles, etc. These vectors can be modulated to achieve delivery of even those drugs which are highly unstable making it difficult to reach the desired sites. This book compilation brings together some of the eminent scientists, working in different dimensions of nanotechnology. They have contributed chapters in their domain of knowledge that we believe would be really useful, not only for the young researchers, but also for the experts looking for some exhaustive compilations.

Chapter 1: It provides a brief exposure to nanotechnology and the imbibition of the drug delivery concept, with relevance to nanotechnology. Some of the pros and cons associated with the implication of nanotechnology have been outlined in this chapter.

Chapter 2: It provides an exhaustive account of the tools and techniques employed for the characterization of nanostructures. This chapter is an overview of the most employed techniques, including dynamic light scattering and laser diffraction for the determination of size distribution, zeta potential and its relationship with the stability and the surface charge of the particles, microscopies (optical microscopy, SEM, TEM, AFM) utilized in morphological analyses, spectroscopies in the infrared or ultraviolet-visible regions, and X-rays diffraction, which help to elucidate the crystalline state, polymorphism and drug-nanosystem interaction and the thermal analyses, which can provide information about the physical state, crystallinity, and stability.

Chapter 3: It deals with liposomes which are spherically shaped micro-vesicles composed of a bilayer of natural or synthetic non-toxic phospholipids, which spontaneously form closed structures when they are hydrated. This chapter accounts for the materials involved in the preparation of liposomes, *i.e.*, techniques for preparation, characterization studies and application of liposomes for drug delivery.

Chapter 4: This chapter discusses dendrimers, which are highly branched synthetic polymeric macromolecules with a distinct three-dimensional spherical shape and nanometric size (1 - 100nm). This chapter reports the role of dendrimers in drug delivery systems for various agents in biology and medical science. Additionally, it covers the important aspectsof the dendrimers in the diagnoses of different types of cancer diagnosis and other therapies.

Chapter 5: It discusses nanomicelles which are colloidal dispersions containing a hydrophobic core and a hydrophilic shell that self-assemble into the nanosize range of 10 -100 nm. Pharmaceutically, nanomicelles serve as outstanding carriers as they can avert or moderate drug degradation by dropping adverse side-effects, thereby augmenting drug permeation through biological barriers with very minimum or no irritation at all, which ultimately enhance bioavailability. In this chapter, a critical history of nanomicelles is provided from its fundamental theory to preclinical and clinical achievements.

Chapter 6: It outlines details about carbon nanotubes (CNTs) which are defined as concentrically rolled graphene sheets but with many chiralities and helicities. Various synthetic approaches like arc discharge method, laser ablation methods, chemical vapour deposition, and pyrolysis are employed to produce CNTs. The present chapter explains the discovery of CNTs, their types, synthetic methods, various reactions possible with the usage of CNTs, applications of CNTs, safety profile, and the challenges involved.

Chapter 7: It describes the application of magnetic nanoparticles in drug delivery. Numerous magnetic nanoparticle-based formulations are already available in the market and even many more are under different stages of development. This chapter describes the progress done in this area due to the well-established synthesis procedures of size and shape-controlled magnetic nanoparticles, as well as their well-recognized surface functionalization approaches.

Dr. Surendra Nimesh
Dr. Nidhi Gupta
Prof. Ramesh Chandra

List of Contributors

Akanksha Mahajan	University Institute of Pharmaceutical Sciences, UGC Centre of Advanced Studies, Panjab University, Chandigarh-60014, India
Ashish Baldi	Department of Pharmaceutical Sciences and Technology, Maharaja Ranjit Singh Punjab Technical University, Bathinda, Punjab, India
Arindam Saha	CSIR-Central Glass and Ceramic Research Institute, Jadavpur, Kolkata-700032, India
Flávia Lidiane Oliveira da Silva	Department of Pharmacy, Faculty of Biological and Health Sciences, Federal University of Jequitinhonha and Mucuri Valleys, Diamantina, MG, Brazil
Gabriel Silva Borges	Department of Pharmaceutics, Faculty of Pharmacy, Federal University of Minas Gerais, Belo Horizonte, MG, Brazil
Guilherme Carneiro	Department of Pharmacy, Faculty of Biological and Health Sciences, Federal University of Jequitinhonha and Mucuri Valleys, Diamantina, MG, Brazil
Gajanand Sharma	University Institute of Pharmaceutical Sciences, UGC Centre of Advanced Studies, Panjab University, Chandigarh-60014, India
Jitender Madan	Department of Pharmaceutics, National Institute of Pharmaceutical Education and Research, Hyderabad, Telangana, India
Kanika Thakur	University Institute of Pharmaceutical Sciences, UGC Centre of Advanced Studies, Panjab University, Chandigarh-60014, India
Kaisar Raza	Department of Pharmacy, School of Chemical Sciences and Pharmacy, Central University of Rajasthan, Ajmer, India UGC-Centre of Excellence in Applications of Nanomaterials, Nanoparticles & Nanocomposites (Biomedical Sciences), Panjab University, Chandigarh-160014, India
Nitika Dehal	University Institute of Pharmaceutical Sciences, UGC Centre of Advanced Studies, Panjab University, Chandigarh-60014, India
Nidhi Gupta	Department of Biotechnology, IIS (Deemed to be University), Gurukul Marg,SFS, Mansarovar, Jaipur 302020, India
Neetu Sehrawat	Department of Pharmacy, School of Chemical Sciences and Pharmacy, Central University of Rajasthan, Bandar Sindri, Ajmer, Rajasthan-305817, India
Neha Bajwa	Department of Pharmaceutical Sciences and Technology, Maharaja Ranjit Singh Punjab Technical University, Bathinda, Punjab, India
Nagarani Thotakura	Department of Pharmacy, School of Chemical Sciences and Pharmacy, Central University of Rajasthan, Bandarsindri, Dist. Ajmer, Rajasthan, India
Osmar Patrício Almeida	Department of Pharmacy, Faculty of Biological and Health Sciences, Federal University of Jequitinhonha and Mucuri Valleys, Diamantina, MG, Brazil
O.P. Katare	University Institute of Pharmaceutical Sciences, UGC Centre of Advanced Studies, Panjab University, Chandigarh-60014, India
P. Sujatha Devi	CSIR-Central Glass and Ceramic Research Institute, Jadavpur, Kolkata-700032, India

Preet Amol Singh	Department of Pharmaceutical Sciences and Technology, Maharaja Ranjit Singh Punjab Technical University, Bathinda, Punjab, India
Rupinder Kaur Sodhi	Department of Pharmacology, Chandigarh College of Pharmacy, Mohali, Punjab, India
Ramesh Chandra	Dr. B.R Ambedkar Centre for Biomedical Research, University of Delhi, Delhi, India Department of Chemistry, University of Delhi, Delhi, India
Satish Sardana	Amity Institute of Pharmacy, Amity University, Gurgaon (Haryana), India
Sarita Rani	Department of Pharmacy, School of Chemical Sciences and Pharmacy, Central University of Rajasthan, Bandar Sindri, Ajmer, Rajasthan-305817, India
Surendra Nimesh	Department of Biotechnology, School of Life Sciences, Central University of Rajasthan, Ajmer 305817, India
Sudhakar Rajoria	Department of Pharmacy, School of Chemical Sciences and Pharmacy, Central University of Rajasthan, Bandar Sindri, Ajmer, Rajasthan-305817, India
Taízia Dutra Silva	Department of Pharmacy, Faculty of Biological and Health Sciences, Federal University of Jequitinhonha and Mucuri Valleys, Diamantina, MG, Brazil
Umesh Gupta	Department of Pharmacy, School of Chemical Sciences and Pharmacy, Central University of Rajasthan, Bandar Sindri, Ajmer, Rajasthan-305817, India
Vaskuri G.S. Sainaga Jyothi	Department of Pharmaceutics, National Institute of Pharmaceutical Education and Research, Hyderabad, Telangana, India
Vipin Kumar	Department of Pharmacy, School of Chemical Sciences and Pharmacy, Central University of Rajasthan, Bandarsindri, Dist. Ajmer, Rajasthan, India

CHAPTER 1

Polymeric Nanoparticles as a Promising Drug Delivery Vehicle

Nidhi Gupta[1], Ramesh Chandra[2] and Surendra Nimesh[3,*]

[1] *Department of Biotechnology, IIS (Deemed to be University), Gurukul Marg, SFS, Mansarovar, Jaipur 302020, India*

[2] *Department of Chemistry, University of Delhi, Mall Road, Delhi 110007, India*

[3] *Department of Biotechnology, School of Life Sciences, Central University of Rajasthan, Ajmer 305817, India*

Abstract: With every era, medical science faces challenges to provide the best health care and efficient drug therapy for the treatment of existing diseases. Current regimes of chemical drugs, though beneficial in a certain sector, have several disadvantages; cytotoxicity and adverse health effects are of primary concern. To accomplish efficient as well as the safe mode of transport for drugs and bioactive molecules, nanotechnology has provided an answer in terms of nanomedicines. Polymeric nanoparticles that can easily be modified to suit the needs not only act as a vehicle to efficiently deliver the drug to the target site but also enhance the bioavailability of the drug. Polymer-drug conjugation delivers the drug specifically to the targeted tumors. The conjugation facilitates increased retention time and enhanced cellular permeability that enables better suppression of the tumors. This chapter gives an insight into the properties of nanoparticles, highlighting the associated advantages and limitations of polymeric nanoparticles as a vehicle of drug delivery to cells.

Keywords: Drug delivery, Gene delivery, Microparticles, Nanocapsules, Nanoparticles, Nanospheres.

INTRODUCTION

The desire to develop efficient drug delivery vectors with required therapeutic need lead to envision of several possible strategies, including the usage of modified or natural polymeric nanoparticles. The bio-availability and distribution of the drug depend upon its physicochemical properties and molecular structure. Due to the uneven distribution of drugs and its accumulation at non-specific sites, the desired efficacy of the drug is reduced at the diseased site that might cause

* **Corresponding author Surendra Nimesh:** Department of Biotechnology, School of Life Sciences, Central University of Rajasthan, Ajmer - 305817, India; E-mail: surendranimesh@gmail.com

toxic and undesirable effects. Henceforth, researchers are always seeking the challenging task of looking for new strategies, to develop delivery systems that may possess maximum therapeutic potential with minimal toxicity. To design and develop better targeting strategies to deliver drugs with more efficiency and specifically, it is essential not only to understand the characteristics of the target but also the mechanism of action of the drugs involved for targeting.

Particle-based drug carriers have been modified to provide better accessibility and efficiency of the drug at the site of action with its increased availability. For desirable therapeutic response, the size of the matrix has to be controlled along with the entrapped drug molecule. According to the size of the particles, the following categorization is done:

i. For particles in size range 50-200 μm: macroparticles
ii. 1-50 μm particles are termed microparticles
iii. Particles with size 1-1000 nm are categorized as nanoparticles

Nanoparticles and microparticles have found their utilization in several genes and drug delivery based researches. The size of microparticles restricts their entry into capillaries and, therefore, are unavailable at tissue sinusoids, though they were found being in circulation. However, these particles get accumulated in the adjacent tissues adjoining the capillaries and were observed to discharge encapsulated drug molecules gradually at the target site. The modified microparticles act as depot system that can be delivered *via* subcutaneous, intraperitoneal, intra-arterial or intravenous administering routes, resulting in a gradual release of the drug and protection from *in vivo* degradation. Engineered polymer with controlled size and swelling and breaking properties can help in the efficient release of drug at the target site. Taking into consideration all these key factors, starch, albumin, polylactic acid, and ethyl cellulose have been deployed to prepare biodegradable microparticles that would be used for chemoembolisation.

The limitation of microparticles was rapid clearance by RES (reticuloendothelial system) defense mechanism. To overcome this rapid removal of the particles from circulation and to have prolonged circulation of particles, a decrease in particle size was sought. The reduction in particle size renders nanoparticles a better carrier vehicle not only for polynucleotides but also for enzymes and proteins, irrespective of the route of administration. Reduced size nanoparticles have increased surface area and henceforth, enhanced adsorption of drug or biological molecules. Nanoparticles not only release the encapsulated or adsorbed drug or a bioactive molecule at the target site but also act as carriers for these molecules. The size of the particles in the nano range can render them to pass easily through capillaries and made them more available for systematic application with significant impact. Particles, either metallic or polymeric, that fall in size range of 1-100 nm are termed nanoparticles. For the first time in science, polymeric nanoparticles were prepared and characterized by Birrenbach and coworkers in 1976 [1]. This study had triggered extensive onset towards research involving designing and developing novel nanoparticles based carrier systems. Thus, these nanoparticles can be employed for the targeted delivery of drugs and biomolecules. On reaching the target site, the entrapped drug or biomolecule in the polymeric nanoparticles can be released from the complex through either one or a combination of the following mechanisms:

i. Hydration of the nanoparticles matrix led to swelling and henceforth, bursting of the particles, or slow-release by diffusion.
ii. The modified polymeric nanoparticles can be targeted for enzymatic degradation at the specific site that resulted in the release of the drug.
iii. Drug molecules entrapped in swollen nanoparticles could be released by chemical cleavage.

One of the most remarkable repercussions of nanotechnology is nanomedicine; the latter term is an umbrella term that conjugates various types of particles in nanoscale size with medical potentials such as liposomes, quantum dots, polymeric micelles, polymer-drug conjugates, dendrimers, inorganic nano particles, biodegradable nanoparticles, and other materials. In nanomedicine, the most potential candidate is nanoparticles that possess wider significant applications in the fields of target-specific drug and gene delivery. The small-sized nanoparticles are capable of better penetration in the tissue and henceforth, having targeted activity at the specific site of action [2]. Polymeric nanoparticles are explored extensively as an efficient delivery vector for drugs, more promising for anticancer drugs. These nanocarriers are modified for the release of drugs at a specific site, for example, the particles can be targeted or release drug based on stimuli response [3].

Polycationic polymeric nanoparticles are extensively worked upon to develop delivery strategies for DNA and siRNA. Polymeric nanoparticles can be categorized, namely, as (i) nanospheres that are spherical shaped particles of nanometer size. These nanoparticles can be modified such that the drug or biomolecule of choice can either be trapped in the spherical nanoparticles or adsorbed on the outer surface or both. (ii) Nanocapsules possess a solid polymeric shell with an inner liquid core. The required molecules can either be adsorbed on the outer surface or entrapped inside the core or both. Other than these two common forms, various other forms of nanoparticles have also been reported, for example, nanorods, nanotubes, cones, spheroids, *etc*. Several natural polycationic polymers, such as chitosan and synthetic ones, including PEI, have been investigated to deliver nucleic acids either in the form of nano-complexes or particles.

Recently, microfluidics-based strategies are researched to prepare nanoparticles that can be tuned to be tailored and reproduce structure. The process is being conducted in a microchannel, in a very controlled manner; small volume of liquid reagents is rapidly mixed to form nanoparticles. These synthesized polymeric and lipid nanoparticles can be used for their applications in nanomedicine [4].

POLYMERIC NANOPARTICLES FOR DNA/SIRNA DELIVERY

Nucleic acids are being extensively investigated as a potent tool for therapeutic gene expression inhibition. DNA and siRNA have similar physicochemical properties, rendering vectors suitable for DNA delivery to be a useful carrier for siRNA as well. Linear and branched, both types of cationic polymers, are efficiently used as DNA transfecting agents. The mechanism behind this complex formation is that the positively charged polymers form polyplexes *via* electrostatic interactions with the negatively charged phosphates of DNA [5], resulting in DNA condensation and protection from being degraded by nucleases. A similar mechanism is utilised for the formation of siRNA-polymer polyplexes or nanoparticles mediated siRNA delivery. Some other polymeric vectors such as micelles, nanoplexes, nanocapsules, and nanogels were also employed for siRNA delivery [6]. A few studies in this arena had shown remarkable strategies for the delivery of specific siRNA for the treatment of several human diseases or silencing of endothelial genes without having any off-target impact on hepatocytic genes [7, 8]. The physico-chemical properties of polyplexes such as surface charge, structure, and size depends on the ratio of the positive charges present on cationic polymers (due to the presence of amino group) to the number of negatively charged phosphate groups of siRNA (*i.e.*, N/P ratio).

ADVANTAGES AND LIMITATIONS OF NANOPARTICLES

Owing to their compact size, nanoparticles can easily cross cellular membranes to facilitate gene or drug distribution in the cells. These are remarkably less cleared by reticuloendothelial system clearance due to their smaller size and could penetrate better into tissues and cells. Polymeric nanoparticles have advantages, and can easily be manipulated because of varying molecular weight, linear and branched, and possess better biostability, are safe and less immunogenic. Even proteins that are beneficial for the study of stem cell research, for example, stromal cell-derived factor 1 (SDF-1) and bone morphogenetic protein 2 (BMP-2) that are involved with mobilization and osteogenesis of mesenchymal stem cells, can be successfully delivered using chitosan-agarose-gelatin nanoparticles [9].

The cost of preparation is quite low and these particles have a high delivery range with respect to the size of the transgene to be delivered. For specific targeted delivery of either plasmids or siRNA, several moieties such as RGD peptides or transferrin can be attached to polymeric nanoparticles [10, 11].

Due to the small size of nanoparticles, usually 10 to 200 nm size ranges, enhanced interaction of nanoparticles with surface biomolecules or within the cell occurs. This size also makes nanoparticles a beneficial delivery vehicle as they can reach inside the tissues, such as tumors with great specificity. This led to improved targeted delivery of genes or drugs to the cancerous cells [12]. But polymeric nanoparticles do have some limitations as well. The polycationic polymers constituting nanoparticles have a charge present on their surface thus; can have strong electrostatic interaction with charged plasma membrane proteins. This leads to instability and, finally, the rupturing of the plasma membrane [13]. In a comparative study between differentially charged polymeric nanoparticles, it was observed polycationic polymers have the highest toxicity, followed by neutral and anionic ones [14]. To overcome the toxicity problem, strategies were developed based on a decrease in surface charge by coating particles with hyaluronic acid or PEG [15, 16]. Of linear and branched PEIs, the former ones are less toxic and henceforth, more suitable for transfection, even at higher N/P ratio [17]. In addition to the above-mentioned limitations, PEI can possess adverse effects as it is a non-biodegradable polymer. Due to this, PEI can be accumulated within the cells and thus, may interfere with important intracellular biochemical processes [18, 19]. The charge of complexes decides the fate of the activation of the complement system. The complement system gets activated if the ratio of positive to negative ions is increased, but its activation is transversely lowered as the PEI/DNA complexes approach neutrality [20, 21]. To reduce the toxicity of complexes, chemical modifications are carried out; in that, small molecular weight PEIs are joined together to generate higher molecular weight PEIs molecules with the help of bi-functional linkages that are degradable. Another polymer that has found its usefulness as an efficient delivery vector for nucleic acids is chitosan. It has applications as a remarkable *in vitro* and *in vivo* siRNA delivery vehicle that is capable of ensuring gene knockdown with minimal toxicity [22, 23]. However, one of the major limitations associated with the usage of chitosan is their low transfection efficiency. The physio-chemical requirements for transfection of nanoparticles are not 100% known and henceforth, must be well explicated before their clinical applications. At times, non-specific stimulation does occur for siRNA delivery *via* nanoparticles. This could be because of the onset of innate immune inflammatory responses that in turn can stimulate type I interferon (IFN) synthesis. RNA duplex interaction with endosomal Toll-like receptors (TLR) leads to INF production that can be started with the delivery of nanoparticle-siRNA in the endosomes [24]. Such problems could be overcome with a few modifications, such as the introduction of 2′-*O*-methyl nucleotides into siRNA duplex strand interrupts TLR-7 interaction and related non-specific effects [25]. Incorporation of pH-sensitive moieties in the nano matrix can ensure the efficient release of siRNA for gene silencing.

Specific binding of siRNA with target RNA depends on high specificity based on Watson-Crick base pairing that constitutes RNA interference *via* initiating nucleolytic activity of the RISC complex. But at times, off-target effects have been observed where non-specific RNAi-induced gene silencing occurred with the introduction of a gene-specific siRNA. This could be due to the occurrence of partial Watson-Crick base pairing, leading to cross-reactivity. In fact, in some cases, pairing consisting of only 11-15 contiguous nucleotides is sufficient to induce gene silencing. 10-200 nm size range of nanoparticles is comparable to proteins and similarly, these nanoparticles can readily interact with surface biomolecules or those present in the cells. The small size of nanoparticles, therefore, provides an added advantage of being able to infiltrate tumor tissues with greater specificity and thus improving the delivery of drug/gene in a more targeted manner [12]. To synthesize functionalized nanoparticles, required changes in the addition of various layers and coatings are made [26].

Liposomes are the simplest of nanoparticles that are widely used in clinics for a long time, spherical in structure with lipid bilayers enclosed in an aqueous compartment [27]. In case of liposomes, the fatty layer is supposed to protect and confine the enclosed drug until bound to the outer membrane of the target cells. The solubility of many amphiphilic drugs could be improved by liposomes as the later consist of a hydrophilic core and hydrophobic phospholipid bilayer coat. The advantages of liposomes include not only prolonged circulation time and reduced systemic toxicity but also enhanced uptake into tumors with a constant discharge of their payload [28].

Dendrimers are branched nanoparticles with an inner core that are made up of different types of polymers such as poly (L-glutamic acid) (PGA), polyamidoamine (PAMAM), poly (ethylene glycol) (PEG), and polyethylenimine (PEI). These polymers have to undergo either convergent or divergent step-growth polymerization [29]. The basic nature of dendrimer is hydrophilicity and henceforth, can be used as a coating agent. The ease of preparation and terminal modifications render dendrimers suitable for targeted delivery or selective imaging of tumors [30, 31]. More than 96% inhibition of growth of tumor cells in animal models was observed with the usage of dendrimers employed to deliver anti-angiogenic angiostatin tissue inhibitor of metalloproteinases gene [32].

Polymer-drug conjugation helps in targeted delivery specifically to tumors with better suppression of the later due to enhanced permeability and retention time. A lot of research had been done to develop polymeric nanoparticles (NPs) that could be biodegraded. These could be prepared from polylactic acid, polyglycolic acid, polylactic-glycolic acid (PLGA), and poly (methyl methacrylate) (PMMA) and served the purposes for both gene and drug delivery. These can be developed as a second generation of carriers to deliver target specific anti-cancer agents and thus form the solid foundation for polymer-bound chemotherapy [33]. The nano-particles have found their use in control as well as treatment of the central nervous system affecting parasitic infections. These particles can cross the blood-brain barrier efficiently and access the infected tissue for drug delivery [34].

One such widely studied, biodegradable, and non-toxic polysaccharide polymer is chitosan, which is biocompatible and protects DNA against DNase ensured degradation [35]. Recent research by Öztürk and coworkers has shown the successful application of poly (lactic-co-glycolic acid) (PLGA) NPs and chitosan (CS)-coated PLGA NPs for oral delivery of clarithromycin against several pathogenic bacteria (D).

Metallic nanoparticles are also pivotal as nontoxic drug carriers for selective delivery, for example, gold NPs. These are modified to be made more beneficial using PEG coatings. In one such study, colloidal gold NPs coated PEG with an incorporated TNF (tumor necrosis factor)-R (PT-cAu-TNF-α) was used to see the effect on tumors. When intravenously administered as thermal therapy in mice models, these NPs showed a significant decline in the growth of tumors [36]. In another study, in MC-38 colon carcinoma, it was observed that colloidal gold NPs bound with thiol derivatized PEG with adsorbed recombinant human TNF on their surface were accumulated specifically in tumors with little or no off-site access to other tissues such as livers, spleens, or other healthy organs [37].

Therefore, it can be concluded that various polymeric nanoparticles can be modified to achieve not only specificity but also enhanced efficiency for the release of drugs or biomolecules at the target site. The increased retention time and reduced clearance by RES provide a lasting effect of the drug moiety at targeted tumors. It can be said at this juncture that in the coming era, polymeric nanoparticles will provide a crucial platform for the delivery of anti-angiogenic and anti-tumorous molecules with enhanced specificity.

CONSENT FOR PUBLICATION

Not applicable.

CONFLICT OF INTEREST

The author declares no conflict of interest, financial or otherwise.

ACKNOWLEDGEMENTS

Declared none.

REFERENCES

[1] Birrenbach G, Speiser PP. Polymerized micelles and their use as adjuvants in immunology. J Pharm Sci 1976; 65(12): 1763-6.
[http://dx.doi.org/10.1002/jps.2600651217] [PMID: 1036442]

[2] Peer D, Karp JM, Hong S, Farokhzad OC, Margalit R, Langer R. Nanocarriers as an emerging platform for cancer therapy. Nat Nanotechnol 2007; 2(12): 751-60.
[http://dx.doi.org/10.1038/nnano.2007.387] [PMID: 18654426]

[3] Nicolas J, Couvreur P. Polymer nanoparticles for the delivery of anticancer drug. Med Sci (Paris) 2017; 33(1): 11-7.
[http://dx.doi.org/10.1051/medsci/20173301003] [PMID: 28120750]

[4] Streck S, Hong L, Boyd BJ, McDowell A. Microfluidics for the production of nanomedicines: Considerations for polymer and lipid-based systems. Pharm Nanotechnol 2019; 7(6): 423-43.
[http://dx.doi.org/10.2174/2211738507666191019154815] [PMID: 31629401]

[5] Agarwal A, Unfer R, Mallapragada SK. Novel cationic pentablock copolymers as non-viral vectors for gene therapy. J Control Rel 2005; 103(1): 245-58.
[http://dx.doi.org/10.1016/j.jconrel.2004.11.022] [PMID: 15710515]

[6] de Martimprey H, Vauthier C, Malvy C, Couvreur P. Polymer nanocarriers for the delivery of small fragments of nucleic acids: oligonucleotides and siRNA. Eur J Pharm Biopharm 2009; 71(3): 490-504.
[http://dx.doi.org/10.1016/j.ejpb.2008.09.024] [PMID: 18977435]

[7] Zhao Y, Gao H, He J, *et al.* Co-delivery of LOX-1 siRNA and statin to endothelial cells and macrophages in the atherosclerotic lesions by a dual-targeting core-shell nanoplatform: A dual cell therapy to regress plaques. J Control Rel 2018; 283: 241-60.
[http://dx.doi.org/10.1016/j.jconrel.2018.05.041] [PMID: 29885417]

[8] Dahlman JE, Barnes C, Khan O, *et al. In vivo* endothelial siRNA delivery using polymeric nanoparticles with low molecular weight. Nat Nanotechnol 2014; 9(8): 648-55.
[http://dx.doi.org/10.1038/nnano.2014.84] [PMID: 24813696]

[9] Wang B, Guo Y, Chen X, *et al.* Nanoparticle-modified chitosan-agarose-gelatin scaffold for sustained release of SDF-1 and BMP-2. Int J Nanomedicine 2018; 13: 7395-408.
[http://dx.doi.org/10.2147/IJN.S180859] [PMID: 30519022]

[10] Schiffelers RM, Ansari A, Xu J, *et al.* Cancer siRNA therapy by tumor selective delivery with ligand-targeted sterically stabilized nanoparticle. Nucleic Acids Res 2004; 32(19): e149.
[http://dx.doi.org/10.1093/nar/gnh140] [PMID: 15520458]

[11] Davis ME. The first targeted delivery of siRNA in humans *via* a self-assembling, cyclodextrin polymer-based nanoparticle: from concept to clinic. Mol Pharm 2009; 6(3): 659-68.
[http://dx.doi.org/10.1021/mp900015y] [PMID: 19267452]

[12] Cuenca AG, Jiang H, Hochwald SN, Delano M, Cance WG, Grobmyer SR. Emerging implications of nanotechnology on cancer diagnostics and therapeutics. Cancer 2006; 107(3): 459-66.
[http://dx.doi.org/10.1002/cncr.22035] [PMID: 16795065]

[13] Fischer D, Li Y, Ahlemeyer B, Krieglstein J, Kissel T. *In vitro* cytotoxicity testing of polycations:

influence of polymer structure on cell viability and hemolysis. Biomaterials 2003; 24(7): 1121-31. [http://dx.doi.org/10.1016/S0142-9612(02)00445-3] [PMID: 12527253]

[14] Jevprasesphant R, Penny J, Jalal R, Attwood D, McKeown NB, D'Emanuele A. The influence of surface modification on the cytotoxicity of PAMAM dendrimers. Int J Pharm 2003; 252(1-2): 263-6. [http://dx.doi.org/10.1016/S0378-5173(02)00623-3] [PMID: 12550802]

[15] Jiang G, Park K, Kim J, *et al.* Hyaluronic acid-polyethyleneimine conjugate for target specific intracellular delivery of siRNA. Biopolymers 2008; 89(7): 635-42. [http://dx.doi.org/10.1002/bip.20978] [PMID: 18322932]

[16] Duan Y, Yang C, Zhang Z, Liu J, Zheng J, Kong D. Poly(ethylene glycol)-grafted polyethylenimine modified with G250 monoclonal antibody for tumor gene therapy. Hum Gene Ther 2010; 21(2): 191-8. [http://dx.doi.org/10.1089/hum.2009.010] [PMID: 19788387]

[17] Wightman L, Kircheis R, Rössler V, *et al.* Different behavior of branched and linear polyethylenimine for gene delivery *in vitro* and *in vivo.* J Gene Med 2001; 3(4): 362-72. [http://dx.doi.org/10.1002/jgm.187] [PMID: 11529666]

[18] Godbey WT, Wu KK, Hirasaki GJ, Mikos AG. Improved packing of poly(ethylenimine)/DNA complexes increases transfection efficiency. Gene Ther 1999; 6(8): 1380-8. [http://dx.doi.org/10.1038/sj.gt.3300976] [PMID: 10467362]

[19] Godbey WT, Wu KK, Mikos AG. Poly(ethylenimine)-mediated gene delivery affects endothelial cell function and viability. Biomaterials 2001; 22(5): 471-80. [http://dx.doi.org/10.1016/S0142-9612(00)00203-9] [PMID: 11214758]

[20] Ferrari S, Moro E, Pettenazzo A, Behr JP, Zacchello F, Scarpa M. ExGen 500 is an efficient vector for gene delivery to lung epithelial cells *in vitro* and *in vivo.* Gene Ther 1997; 4(10): 1100-6. [http://dx.doi.org/10.1038/sj.gt.3300503] [PMID: 9415317]

[21] Plank C, Mechtler K, Szoka FC Jr, Wagner E. Activation of the complement system by synthetic DNA complexes: a potential barrier for intravenous gene delivery. Hum Gene Ther 1996; 7(12): 1437-46. [http://dx.doi.org/10.1089/hum.1996.7.12-1437] [PMID: 8844203]

[22] Howard KA, Rahbek UL, Liu X, *et al.* RNA interference *in vitro* and *in vivo* using a novel chitosan/siRNA nanoparticle system. Mol Ther 2006; 14(4): 476-84. [http://dx.doi.org/10.1016/j.ymthe.2006.04.010] [PMID: 16829204]

[23] Pillé JY, Li H, Blot E, *et al.* Intravenous delivery of anti-RhoA small interfering RNA loaded in nanoparticles of chitosan in mice: safety and efficacy in xenografted aggressive breast cancer. Hum Gene Ther 2006; 17(10): 1019-26. [http://dx.doi.org/10.1089/hum.2006.17.1019] [PMID: 17007568]

[24] Marques JT, Devosse T, Wang D, *et al.* A structural basis for discriminating between self and nonself double-stranded RNAs in mammalian cells. Nat Biotechnol 2006; 24(5): 559-65. [http://dx.doi.org/10.1038/nbt1205] [PMID: 16648842]

[25] Judge AD, Bola G, Lee AC, MacLachlan I. Design of noninflammatory synthetic siRNA mediating potent gene silencing *in vivo.* Mol Ther 2006; 13(3): 494-505. [http://dx.doi.org/10.1016/j.ymthe.2005.11.002] [PMID: 16343994]

[26] Groneberg DA, Giersig M, Welte T, Pison U. Nanoparticle-based diagnosis and therapy. Curr Drug Targets 2006; 7(6): 643-8. [http://dx.doi.org/10.2174/138945006777435245] [PMID: 16787165]

[27] Rawat M, Singh D, Saraf S, Saraf S. Nanocarriers: promising vehicle for bioactive drugs. Biol Pharm Bull 2006; 29(9): 1790-8. [http://dx.doi.org/10.1248/bpb.29.1790] [PMID: 16946487]

[28] Park JW. Liposome-based drug delivery in breast cancer treatment. Breast Cancer Res 2002; 4(3): 95-9.

[http://dx.doi.org/10.1186/bcr432] [PMID: 12052251]

[29] Aulenta F, Drew MG, Foster A, *et al.* Synthesis and characterization of fluorescent poly(aromatic amide) dendrimers. J Org Chem 2005; 70(1): 63-78. [http://dx.doi.org/10.1021/jo048799a] [PMID: 15624907]

[30] Quintana A, Raczka E, Piehler L, *et al.* Design and function of a dendrimer-based therapeutic nanodevice targeted to tumor cells through the folate receptor. Pharm Res 2002; 19(9): 1310-6. [http://dx.doi.org/10.1023/A:1020398624602] [PMID: 12403067]

[31] Padilla De Jesús OL, Ihre HR, Gagne L, Fréchet JM, Szoka FC Jr. Polyester dendritic systems for drug delivery applications: *in vitro* and *in vivo* evaluation. Bioconjug Chem 2002; 13(3): 453-61. [http://dx.doi.org/10.1021/bc010103m] [PMID: 12009933]

[32] Vincent L, Varet J, Pille JY, *et al.* Efficacy of dendrimer-mediated angiostatin and TIMP-2 gene delivery on inhibition of tumor growth and angiogenesis: *in vitro* and *in vivo* studies. Int J Cancer 2003; 105(3): 419-29. [http://dx.doi.org/10.1002/ijc.11105] [PMID: 12704680]

[33] Vicent MJ, Duncan R. Polymer conjugates: nanosized medicines for treating cancer. Trends Biotechnol 2006; 24(1): 39-47. [http://dx.doi.org/10.1016/j.tibtech.2005.11.006] [PMID: 16307811]

[34] Gondim BLC, da Silva Catarino J, de Sousa MAD, *et al.* Nanoparticle-mediated drug delivery: Blood-brain barrier as the main obstacle to treating infectious diseases in CNS. Curr Pharm Des 2019; 25(37): 3983-96. [http://dx.doi.org/10.2174/1381612825666191014171354] [PMID: 31612822]

[35] Mansouri S, Lavigne P, Corsi K, Benderdour M, Beaumont E, Fernandes JC. Chitosan-DNA nanoparticles as non-viral vectors in gene therapy: strategies to improve transfection efficacy. Eur J Pharm Biopharm 2004; 57(1): 1-8. [http://dx.doi.org/10.1016/S0939-6411(03)00155-3] [PMID: 14729076]

[36] Visaria RK, Griffin RJ, Williams BW, *et al.* Enhancement of tumor thermal therapy using gold nanoparticle-assisted tumor necrosis factor-alpha delivery. Mol Cancer Ther 2006; 5(4): 1014-20. [http://dx.doi.org/10.1158/1535-7163.MCT-05-0381] [PMID: 16648573]

[37] Paciotti GF, Myer L, Weinreich D, *et al.* Colloidal gold: a novel nanoparticle vector for tumor directed drug delivery. Drug Deliv 2004; 11(3): 169-83. [http://dx.doi.org/10.1080/10717540490433895] [PMID: 15204636]

CHAPTER 2

Characterization of Nanostructured Systems for Bioactive Compound Delivery

Gabriel Silva Borges[1], Flávia Lidiane Oliveira da Silva[2], Osmar Patrício Almeida[2], Taízia Dutra Silva[2] and **Guilherme Carneiro[2,*]**

[1] *Department of Pharmaceutics, Faculty of Pharmacy, Federal University of Minas Gerais, Belo Horizonte, MG, Brazil*

[2] *Department of Pharmacy, Faculty of Biological and Health Sciences, Federal University of Jequitinhonha and Mucuri Valleys, Diamantina, MG, Brazil*

Abstract: Characterization of nanostructured systems is an important aspect to support the choice of the better formulation composition and the best production conditions throughout a development process. Several methods can be used alone or combined for the determination of physical (*e.g.*, mechanical, electrical, electronic, magnetic, thermal and optical), chemical or biological properties of a nanomaterial. This chapter is an overview of the most employed techniques, including dynamic light scattering and laser diffraction for the determination of size distribution; zeta potential and its relationship with stability and the surface charge of the particles; microscopies (optical microscopy, SEM, TEM, AFM) utilized in morphological analyses; spectroscopies in the infrared or ultraviolet-visible regions, and X-rays diffraction, which help to elucidate the crystalline state, polymorphism and drug-nanosystem interaction; and thermal analyses, which can provide information about the physical state, crystallinity, and stability. Further complementary information can be obtained from many other methods, such as nuclear magnetic resonance or Raman spectroscopy, but they are beyond the scope of this chapter. The careful choice of the characterization techniques to be used is certainly a decisive step in the successful and rational development of a nanocarrier formulation.

Keywords: AFM, Confocal microscopy, Dynamic Light scattering, Fluorescence microscopy, Nanoparticles, SEM, TEM, XRD.

---

* **Corresponding author Guilherme Carneiro:** Department of Pharmacy, Faculty of Biological and Health Sciences, Federal University of Jequitinhonha and Mucuri Valleys, Diamantina, MG, Brazil;
E-mail: guilherme.carneiro@ufvjm.edu.br

Surendra Nimesh, Nidhi Gupta and Ramesh Chandra (Eds.)

INTRODUCTION

An important part of the development process of nanostructured systems for bioactive compounds delivery is the determination of relevant physicochemical properties to be measured in order to better understand the developed system. Thus, the characterization of nanosystems has paramount importance, being necessary for product development, both to choose a better formulation composition and to define the best production conditions [1].

Nanomaterials may assume several forms, such as free-standing nano-objects (like nanoparticles and carbon nanotubes), materials with nanosized holes (including porous films), and other types of materials with a variety of different nanoscale structures (such as nanometer-thick layered structures) [2]. The size distribution, visual aspect, incorporation efficiency of the active molecule to the system, and zeta potential (ZP) have been considered the most important parameters for the characterization of nanostructured objects [3, 4].

Characterization of nanosystems may include the evaluation of physical (*e.g.*, mechanical, electrical, electronic, magnetic, thermal, and optical), chemical, or biological (*e.g.*, bioactivity, immunogenicity) properties. Often, samples will be characterized in two or more of these properties. Nanomaterials characterization can provide a broader view of physical parameters like size distribution, shape, an organization in phases and structures, polymorphism, or more detailed information about the chemical composition, like the atomic structure, chemical composition of the surface and the presence of coating, as in the case of protein deposition over the nanoparticles' surface. Thus, characterization methods should provide information about the structures of nanomaterials at various levels of sophistication: crystal structure, microstructure, atom-level structure, and electronic structure [5].

The nanomaterial can normally be considered a set of four entities: surface, coating (or film), interface and bulk (Fig. **1**). They should all be proper characterized to better describe it. Nanoparticles and other nanosystems exhibit different physical and chemical properties on the surface, compared to the bulk. The scale reduction of a nanomaterial brings a higher surface area per mass unit, which implies a surface with more atoms. Surface atoms have less neighbor atoms than the bulk ones, having less bonds and interactions, being usually less stable. Therefore, the surface of nanomaterials is more reactive than the bulk condition [6, 7].

Nanomaterials are commonly made of metallic, semiconducting or organic molecules and are generally coated with polymers or other biomolecules. Nanomaterials are characterized by a relatively large surface area per unit mass and the sphere is the form with the smallest surface area per unit mass. This way, the surface of a solid depends on its shape as well as its size. Still, the final size and structure of nanoparticles depend on the salt and surfactant additives, reactant concentrations, reaction temperatures, and solvent conditions used during the synthesis [8].

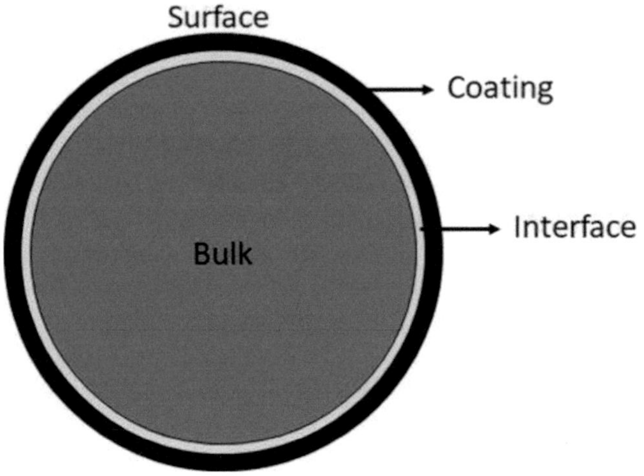

Fig. (1). Schematic representation of a nanoparticle divided into four parts: surface, coating, interface and bulk.

Characterization will evaluate the response of a material or system to a stimulus that can be an incident beam of radiation, particles (such as electrons, or neutrons) or ions. Physical and mechanical interactions may also occur in the evaluation of nanoparticle topography and in profilometry. Therefore, almost all methods will interfere somehow with the sample, causing some type of disturbance, especially in its surface. Not necessarily, the method will induce destructible disturbance.

On the other hand, the determination of some properties of nanoparticles and other nanocarriers can be misleading at times. Depending on the sample, the obtained experimental data can be reported without complete comprehension and with some associated errors. Polydisperse samples are usually difficult to have their properties measured due to size variations and thus can lead to experimental errors. Due to the fact, there is a lack of references to the metrology of nanoscaled materials, there is not even a consistently preferred measurement for a given parameter. If the particle size is measured using the dynamic light scattering (DLS) method, it takes into consideration the ligand shell, while transmission electron microscopy (TEM) ignores it. Which is right? Which is relevant? Are both needed? Are both stable over time? These are questions that should be answered by the researchers prior to conduct certain studies more efficiently [9].

Nanoparticles are thus highly complex systems theoretically composed of core (bulk), shells, dynamic shells, ligands, ligand functional groups, surface charge, surface adsorbates, and organized in a structural geometry; all of these properties should be previously planned by the formulating researcher. However, more species can coexist: side products, decomposition products, extra ligands, reactants, catalysts, and salts. Most of the characterization methods usually employed do not measure these other species and their influence is not clearly reported in the literature. Thus, many methods assume material purity without prior demonstrating it [9].

Nanoparticles dispersed in complex media such as biological, which contain potential interferents, should be measured using an instrument method with high selectivity and sensitivity. Besides, in such media, nanoparticles may undergo structural transformations: they can dissolve, agglomerate, sediment, or even be coated by the deposition of external molecules, resulting in instability and further difficulties to determine the properties of the samples and predict their *in vivo* biological behavior [2, 10].

Nanostructured materials are usually synthesized, shipped, processed, cleaned, tested, or dispersed in many solution types. Interactions between nanoparticles and solvent can have some intended and unintended consequences. If nanoparticles are characterized in a dried form, the removal of the solvent and co-solvents should be processed, avoiding phenomena of aggregation or deposition of salts or other solution species on the nanoparticles' surface [2, 10].

Considering the small size of nanoparticles, most of the constituent atoms are placed on the surface or next to a surface atom. If the nature of the surface is not known, the particles are not well characterized. For instance, surface coatings, such as those associated with deposition of substances present in biological media, are not identified by many of the commonly used characterization techniques. In addition, attention should be paid to the synthesis route, handling, and processing since variation on these processes can produce totally distinct types of nanoparticles. Finally, nanoparticles are dynamic; they change shape and size with time and respond to the environment. They can grow by agglomeration or just individually, and can be dissolved in biological media or adsorb molecules from it (it is stated single-phase nanoparticles are actually coated by molecules from the environment). Nanoparticles can change by exposure to light or as a function of temperature, and the time for these changes varies from a fraction of seconds to years [2, 11, 12].

There are numerous useful methods to characterize nanosystems, among them, the main and most applied in practice are: particle size and ZP analysis;

thermogravimetric analysis (TGA), differential thermal analysis (DTA), differential scanning calorimetry (DSC), X-ray diffraction (XRD and SAXS), microscopies, including fluorescence techniques, infrared (IR) and ultraviolet-visible (UV-vis) spectroscopies. Classical analytical techniques may also be applied to characterize encapsulation efficiency in the nanosystems, among others: high performance liquid chromatography (HPLC) coupled or not to other techniques such as mass spectrometry (HPLC/MS) and nuclear magnetic resonance (HPLC/NMR), gas chromatography (GC or GC/MS), IR, and UV-vis spectroscopies [3, 13]. In this chapter, some of the most employed characterization techniques in the process of nanosystems development will be presented, as well as the applicability of these methods for investigation of their physicochemical properties.

SIZE ANALYSIS

Particle size determination is considered to be necessary for virtually all processes and areas involving the development of nanometric particles [14]. Properties of nanostructured systems associated with their nanometric size, high structural integrity [13], stability [15], ease of fabrication [16], compatibility properties, the possibility of sustained release [1, 15] and surface functionalization [17] make them a versatile and promising therapeutic delivery vehicle [1].

DLS is the main technique used to determine the diameter of materials composed of different types of particles dispersed in a liquid medium. These particles can be organic (made of polymers, lipids, carbohydrates, proteins and surfactants), inorganic (made of metals, such as gold or silver nanoparticles, or those formed by oxides of transition metals), or hybrid (formed by an inorganic nucleus covered by organic molecules). Determination of size distribution (polydispersity) is necessary because particles of different sizes have different properties such as reactivity, dissolution rate, stability and efficiency in drug loading (DL) and delivery. Therefore, understanding the use of the proper equipment and the application of methods to measure these parameters is essential to achieve a reduction of the time used to develop a nanostructured formulation [18].

Particle size, together with surface charge, are the two primary factors responsible for a number of biological effects and responses generated by nanoparticles, including cellular uptake [19], toxicity [20], dissolution [21] and release profile of incorporated drugs [22].

The measurements of DLS are very simple, easy and reproducible tools to characterize the particles and analysis the quality of their production [14]. The instruments are usually robust, compact and simple to operate, offering easy-t-

-use digital interfaces and enabling extensive data analysis. In addition, the techniques are non-invasive, requiring minimal preparation of the sample and no need for calibration at each analysis [23].

Dynamic Light Scattering

DLS, also known as photon correlation spectroscopy, or quasi-elastic light scattering, is the most widely used technique to evaluate the size of suspended particles, both in research and in industry. This technique can be applied to particles with a size in the range of 0.3 nm to 10 μm and is based on the principle that when a particle is irradiated by a light source, such as a laser, the optical light path is disturbed and the light is scattered in all directions. The DLS technique then consists of analyzing the fluctuations of the intensity of scattered light at a given angle. This analysis provides information about the particle motion, which is the cause of fluctuations in intensity [24].

When small particles are dispersed in a solvent, they assume a movement pattern known as Brownian motion. This is the random motion of microscopic particles immersed in a solvent caused by the shocks of the solvent molecules against these particles as first defined by Einstein [25].

If the particle dispersion is irradiated by a monochromatic incident light laser, this light will be scattered by the particles in all directions. When the particle size is smaller than $\lambda/10$ (λ is the wavelength of the incident light), the radiation will be dispersed in several directions without modification in the energy (elastic scattering) or wavelength of the incident light, occurring oscillations only in the amplitude. This phenomenon is known as ***Rayleigh scattering*** and is not angle-dependent. However, when the particle size is larger than $\lambda/10$, the light is unequally scattered in energy (inelastic scattering) and is angle-dependent. The scattered light has the most intense energy in the same direction as the original beam. This anisotropic phenomenon is known as ***Mie scattering*** [14] (Fig. **2**).

As the particles are not stationary, but in Brownian motion, the distance between them is constantly changing, causing constructive or destructive interferences in the diffracted light. When the scattered light is projected onto a detector that collects the signal pattern, the diffracted light can arrive out of phase, due to destructive interferences (dark spots), or in phase, due to constructive interferences (bright spots). Therefore, the intensity of the signal fluctuates over time. The rate of these intensity variations depends on the speed of particles (smaller particles diffuse faster), thus, size distribution information is obtained by analyzing the fluctuations in the intensity of the scattered light [26]. Equation 1 relates the signal intensity with the radius of the particle:

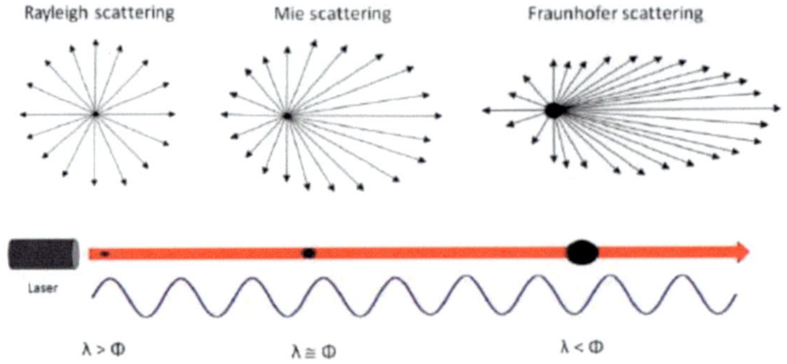

Fig. (2). Light scattering as a function of particle size and wavelength ratio.

$$I = I_0 \frac{1+cos^2\theta}{2R^2} \left(\frac{2\pi}{\lambda}\right)^4 \left(\frac{n^2-1}{n^2+2}\right)^2 \left(\frac{d}{2}\right)^6 \quad (1)$$

where I_0 = intensity of the incident radiation and λ = wavelength of the incident radiation, R = distance to the scattering center, θ = angle at which the light is scattered, n = refractive index, and d = particle diameter.

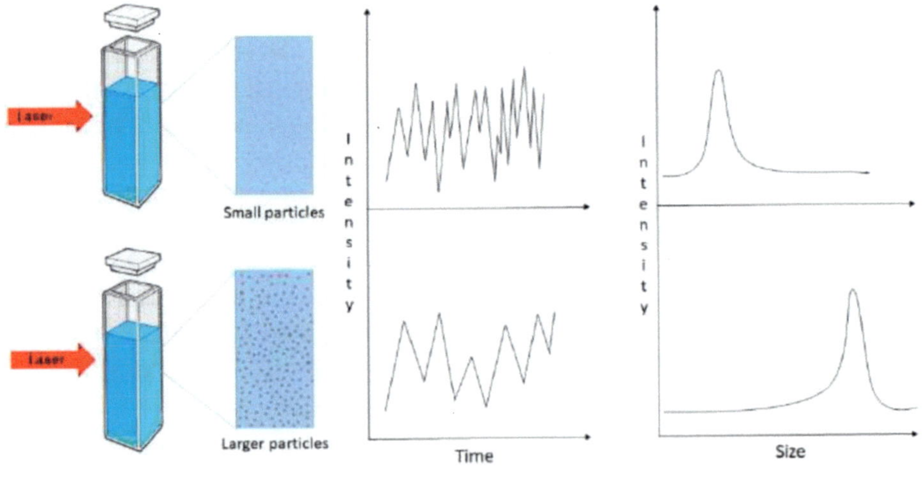

Fig. (3). Correlation between light intensity changes and particle size.

If the scattering centers are independent and are randomly distributed in the dispersion media, the probability of constructive and destructive interferences between the scattered radiation waves will be the same since, due to Brownian motion, the distance between the particles (scattering centers) varies with time. In this case, smaller particles move faster when compared to larger ones (Fig. **3**). Therefore, smaller particles cause higher rates of fluctuation in the intensity, and larger particles generate less fluctuation. Thus, the measures of fluctuations in the intensity of scattered light as a function of time bring with them information about the motion speed of these scattering centers, *i.e.*, this correlation can define the value referring to the translational diffusion coefficient of the particles (D_T) [14, 21, 25].

Assuming that the particles are spherically symmetrical and are in Brownian motion (for example, globular proteins, other polymers and lipid vesicles in suspension), the Stokes-Einstein equation can be used to calculate the hydrodynamic diameter (D_H) of the particles, which is inversely proportional to the D_T (Equation 2).

$$D_T = \frac{k_B T}{3\pi\eta D_H} \quad \textbf{(2)}$$

where k_B = Boltzmann constant (1,38064852 x 10^{-23} J/K), T = temperature (K) and η = absolute viscosity of the solvent and D_H = hydrodynamic diameter.

To define D_H, the measurements consider the association of the nanoparticle with counterions and hydration layers (Box **1**). It is then expressed as the mean diameter (Z-average) of the particles, which must be in thermal equilibrium with the solvent. The Rayleigh and Stokes-Einstein scattering equations indicate that DLS results depend on a few variables, including solvent viscosity, instrument, temperature, and refractive index of the material. However, DLS provides only an indicative size of the colloid since D_H varies with the variation of the particle corona, especially the soft corona. Another limitation is that the DLS measurements provide information in D_H, which assumes that the particles have a spherical shape. Nevertheless, if the particle has a shape too different from a sphere, such as cylinders, nanostars, nanorods, the obtained diameter would be too different from the real dimensions of the object [14].

Several studies cited in the literature have used DLS as the method of choice for the determination of Z-average and polydispersity or polydispersion index (PDI) [15, 27, 28]. In previously published studies by our group, involving the incorporation of retinoic acid [29 - 33], paclitaxel [34], and benznidazole [35] in nanostructured lipid systems, one can see the application of the DLS technique for the determination of mean particle size and PDI.

There are many examples in the literature showing the differences in measuring particle size by DLS or microscopy methods. Just to exemplify, in the work of Li *et al.* [36], the size distribution of doxorubicin-loaded keratin nanoparticles was determined by DLS and scanning electron microscopy (SEM). The values obtained for the same formulation were around 215 and 150 nm for DLS and SEM, respectively. Clearly, it is expected that the value obtained in DLS is somewhat higher since it is associated with the hydrodynamic diameter, unlike microscopy, which determines particle diameter in a dry sample [14].

The great advantage of using DLS in many research laboratories as well as in industries is the applicability in production stages without destruction or alteration of the product. In addition, DLS commercial equipment can be used for diverse systems such as polymers, proteins and particles [37].

Box 1. Protein corona formation in nanostructured systems: impact on particle size analyses.

Particles in colloidal systems develop random translational movements and rotate very fast. When the particle concentration is increased, the number of collisions between them also rises, since the average pathlength traversed by them is reduced. This way, by changing the concentration of the particles in the sample, the interparticle interactions also change.

With the size reduction to the nanometer scale, the surface of the dispersed particles is also changed, including surface charges. Thus, the adsorption of other components on the surface of the nanoparticles is facilitated due to the higher surface area. A classic example of this phenomenon is the deposition of proteins from a biological medium, creating the characteristic protein corona. This way, the particle has its surface hydrated and surrounded by molecules which are not ingredients of the particle itself. The corona has components attached by strong or weak interactions: the hard corona, a stable inner layer tightly attached to the surface particles [38] and the outmost layer formed. The soft corona, a relatively loose layer composed of molecules of different charges and size on hard corona top [39].

Consequently, what actually disperses the light is the nuclei containing the nanoparticles with their surface layer hydrated/solvated, with different compositions. Therefore, in DLS, the size (hydrodynamic diameter) is determined for nanoparticles with different chemical compositions on the surface when compared to the ones originally synthesized [40] (Fig. **4**).

The hydrodynamic radius (R_H) is the radius of a hypothetical sphere that moves with the same speed as the sample particles. Such hard spheres do not exist in colloidal dispersions, since the composition of corona, especially the soft corona, is dynamic and changes with the ionic strength, types and size of the molecules present in the dispersion media, and types of solvents.

Equipment

Several instruments for DLS analyses are available from several suppliers. It is not the scope of this text to mention this but to highlight the components common to them. Basically, these instruments generally have three main components: light source, sample, and laser detector (Fig. **5**).

Light Source

Usually a 4 mW He-Ne or with variable wavelengths lasers (depending on each equipment, most common at 633 and 638 nm - red laser, or 532 nm - green laser); 60 mW diode lasers at 830 nm are also used [14]. Laser sources provide a stable beam of coherent monochromatic light. There is an attenuator available to change the power of the laser, enabling analyses that do not destroy or alter the samples, so it is applicable for industrial purposes of development and production of nanoparticles.

Sample

Clean square cuvettes made of scratch-free glass or optically translucent disposable plastic are used (3x3 mm, 5x5 mm or 10x10 mm) [14]. These cuvettes are inserted into a dark, isolated chamber to perform a measurement (Fig. **5**). Inside this chamber, which must be properly closed to perform the analyses, it is also possible to control the temperature of the sample in the range of 2 to 90 °C. Plastic cuvettes with inbuilt electrodes capable of measuring DLS and ZP at the same time are also available [23].

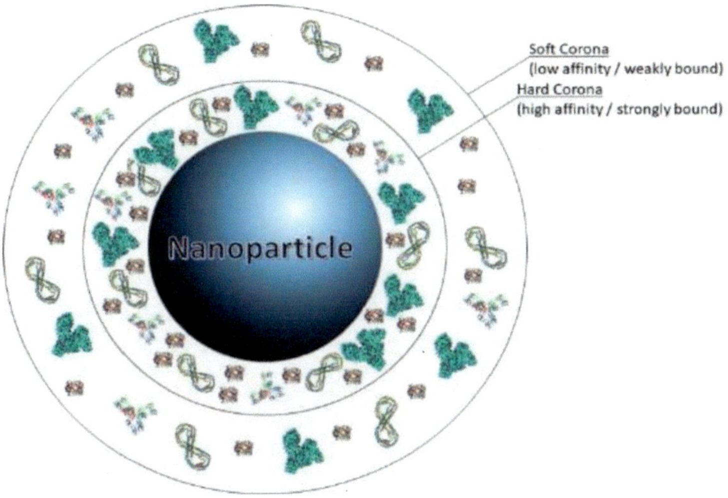

Fig. (4). Nanoparticles core with hydrated/solvated surfaces and different attached compounds.

Detector

Newer DLS instruments are equipped with avalanche photodiode detectors (APD) efficient to detect the red wavelength 633 nm laser. In the latest versions, detectors can be placed at the angles of 13° or 173° to detect backscattering, although in some older versions, the only angle available is 90° (Fig. **5**). By enabling the backscattering detection when the detectors are placed at 173°, the incident beam does not have to pass through the entire sample, avoiding multiple scattering, where the scattered light from one particle is re-scattered by other particles. Thus, the excess of dispersed light can be excluded. This helps to unmask low-intensity scattered light signals from smaller particles. The area of the illuminated sample within the cuvette is also increased (about 8-fold) compared to the 90° arrangement. In this backscatter arrangement (non-invasive backscatter system – NIBS), if necessary, the lens can be focused in the middle or close to the walls of the cuvette if the sample is too diluted or concentrated, respectively [14].

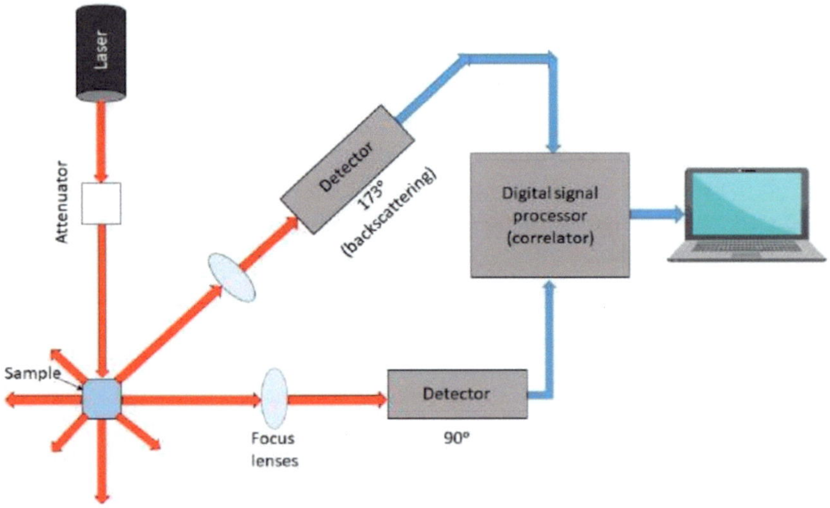

Fig. (5). Schematic representation of DLS equipment.

In addition to the above components, the DLS instruments also have a goniometer (to measure the scattered light angles), an autocorrelator (which correlates the variation rate of the dispersion intensity over time) and a processor for analysis of the signal data.

Performing the Technique

Sources of error and variations may arise during sampling, sample preparation, instrument configuration and instrument operation. Some of them may be user-dependent, so when performing the DLS technique, it is necessary to consider some important factors to obtain reliable data, which truly represent the characteristics of the particles present in the sample.

Samples should be clear, homogeneous and free of mist, and care should be taken to avoid contamination by particles larger than the particles under analysis (*e.g.*, dust and lipids). Some solvents, as toluene, scatter light and cause interference, creating background noise while others, such as DMSO, changes the viscosity drastically with slight variation in temperature. Deionized water is not a good solvent either as the absence of ions fails to block long-distance interactions between particles. Checking precipitates is also important since it confirms the presence of bigger particles due to poor dispersion, polydispersity, alterations in pH, and poor sonication process (to avoid agglomerates). Sonication deserves attention to avoid protein denaturation and can be sluggish to obtain a homogeneous dispersion of nanoparticles [23, 41].

In higher nanoparticle concentrations, the scattered light from one particle interacts with others, losing intensity before reaching the detector, which is called multi-scattering phenomenon. Therefore, the size measured will be smaller, underestimated. Moreover, particle agglomeration is favored in high concentrations. On the other hand, if the sample is too diluted, the generated signal from the scattered light might be too low to analyze. To obtain the optimal concentration is important, but it depends on the instrument and the sample conditions. Factors such as scattering volume, scattering angle, laser power, detector sensitivity, molecular weight, compactness, and shape of the material directly influence the required concentration for the measurements [23].

Colored and fluorescent samples should be avoided to have their diameter measured by DLS. Since it is possible for the sample to absorb the light at the wavelength of the incident beam, the intensity of the scattered light will be lower, and a smaller size will be determined, underestimating the value. In the case of fluorescent light, it is non-coherent and recorded as noise, resulting in low-quality data and broadening of peaks [24].

Analyzing the Results

The DLS technique gives us results regarding the intensity of the light scattered as a function of the particle size distribution and uses specific algorithms to convert them in the contribution by volume and number.

In order to better understand the size distribution as a function of intensity, number and volume, consider as an example a sample containing nanoparticles with diameters of 5 and 50 nm in equal quantities. Distribution by ***number*** would generate two peaks with same relative percentage (1:1) centered at each size value, 5 and 50 nm. If these data are considered as a function of ***volume***, the peak of 50 nm would have a contribution 1,000 (10^3) times higher than the peak of 5 nm, because the volume of the sphere is proportional to its radius (r) raised to the cube (r^3). Finally, in the distribution by ***intensity***, the information directly provided by the DLS technique, would show a contribution 1,000,000 (10^6) times higher for the peak of 50 nm, since, by the Rayleigh equation (Box **2**), the intensity of the scattered light varies with r raised to the sixth power (r^6) (Fig. **6**) [42].

Thus, the more monodisperse the system is- with a narrow distribution range of the sample size – the more reliable the particle diameter result (the lower the standard deviation). In polydisperse systems, the particle size distribution will be defined using algorithms present in the equipment software to aid in the analysis, such as Contin [43], already present in the software. Anyway, analysis of polydisperse samples by utilizing the intensity distribution might not correspond effectively to the number of particles present in each distribution.

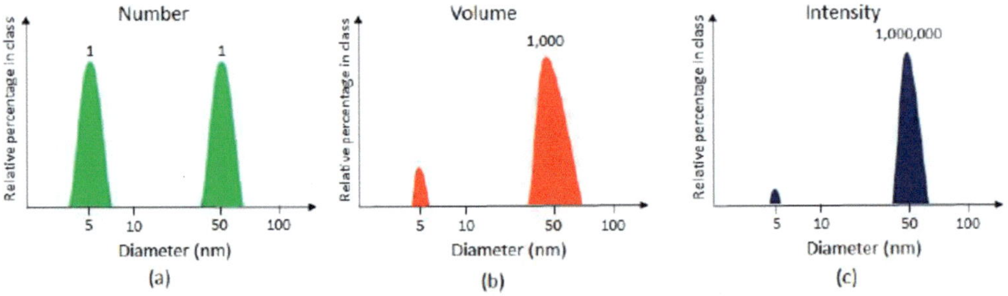

Fig. (6). Particle size distribution by number (**a**), volume (**b**) and intensity (**c**).

PDI can be obtained from the width of size distribution. If a Gaussian distribution is assumed, PDI is the relative variance, associated with the standard deviation (σ) and to the R_H, as demonstrated in equation 3 [44]:

$$PDI = \frac{\sigma^2}{(R_H)^2} \tag{3}$$

Therefore, high PDI values indicate heterogeneity of the suspended nanoparticles diameter and low PDI values suggest that the particle diameter distribution has a narrow range, indicating a tendency to monodispersion. If the sample is spherical, reasonably narrow and monomodal, PDI should be below 0.1. Samples with a PDI up to 0.3 are considered to have a narrow size distribution and Z-average size and PDI should be used for comparative purposes to samples measured in the same conditions. In broader distributions, with PDI over 0.7, the sample is so polydisperse that Z-average values are not reliable and a distribution analysis should rather be performed [23].

Laser Diffraction

Laser diffraction (LD), also known as small angle static light scattering, has been applied for size determination of particles ranging from hundreds of nanometers up to several millimeters, and mixtures of those. This technique uses the fact that depending on the particle size, the pattern of diffracted light should be different.

That is what happens when a laser beam (light source) hits a particle and is detected by a ring detector system [45].

However, besides diffraction, other phenomena are observed: reflection, refraction, absorption, or re-radiation (as heat or light) and the detector cannot distinguish these more complex scattering patterns, which spread light in every direction [46]. The scattering patterns depend on the ratio of particle size to the incident light wavelength and, thus, three shapes of scattering can be distinguished: Fraunhofer, Mie and Rayleigh scattering [47] (Box **2**).

Box 2. Shapes of scattering patterns.

According to the ratio of particle size to the incident light wavelength, three types of scattering can be shown. If the particle is much larger than the wavelength, it will show a Fraunhofer scattering pattern, if the particle is much smaller (>10 times), it should show the Rayleigh scattering. In the intermediate region, particles show Mie scattering (Fig. **2**).

The pattern showed by smaller particles is the Rayleigh scattering, characterized by having the same intensity of forward scattered and backscattered light. The intensity of forward scattered light increases as the particle size increases. Thus, in Mie scattering, the ratio of forward scattered to backscattered light intensity is higher, and even higher in Fraunhofer scattering, which is associated with strong forward scattering and weak backscattering intensity.

Rayleigh scattering occurs when the particles have a size less than 1/10 of the incident light wavelength. Above this ratio, the scattered light should have more intense energy in the same direction as the original beam, characterizing Mie scattering [14].

Considering that the scattering phenomenon is complex, the application of the Mie formula (Equation 4) is important to obtain the correct analysis. This formula has, among many others, three influence factors: the particle radius (A), the angle of scatter (W) and the optical parameters (m), the imaginary and real refractive index (expressed as complex refractive index).

$$I(w) = E\begin{Bmatrix} k^2A^4[J1]^2W^{-1} + [K_1W]^1[K_2W]^3[K_3W]^5 \\ +k^4A^6(m-1)^2W^6/8\pi^2 \end{Bmatrix} \quad \textbf{(4)}$$

where I = intensity of scattered light, E = flux per unit area of the incident light, k and K = constants, J1 = first order Bessel function of the first kind. Fraunhofer term is highlighted in blue and Rayleigh term in green.

Therefore, the particle size can be calculated from the scattering angle and the optical parameters. In Fraunhofer approximation, when the particle size is 5-6 times higher than the wavelength of the incident beam, the angle of scatter is very small and then the optical parameters are not required since all the mathematical term is cancelled out. Considering that, in practice, the wavelengths range from 633 to 800 nm, and more usually from 633 to 750 nm, this approximation is applied to particles above 3.8 to 4.5 µm. For particles under this size, all Mie formulas should be considered, and thus the complex refractive index should be included.

Depending on the wavelength of the light source, the lower size limit of detection is around 400 nm; thus, in order to analyze small particles up to 20 nm, additional techniques are associated with enhancing the measuring range for submicron particles. It is possible to gather more information on the scattering pattern by detecting light intensities not only in the forward scatter, but also in other directions, to detect side scattering and backscattering. Therefore, it is possible to determine if the particle generates Rayleigh, Mie or Fraunhofer scattering [46].

In addition, many manufacturers use two or more different wavelengths and the measure is performed twice (with horizontal and vertical polarized light) to analyze the change in the scattering pattern, obtaining a good analysis result. Therefore, LD analysis involves not only LD measurements, but the combination with other techniques through determined algorithms [47].

Thus, LD, as well as DLS, uses the principles of light scattering to determine particle size. In LD, the size distribution estimation is defined by collecting the scattered light intensity as a function of the scattering angle (θ) [45]. This angle is generated when the incident beam collides with a suspended sample with the aid of a series of detectors placed at different spatial locations. Thus, it covers a defined extent of dispersion angles of incident light (Fig. **7**).

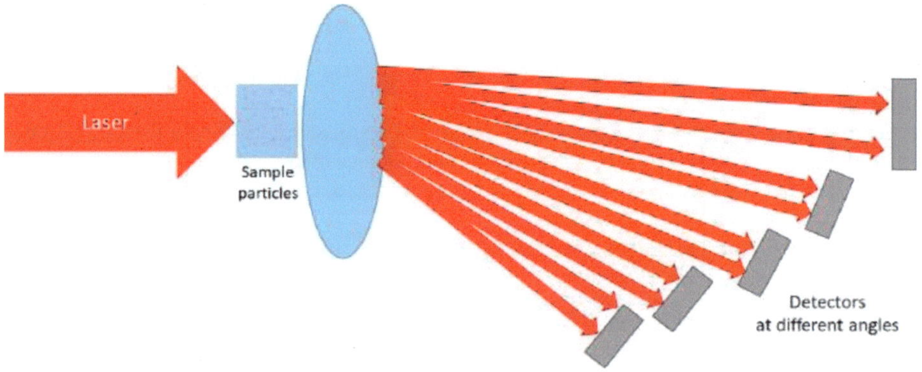

Fig. (7). Spatial arrangement of LD detectors.

As an example, one can imagine two spherical particles of equal size exposed to a beam of parallel monochromatic light (Fig. **8**). The light, when scattered at equal angles, remains parallel and will be focused by an adjustable lens on a specific point in the plane of detection. Each particle size will have its own characteristic dispersion patterns, since particles with equivalent size will disperse light at the same deviation angle, as a fingerprint related to particle size [46].

One of the detectors will be aligned with the specific light scattering angle generated by particles of equal size. This signal point is collected by detectors scattered at different angles, added and combined with other possible points generated by light scattered at other angles by particles of distinct sizes. Thus, it is created a scattering pattern that is independent of the particle position and its motion in the optical field. This happens since, at any point where the particle is passing through the light beam, the angle of light deviation will be the same [46].

This pattern directly correlates the intensity with the light angle deviated by particles of different sizes contained in the sample, generating results that will be expressed as a function of the size distribution (Fig. **9**).

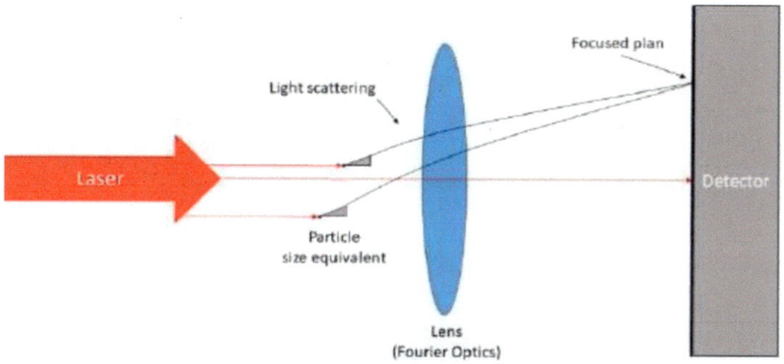

Fig. (8). Parallel scattering of light by particles of the same size and after focusing to a detector by an adjustable lens.

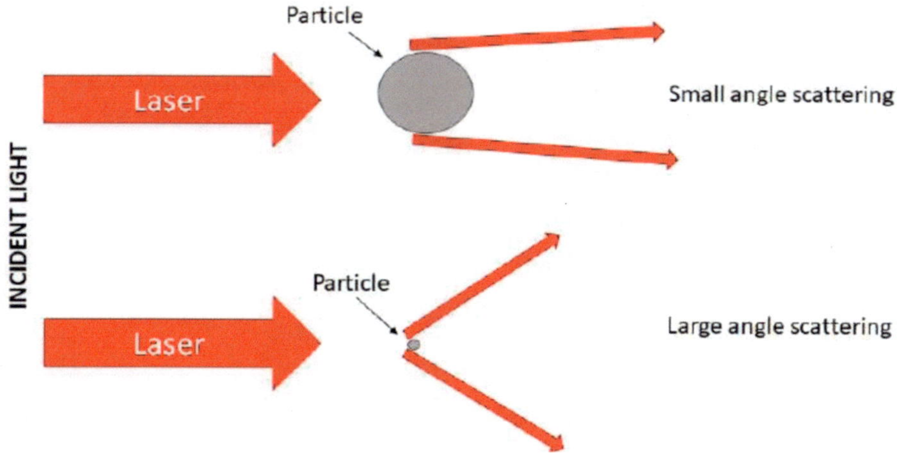

Fig. (9). Angle of deviations (θ) of the light from different particle sizes.

The success of this technique is based on the fact that it can be applied to various types of particle systems, since the equipment has specific compartments, known as sample dispersion units, enabling analysis of materials in the form of dry powders, suspensions, emulsions or even in aerosol [48], in the size range of 10 nm to 3500 μm [45]. The experimental procedure is rapid and there is a variety of commercially available instruments.

However, in general, LD equipment is high-cost, not very compact, and does not have the necessary resources for the determination of ZP, which is usually present in DLS systems. Therefore, the versatility of DLS equipment is higher than LD equipment and should also be considered. For multimodal samples, LD is more useful to evaluate if the nanoparticulated system contains larger particles than to determine the size distribution.

When comparing DLS and LD techniques, the results obtained are not always the same and often differ between 10 and 20%, depending on the experimental conditions and the type of nanoparticle. While the lower limit of detection is smaller for DLS, the upper limit of detection is higher in LD. Therefore, for smaller nanoparticles (for example, ≤ 50 nm), DLS provides better data, whereas for larger particles (*e.g.* ≥ 1 μm) LD is more adequate [14].

Equipment

Laser diffraction instruments are basically composed of one or more light sources, typically a low power laser (red laser – 4 mW He-Ne, 632.8 nm; blue LED laser – 10 mW, 470 nm), an arrangement of the convergent lens (Reverse Fourier Optics), a sample holder compartment, detectors positioned at variable angles, a correlator, and a computer to analyze the results (Fig. **10**) [45].

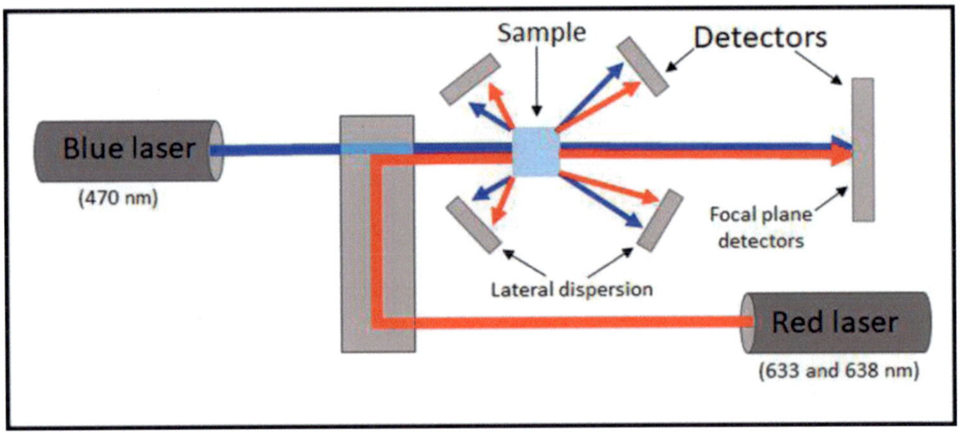

Fig. 10. Schematic representation of LD equipment.

Performing the Technique

The equipment available in the market is relatively simple to handle; however, it is necessary that the operator has attention in some steps of the procedure.

Sample Preparation

Usually, the most modern equipment will automatically define the correct sample concentration by providing the feedback in real time when the ideal condition is reached in the dispersion units; if it is not the case, the sample should be prepared by dispersing it in the correct concentration [45].

Selecting the Dispersant

One of the crucial steps to obtain reliable results in particle size distribution is the appropriate choice of the dispersing medium. That happens because the refractive index of the medium must be known and the sample should present its real individual size and not dissolve in the solvent. The dispersant may be any liquid which is clear and optically uniform at the wavelength used in the analysis, as long as it does not interact with the sample by changing its original size. The most commonly used dispersants in the analysis routine are water (refractive index 1.33), ethanol (refractive index 1.36), isopropyl alcohol (refractive index 1.39), acetone (refractive index 1.36), butanone (refractive index 1.38) or hexane (refractive index 1.38) [41].

Choosing the Dispersion Unit

Each sample has a dispersion unit suitable for properly disperse each type of sample (liquid dispersions or aerosols) in an ideal dispersing medium. The objective of the sample dispersion units is to ensure that the sample is pre-treated by being properly dispersed prior to interact with the light beam [45].

The equipment available on the market may have fixed dispersion units (in older models), or have dispersion units that are easy to replace (more modern and versatile models), increasing the possibility of analyzing various materials. These models generally have individual dispersion units, compatible with the analysis software. They can be purchased individually according to the needs of the formulator regarding the samples that will be characterized. Some dispersion units may have a sonic probe (40 W) for fast dispersion of agglomerates, helping to generate results that represent the real particle size distribution [45, 49].

After prepared, the sample is taken to the optical field by pumping, if particles are dispersed in liquid; by vibration and flow, if dry powder, and by spray, for aerosol particles. In the optical field, the sample will collide with the light beam and will be deviated towards to the detector. If there are errors in the preparation, like if the sample is not representative, poorly dispersed, or exhibits agglomerates, the obtained results will not be trustworthy [45].

Therefore, when the sample is properly prepared, the obtained results are more likely to be representative of the real size distribution of the sample.

Analyzing the Results

The oscillations of light intensity as a function of the angle θ will be correlated in the equipment software by specific mathematical algorithms, varying according to each manufacturer [50].

The results will be presented as a histogram of particle size distribution, *i.e.*, frequency of particles in function of size (Fig. **11**). The histogram divides the results in discrete size ranges along the horizontal axis. The number and size of ranges are determined in the equipment software. The height of the vertical bars corresponds to the volume percentage of the particles in each range. The curve of particle distribution is therefore constructed by connecting the tops of the bars with the median point of each size range. The shape of the graph may then follow a Gaussian trend or may present an asymmetrical shape [45].

The data can also be presented by a cumulative particle size curve (Fig. **12**). In this case, the horizontal axis scale represents the mean diameter of the particles and the vertical axis represents the cumulative frequency, varying from 0 to 100%. This form of presenting the results enables the formulator to obtain information about the percentiles of particles above or below a specific diameter value, allowing optimization in the development and characterization process [45].

Knowing the forms of presenting the results inherent to each instrument, as well as the editing and adjustment tools of the software, is of great value to the formulator. Through the careful analysis of the results, LD can also be a useful tool to conclude when the formulation developed meets the size and polydispersity specifications.

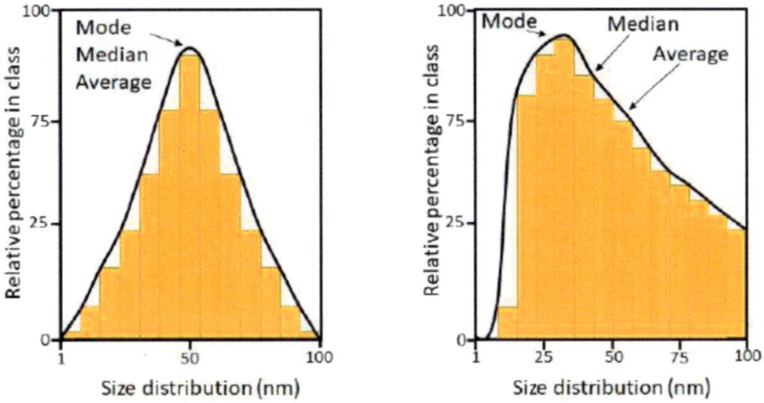

Fig. (11). Particle size distribution histogram.

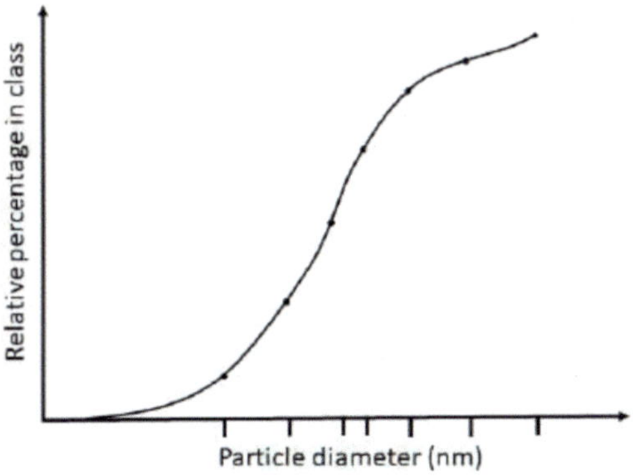

Fig. (12). Cumulative particle size curve.

Zeta Potential

ZP is a very useful tool to understand and characterize colloidal suspensions, being directly associated with colloidal system stability (Box **3**). It is expressed as a measurement of the electrostatic attraction or repulsion between ions of the dispersing medium and the particles. The interaction between particles and polar liquids is governed not only by their surface electrical potential but by the effective relationship between the particle (and associated ions) to its colloid nucleus, which is present in charged layers over the surface and in the surroundings of the particle [51].

Box 3. Stability of colloidal systems and DVLO theory.

Particles suspended in a liquid medium exhibit Brownian motion (refer to section 2.1 for more detailed explication), which generates constant collisions between them.

The DVLO theory (developed by Derjaguin, Verwey, Landau and Overbeek in 1940) suggests that colloidal system stability is dependent on the particles total interaction energy which in turn depends on some competing contributions (Equation 5):

$$W(D) = W(D)_A + W(D)_R + W(D)_S \quad \textbf{(5)}$$

where $W(D)_A$ = attractive interaction energy due to van der Waals interaction, $W(D)_R$ = repulsive interaction energy due to electric repulsion, and $W(D)_S$ = energy due to the solvent, which has a marginal contribution. Among other factors, $W(D)_R$ also depends on ZP.

Thus, an energy barrier resulting from the repulsive force prevents two particles from approaching and aggregating (Fig. **13a**). But if, due to Brownian motion, the particles collide with enough energy to overcome this barrier, the attractive forces overcome the repulsion generated by the surface charges. This collision will pull the particles into contact by sticking together in strongly and irreversibly attached aggregates. Therefore, the system is less stable [52, 53].

If ZP is reduced (in high salt concentrations), reversible adhesion between particles can take place (Fig. **13b**). Despite these flocs are weak, they are not broken up by the Brownian motion, but they may be dispersed if a vigorous agitation or another kind of energy is applied.

On the other hand, if the particles have a charge that generates sufficient repulsion to overcome the attraction forces, the dispersion will resist flocculation and the colloidal system will be stable. However, if the system does not exhibit sufficient repulsion energy, flocculation will eventually occur.

From the DVLO theory, the conclusion is that the stability condition of a colloid system is not dependent only on ZP, since it only provides information about the electrostatic repulsive forces, but also on van der Waals and other interactions. Therefore, it is not uncommon to find stable colloids with low ZP and unstable colloids with high ZP values [52 - 55].

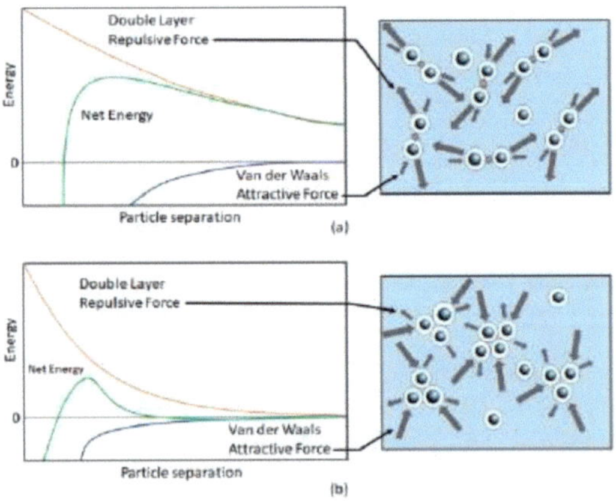

Fig. (13). Interaction forces between two dispersed particles. (**a**) Repulsive forces preventing two particles from approaching and aggregating. (**b**) Attractive forces favoring these phenomena.

A charged particle dispersed in a polar liquid has a layer of opposite charged ions attached to its surface, named the Stern layer, and a second diffuse outer layer composed of moderately associated ions. These two layers are called a double electric layer. As the particle moves (due to gravity, kinetic energy and/or under an applied electric field), there is a limit between the ions in the diffuse layer (outer layer) that moves with the particle and the ions that remain interacting with the dispersing medium. The electrostatic potential at this boundary of the slip plane (an imaginary plane that separates immobile ions on the surface from the mobile ions in solution) can be defined as the ZP [56].

The potential of the particle surface is known as the Nernst potential (ψ_0) and it is difficult to be measured. The electrostatic field decreases in dispersion as the distance from the particle surface increases, and thus the surface potential at a distance x from the Stern layer can be defined as by Equation 6 [54]:

$$\psi = \psi_d e^{-\kappa x} \tag{6}$$

where ψ = surface potential at distance x from the Stern layer, ψ_d = surface potential at Stern layer, κ = Debye-Hückel parameter, x = distance. When the slipping plane is in close proximity to the Stern layer, $\psi_d \approx$ ZP.

Considering a negatively charged particle (Fig. **14**), some of the positive ions present in the dispersant will form a tightly attached layer around the surface of the particle. This layer of counter-ions is the Stern layer. Additional positive ions are also attracted to the negative colloid center, but they are repelled by this positive Stern layer as well as by other positive ions that are also trying to approach the negative particle. This dynamic balance results in the formation of a diffusive layer of counter ions. This layer has a high concentration near the surface that gradually decreases with the distance until the point that reaches equilibrium with the ions of the dispersing solution [24].

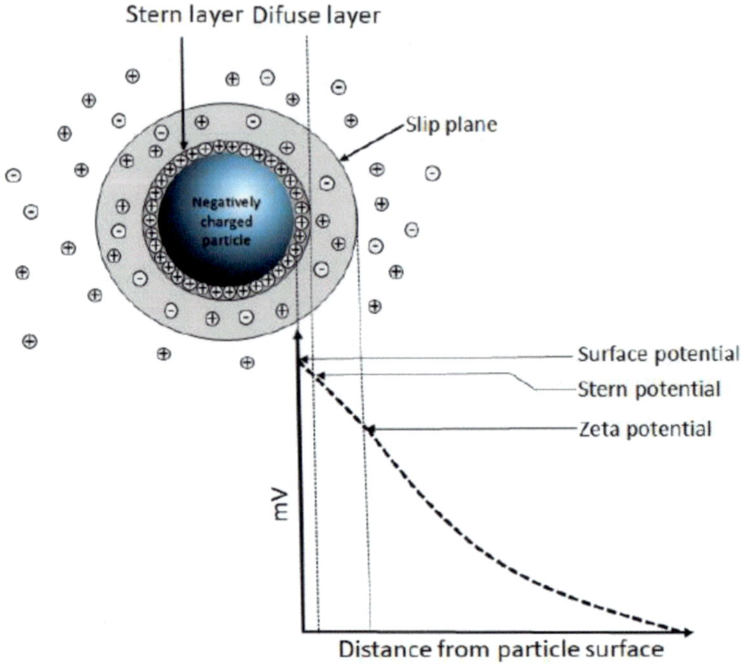

Fig. (14). Representation of the negative charges in a particle surface.

ZP is, therefore, the difference between the electric potential at the shear surface of the particle and the electrical potential of the solution [51]. This is calculated by tracking the movement of suspended charged particles by the influence of an electric field. When an electric field is applied through an electrolytic medium, the charged particles suspended in the medium are attracted to the oppositely charged electrode (electrophoresis). Viscosity forces act on these moving particles opposing their movement. When equilibrium is reached between these two opposing forces, the particles move with constant velocity.

This velocity depends on the strength of the electric field or voltage gradient, the dielectric constant of the medium, the viscosity of the medium, and the ZP. In the measurements of ZP, an electric field is applied throughout the sample and the movement of the particles (electrophoretic mobility - U_e) is measured by the light scattering of the particles applying the DLS technique (see item 2.1) [57].

The Henry equation is applied to calculate the ZP (z) from U_e (Equation 7):

$$U_e = \frac{2\varepsilon z f(ka)}{3\eta} \qquad \textbf{(7)}$$

where ε = dielectric constant, η = absolute zero-shear viscosity of the medium, f(κa) = Henry function, and κa = measure of the ratio of the particle radius to the Debye length.

The parameter κ is the Debye-Hückel parameter, and it depends on the ionic strength. Hence, with increased ionic concentration by the addition of electrolytes, the double layer gets compressed and the ZP decreases. The κ units, called the Debye length, are the reciprocal lengths and κ^{-1} is often considered as the measurement of the electric double layer thickness. The parameter 'a' refers to the radius of the particle, and, therefore, κa measures the ratio of particle radius to the thickness of the electrical double layer (Fig. **15**) [58].

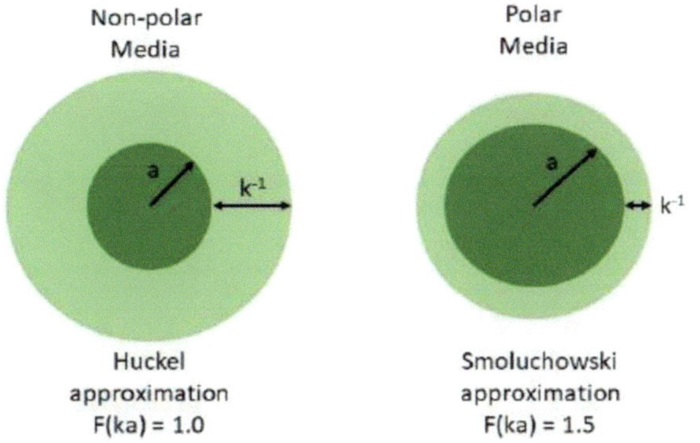

Fig. (15). Huckel and Smoluchowski approximation used to convert electrophoretic mobility in zeta potential.

Electrophoretic determinations of ZP are most commonly made in aqueous media and moderate electrolyte concentrations. f (κa) in this case is 1.5, and this is referred to as the Smoluchowski approximation. Therefore, the calculation of ZP from the mobility is direct for systems that fit the Smoluchowski model, *i.e.*, particles larger than 200 nm dispersed in electrolytes containing a concentration higher than 10^{-3} M. For smaller particles in a low dielectric constant medium (*e.g.*, non-aqueous media), f (κa) is 1.0 and is equally simple for the calculation of ZP [58].

Therefore, U_e (generally expressed in volts/cm) is the velocity (in m/sec) for unit field strength and is considered as a relative measure of how fast a particle moves in an electric field. Its measurement can give a detailed view on the conditions of the dispersion, whether they tend to remain stable, suffer aggregation or flocculation, being important for the development of dispersions, as emulsions and suspensions [23].

Equipment

ZP measuring instruments generally have six main components (Fig. **16**). The first one is a laser source, used to provide irradiation of polarized light on the particles contained in the sample. The incident laser beam passes through the center of the sample cell (second component), which is contained in a chamber in the absence of light and with controlled temperature. In modern equipment, the sample cell is a cuvette, which is inserted in the dark chamber to proceed with the ZP measurement. After striking the particles, light is scattered and collected by a detector (third component) positioned generally at the angle of 13° [58].

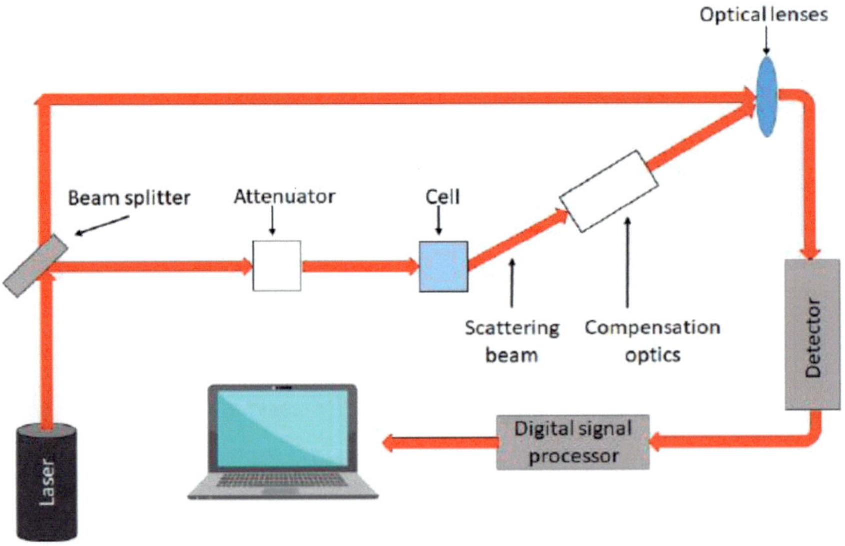

Fig. (16). Schematic representation of ZP equipment.

When an electric field is applied to the sample cell, the particles present there move towards the opposite charged electrode and cause fluctuations in the detected intensity in a certain frequency, proportional to the velocity of the particles. This information is passed to the fourth component, a digital signal processor, and subsequently, the information is analyzed in the software, the fifth element of the instrumental part. In order to successfully detect the intensity of light dispersed by the particle during movement, it must be within a specific range. To maintain the signal within this ideal range, the equipment uses an attenuator (sixth component), which acts to adjust the light intensity, making the signal possible to be detected [56].

It is worth noting that the sample cell, usually a cuvette, is an external important component of the equipment. It has a U-shaped channel, which has two ending metallic plates capable of conducting an electric potential to the solution. These cells are made of acrylic material with a known refractive index, which must be informed to the software before ZP analysis [23].

Performing the Technique

The most commonly used method for measuring ZP is based on the technique of capillary electrophoresis, where ZP is defined as a function of the Ue of a nanoparticle dispersed in a sample [58].

Initially, prior to ZP characterization of a sample, the analyst should consider choosing the appropriate cuvette for ZP analysis (Fig. **17**) and set in the software the parameters related to the specificity of each sample, such as dispersing medium and temperature. Analysis can be performed in a manual single analysis mode, or when analytical routines are used in the same conditions. For that, Standard Operating Procedures (SOP) are usually created. These SOP measurements make the analysis faster and minimize the risk of errors in the settings; therefore, they are very useful in quality control routine analysis [23].

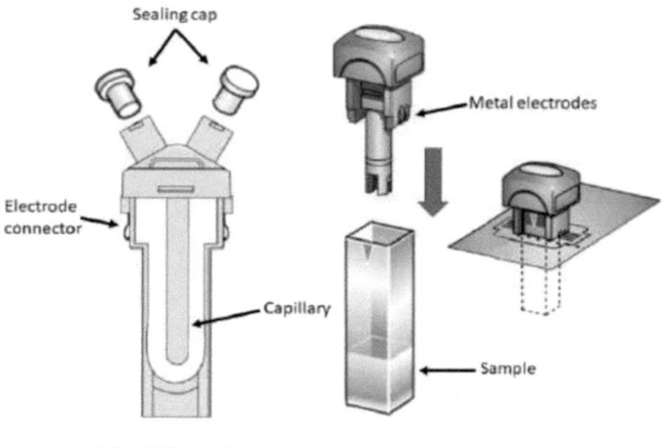

Fig. (17). Different cuvettes used for ZP analysis.

Once the ideal conditions are defined, an electric potential is applied to the sample by electrodes at the ends of the capillary. The charged particles will migrate to the opposite charge electrode at a rate that depends on the magnitude of the charge, the system viscosity and the electrolyte concentration present in the dispersant [23].

Since U_e of the particle is obtained and it is related to the known electrical potential applied, using Henry's equation is possible to obtain ZP of nanoparticles, with the help of two other known constants of the sample: viscosity and dielectric constant [56].

The characteristics of the particle surface and the interaction of charge with the dispersing medium are also fundamental in the ZP measurements and can influence stability of nanosystems, due to the possibility of agglomeration or even interaction with other components if biological samples are used (these influences are described in more detail in Box **3**.

Analyzing the Results

ZP measurements provide accurate analysis of the electronic state of nanoparticles surface and the data obtained can be used to predict the stability of formulations containing these nanoparticles, among other information [51]. However, these ZP measurements are complex because they depend on a number of factors, the main ones being described in Box **4**.

The instability of the formulation may arise from the interaction between weakly charged or uncharged nanoparticles, leading to a tendency to form aggregates. Formulations having absolute values of ZP between 0-10 mV, 10-20 mV, 20-30 mV and $\geq$ 30 mV are considered highly unstable, relatively stable, moderately stable and highly stable, respectively [54].

Therefore, as a general rule, ZP values higher than +30 mV and lower than -30 mV are indicative of stable conditions, while values between -30 mV and +30 mV indicate unstable conditions, favoring aggregation, flocculation and coagulation [51, 54].

However, despite the rule, the reality regarding the study of nanosystems stability is more complex than just the interpretation of ZP values, also considering the sum of the van der Waals attractive forces and repulsive electrostatic forces generated by the double layer effects [58]. On the other hand, the presence of a long hydrophilic chain, usually a polyethylene glycol chain (PEG) on the surface of the nanoparticle, which provides steric stabilization and makes monodispersed particles, can also enhance the stability without changing the superficial charge and ZP.

Box 4. Factors influencing ZP measurements pH.

If the sample is dispersed in an aqueous medium, which commonly happens, pH is one of the most important factors influencing ZP values. A ZP value without the description of pH becomes a practically meaningless number [14, 23].

To better understand the influence of pH, a suspension particle with a negative zeta potential can be taken as an example. When an alkalizing agent is added to the suspension, the particles tend to acquire an even more negative charge. On the other hand, if an acidulating agent is then added to this suspension, there will be a moment at which the negative charge will be neutralized, known as the isoelectric point. In the isoelectric point, the ZP of the system is close to zero, a condition of less stability, favoring the formation of aggregates. Any addition of more acid may cause a positive charge accumulation on the surface of the particles, thereby reversing the ZP values.

Therefore, a zeta potential *versus* pH curve starts from positive values at low pH, decreasing to negative values at high pH. The point when the curve passes through zero ZP is the isoelectric point (Fig. **18**) [23, 58].

Concentration of Formulation Components

The effect of the concentration of a formulation component on the ZP can provide information that assists the development of a product with adequate stability. The formulation components, mainly in the development stages, are tested in variable proportions to obtain ZP values proper for stable systems. For instance, different concentrations of lipophilic amines in the composition of solid lipid nanoparticles (SLN) yielded nanoparticles with different ZP, measured in the same conditions. In the same way, the addition of a coating can neutralize the nanoparticle charges and then reduce ZP [29, 59 - 61].

Therefore, the utilization of cationic or anionic ingredients in the surface of nanoparticles, such as surfactants or some polymers, can make the ZP more positive or negative, respectively, if it is measured in the same conditions.

Concentration of Particles in the Sample

Samples with a lower concentration of suspended particles favor the passage of the light beam, making more accurate the measurement of U_e [14]. However, ZP is highly sensitive to changes in the constitution of the formulation. For example, making a dilution for the measurements while improving detection of U_e, could alter the pH or ionic strength of the medium, and the results obtained would not be the same as those of the original suspension of nanoparticles. Therefore, a dilution should be carried out only if necessary and with caution, paying attention to the ideal conditions of the sample type and the limits of the equipment chosen for ZP analysis [23].

Conductivity

κ^{-1} values (thickness of the double layer) depend on the concentration of ions in solution and can be calculated from the ionic strength of the medium. The higher the ionic strength, the more compressed the double layer. The valence of the ions will also influence the thickness of the double layer. A trivalent ion such as Al^{3+} should compress the double layer more than a monovalent ion such as Na^{+}.

Inorganic ions can interact with charged surfaces by a non-specific or specific ion adsorption. The specific adsorption of ions onto a particle surface changes the value of the isoelectric point and can also drastically change ZP values. In some cases, this specific ion adsorption can lead to charge reversal of the particle surface [58].

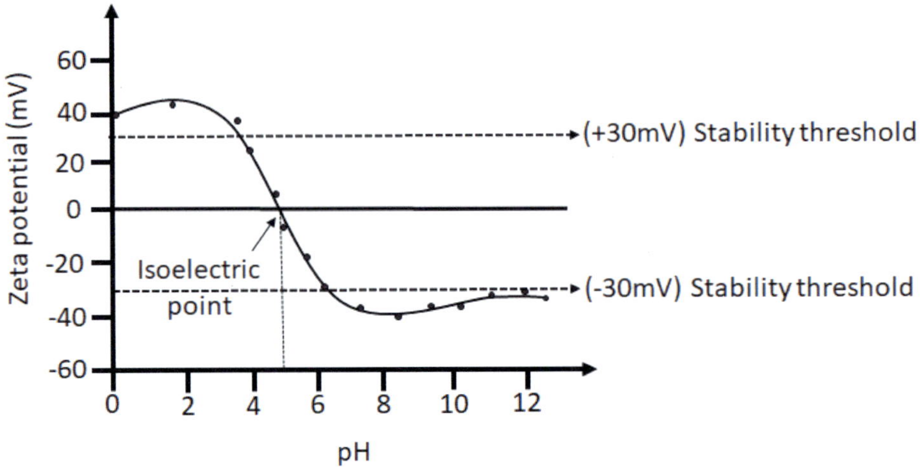

Fig. 18. Influence of pH in ZP values.

MORPHOLOGY

Optical Microscopy

The human eye is able to directly observe objects or details bigger than 70 µm, which corresponds to the minimum distance of separation between two points that are perceived as separate entities. The function of the microscope, in turn, is to produce an enlarged image of the object under observation, containing the desired structural information that can be perceived by the eye-brain system. Some characteristics should be considered when selecting a microscopy system for morphological observation, such as resolution, magnification and contrast [62].

The distance between the object and the detector that records the final image, the aperture size of the imaging system and the nature of light are the three basic factors that limit the resolution of an optical microscope [63]. This resolution is about 0.2-0.3 µm and such a limit can be calculated using the Rayleigh criterion [64]. This criterion is given by Equation 8:

$$R = \frac{0.61\,\lambda}{NA} \tag{8}$$

where R = maximum resolution, λ = wavelength of light in the sample and NA = numerical aperture of the lens.

Resolution is also limited by light diffraction, so an alternative to improve resolution is to decrease the wavelength of the radiation. A simpler way is to use immersion oil, which reduces the wavelength of light to be focused and diffracted [65].

The magnification required to detect an object of diameter D must be such that the magnified diameter of the object (M x D) is equal to the resolution of the object in the eye. Therefore, in Equation 9:

$$M = \frac{\Delta R}{D} \quad \textbf{(9)}$$

where M = magnification, ΔR = resolution of the object in the eye and D = object diameter.

Considering the maximum eye resolution (70 μm) and the resolution limit of the optical microscope, we have that this equipment allows us a magnification around 300 times higher in comparison to the naked eye [65]. Moreover, the way light interacts with the object influences the formation of the image and its contrast [62].

The optical microscope can be used in two basic modes: transmission or reflection; the first being used for transparent samples and the second for opaque ones [62, 65]. The composition is basically the same for the two modes and is outlined in Fig. (**19**). The main constituents are the light source, the condenser, the objective lenses, and the ocular lenses.

The light beam passes through the first lens system towards the condenser, which will concentrate the light uniformly in the illuminated area of the sample. A field diaphragm is placed before the condenser to control the size of the illuminated area [62]. The general rule is to open the field diaphragm sufficiently that only the area of the object under observation is illuminated. This is because if the aperture is too wide, the sample contractor will be reduced due to stray reflections inside the microscope. The condenser diaphragm, in turn, controls the angle of the cone of light that illuminates the object and also assists in reducing stray reflections [63].

The objective lenses collect light from the sample and form the primary image, being the main responsible for the resolution and magnification of the image. The primary image represents a real and magnified image of the object under observation but inverted. Ocular lenses are responsible for collecting the diverging rays and producing the final image. The final magnification of the image is given by the product of the objective lens magnification by one of the ocular lenses [62]. For instance, an object analyzed with 10x magnification in the objective lens and other 10x in the ocular lens results in a final image with a magnification of 100x.

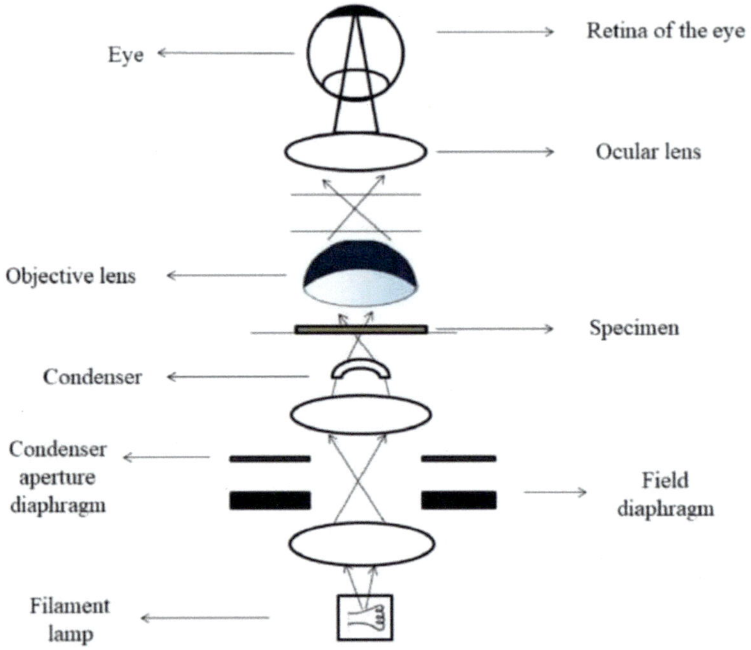

Fig. (19). Schematic representation of an optical microscope.

Types of Optical Microscopy

Light Field Microscopy

This type of microscopy is suitable only for objects that cause amplitude reduction of the incident light by absorption and diffusion, or systems that have quite different refractive indexes. The light deviation is so intense that it prevents light from entering the lens and changes the gray level or color contrast. This way, the object appears dark against a bright field (BF). It is used, for example, in the analysis of dispersion and distribution of pigment particles [62, 66].

Dark Field Microscopy

Dark field illumination greatly improves the ability of the optical microscope to detect small structures or particles, often below the theoretical resolution limit [66]. In this type of microscopy, a diaphragm with an annular aperture is placed below the condenser. This device prevents the direct light beam traversing the sample from being collected by the objective lens, thus only the beams deflected by diffraction and diffusion form the image. If the sample is homogeneous, the light beams do not deviate, do not enter the lens and the field of view of the microscope will remain dark. In heterogeneous samples, there will be diffraction and diffusion of light and some of these deflected beams will be collected by the lens to form the image. Thus, heterogeneities in the sample will appear bright on a dark background, making this technique advantageous for detecting heterogeneities of very small dimensions [62].

Phase Contrast Microscopy

This technique was developed to generate contrast in the images of phase objects, which are objects that do not absorb light. A phase ring is placed in the back focal plane of the objective and allows the beam passage directly from the light to be delayed or advanced in relation to the beams that undergo diffraction, reducing its luminous intensity. Thus, if the sample is homogeneous, there will be no light diffraction, all the beams pass through the phase ring and will be collected by the objective lens, resulting in an image with uniform luminosity. In heterogeneous samples, the diffracted beams pass out of the ring while the direct beam passes through the ring to be delayed or advanced and to match the diffracted beams. The interference of the two groups of beams gives rise to an image with adequate contrast [62].

Polarized Light Microscopy

Polarized light microscopy explores the anisotropic optical properties of a sample, such as refraction and absorption. The anisotropy of the refractive index is called birefringence, whereas the anisotropy of the absorption coefficient is called dichroism [67]. The inclusion of a polarizer and a compensator before the condenser and the inclusion of an analyzer, and a graduated circular rotary stage before the ocular lens, transform an optical microscope into a polarized light microscope [67, 68].

Birefringent or optically anisotropic structures appear bright against a dark background. By rotating the sample between the crossed polarizers, it is possible to recognize birefringent components and determine their axis orientations. Birefringence occurs when there is a molecular order, a non-random molecular orientation, as in crystals or in aligned polymers [68].

Applications

Optical microscopy applications may include analysis of biological tissues, liquids, dispersions, powders, fibers, and polymer films. A common application is the observation of defects in polymer films. The cross-sectional observation of these films provides information on the number and thickness of the layers that make up the film, as well as information about the presence (or absence) of entrapped material within the film. Phase contrast microscopy is extremely useful in observing polymer blends as it allows for improved visibility of low contrast objects. Materials that exhibit birefringence, such as drug crystals, can be studied using polarized light microscopy to provide information on their morphology [69].

Osteoselective and biocompatible polymer films were developed by Gulsuner *et al.* [70] and optical microscopy was one of the techniques used to characterize these films' surface. Through this technique, the authors could observe differences in the topography of the films, such as the presence of roughness, when different polymers were used in the production.

Joksimovik *et al.* [71] studied the influence of the presence of DNA molecules in the formation of silica nanoparticles structured hybrid films. They used polarized light microscopy to observe the influence of DNA on the structure of a dried drop of a nanoparticle colloidal solution. The images showed that the DNA significantly influenced the mechanical properties of the film, reducing the fragility and increasing the elasticity.

Cubic nanoparticles composed of monoolein and chitosan to load clotrimazole were developed and characterized by Verma and Ahuja [72] and polarized light microscopy was used as one of the characterization techniques. The anisotropic structures exhibited typical birefringence, while the isotropic structures appeared as dark background, confirming the formation of cubic liquid crystalline phases. The low birefringence indicates the existence of some unordered structures and may be due to the chitosan coating of the nanoparticles.

Dark field microscopy was used by Xiao *et al.* [73] to analyze the three-dimensional orientation of gold nanorods with an axial diameter of 25 nm and a longitudinal length of 60 nm. The orientation angles of the gold nanorods were determined after analysis of the field distribution pattern of each oscillation dipole in the defocused dark field image.

Wang *et al.* [74] studied the evolution of gold nanoparticles clusters in human alveolar basal epithelial adenocarcinoma cells (A549) through dark field light microscopy and chromatic analysis approach. The analysis of the brightness/hue values of the dark field images helps in the estimation of the number of nanoparticle clusters. The study collaborates to understand the kinetic profile of endocytosis of nanoparticles under physiological conditions.

The optimization of the spray coating of polymer films was carried out by Bose *et al.* [75] with the aim of producing films with uniform thickness (500 nm to 1 μm) and with low roughness. Light field optical microscopy was used to observe the surface texture of the films.

Scanning Electron Microscopy

SEM is one of the most versatile instruments available for analysis of microstructure morphology and characterization of chemical composition [76]. It allows the observation and characterization of organic and inorganic materials, on a micrometric and nanometric scale. It offers much of the ease of use and image interpretation of a conventional light microscope while providing greater depth of field and image enlargement [77].

Principles of the Technique

In electron microscopy, the light source of the optical microscope is replaced by a beam of high-energy electrons. This electron beam is focused on a minimum point of the sample and used to "sweep" its surface [76]. This electron point is moved at a well-defined speed and based on a pattern of lines named raster. Line by line, the surface of the sample is analyzed, each line being constructed by a large number of elements that will be converted into an image [78].

The interaction between the electron beam and the sample produces signals that will be used in the image formation. These signals are subsequently converted into a visual signal in a cathode ray tube. There are two main interactions between the electron beam and the sample: elastic and inelastic. The elastic dispersion is the result of the incident electron deflection by the specimen atomic nucleus or by electrons of similar energy. Inelastic dispersion results from the transfer of energy from the electron beam to atoms in the sample [76]. There are several electronic signals emitted, including Auger electrons, characteristic X-rays, transmission electrons, absorption electrons, backscattered electrons (BSE), secondary electrons (SE), among others [79] (Fig. **20**).

For surface morphology characterization, the two main types of detected signals are SE and BSE. BSE is primary electrons that have been ejected from the sample and spread elastically at an angle higher than 90°. BSE collects information on the chemical composition of the sample and produce a signal that is useful for image formation, as they show contrast due to variations in chemical composition. SE, in turn, is generated from the excitation of electrons of the sample after being struck by the electron beam. They provide topographic information with a high resolution, allowing the visualization of the texture and surface roughness of the material and have an optimal signal-to-noise ratio [65, 76, 79]. The three-dimensional appearance of the image is a result of the large depth of field of SEM as well as the relief effect produced by SE shadows and the contrast produced by BSE [77].

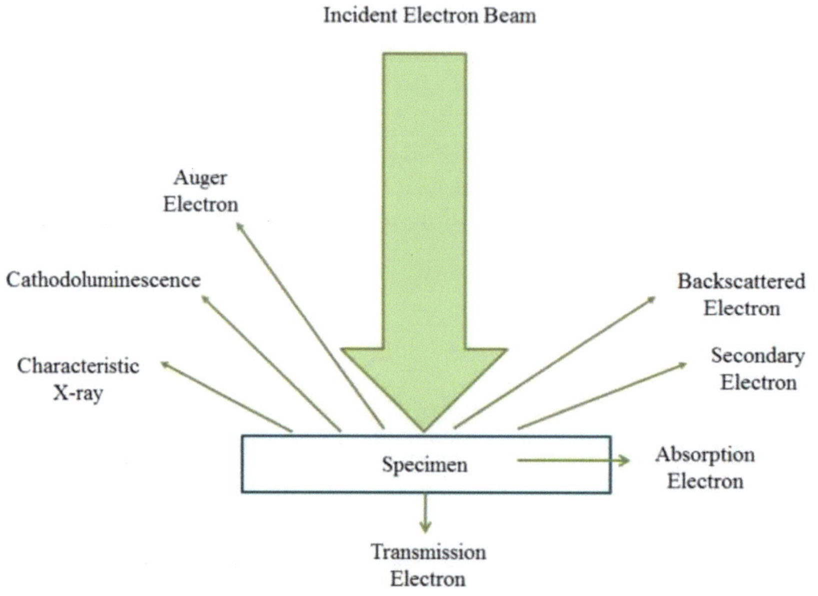

Fig. (20). Representation of the transmitted electronic signals resulting from the interaction between the incident electron beam and the sample.

Regarding the determination of mean diameter by SEM, this is possible by counting and measuring a sufficiently large number of random particles in the sample. The data of many images of particles and multiple samples of the material under analysis can be used to obtain statistical results, the mean and standard deviation of the diameter measurement. Image processing software, such as ImageJ, can be used for processing, analysis, and image measurements [80].

A great advantage of SEM is the ease of sample preparation. There is no need for dilution, solid samples can be observed directly, and if the specimen is conductive, it requires no special preparation before analysis. If the material is hygroscopic or undergoes structural modification with moisture, drying techniques may be required. When the sample is composed of insulating materials, they do not provide a path to the current and can acquire a positive or negative electrostatic charge when exposed to the electron beam. Negative charges represent a major problem, since they repel the incident electrons, deflecting the electron beam and thus causing image distortion, variations in brightness and, consequently, impaired resolution. An alternative to overcome this drawback is to coat the surface of non-conducting samples with a thin metal film, commonly gold; this process is called metallization [65, 78].

Besides the easy preparation of the samples, other advantages of SEM are: the great depth of field, which allows the collection of a considerable amount of structural data; ability to change the orientation of the sample during observation (rotation and inclination); availability of a wide range of magnification and non-mandatory mechanical crushing that will cause sample destruction [78, 81]. The latter allows the morphological and surface information to be maintained and is crucial to confirm if the pores are open to the outer surface in nanoporous materials [81].

Equipment

The two main components of SEM are the electron column and the control console, as shown in Fig. (**21**). The electron column consists of an electron gun and two or more electronic lenses, in addition to vacuum pumps at its base. The control console consists of a cathode ray tube viewing screen and a computer [77]. The electron gun, located at the top of the column, is responsible for producing and accelerating the electrons to an energy level from 0.1 to 30 KeV [76].

Electron sources can be divided into two categories: thermionic sources and field emission sources. Among the thermionic sources, there are the filaments of tungsten and lanthanum hexaboride (LaB_6). Tungsten filaments have both reliability and low-cost advantages; however, they are operated at elevated temperatures and provide low magnification compared to other types of sources [76]. The LaB_6 emitters provide that the same amount of electrons can be emitted at lower heating temperature, exhibiting a 5-10 times higher brightness and having 10 times longer lifetime compared to tungsten filaments [82]. The thermionic sources are relatively cheaper and require a lower vacuum, however, the lifetime is short and the brightness is lower compared to field emission sources. In addition, large energy spread restricts their application [76, 82].

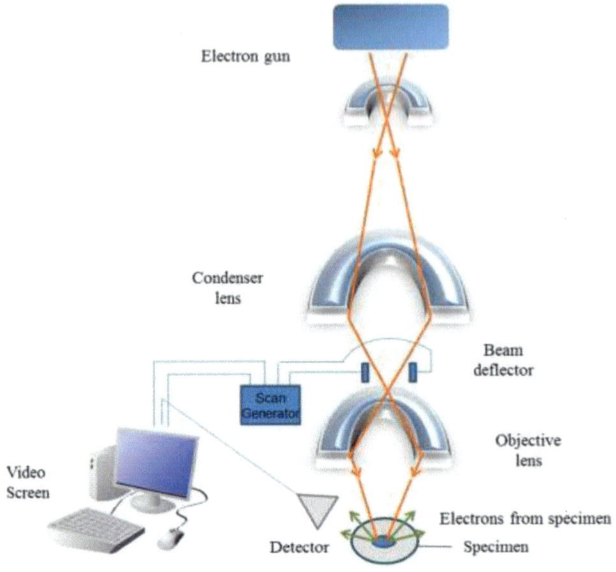

Fig. (21). Schematic representation of a scanning electron microscope.

Regarding field emission sources, these can be divided into two types: cold field emission, which operates at ambient temperature, and thermal field emission, which operates at high temperatures. The latter includes the Schottky field emission source, which has greater stability and the operation is easier than the other thermal field emitters. The best advantage of field emission sources is the high brightness, which is fundamental to improve the resolution, so these sources obtain resolutions in the order of 1-2 nm [82]. All field emission sources require an ultra-high vacuum to stabilize the electron emission and contamination of the electron guns and other components [76].

Electromagnetic lenses are used to focus and define the electron beam at a small point and are necessary since the diameter of the beam produced is too large to obtain high-resolution images. The condenser lens controls the electron beam and the objective lenses are used to focus the electron beam at a point on the sample surface and provide a new demagnification [76]. Most SEM can produce an electron beam in the sample with a spot size of less than 10 nm, allowing obtaining images with adequate resolution [77]. Electron microscopes require a vacuum for proper operation. This vacuum is used both to permit electron source operation and to minimize the dispersion of electrons, except the ones from the sample [83]. In order to reach the necessary vacuum, more than one type of vacuum pump is employed; usually, mechanical and diffusion pumps are used [76].

The SEM operator can control several parameters, among them: the electron acceleration voltage, the working distance (distance between the sample and the objective lens), and sometimes the aperture diameter of the objective lens. The choice of these variables influences the resolution and image quality. The acceleration voltage determines the kinetic energy of the primary electrons and their depth of penetration into the sample. It thus influences the depth of information that BSE provides. Since SE is generated within a very low energy level, the image provided by this signal is considered independent of the choice of kinetic energy. Working distance should be small so that a good resolution can be obtained [65].

Environmental SEM

Environmental scanning electron microscope (ESEM) allows the analysis of several samples, wet or dry, since the environment around the specimen does not need to be in high vacuum [77]. ESEM is widely used to obtain surface images of samples in a vapor environment with nanometer-scale resolution [83].

Low vacuum analyses are possible due to the design of the electronic optical column. In ESEM, the optical column is divided into different vapor pressure zones. These regions are separated by pressure-limiting apertures and allow a differential vacuum [82]. The sample is maintained at a low pressure to prevent dehydration while the column is held in a high vacuum to minimize the scattering of the electron beam [84].

The interaction between the primary electrons and the sample is similar to what happens in conventional SEM. The electrons are accelerated by a detection field and collide with the gas atoms present in the sample chamber, producing additional electrons. This electron cascade is responsible for strong signal amplification [84]. In addition to facilitating signal detection, the gas acts as a conductor of electrical charge, preventing the sample from becoming charged. The collision between the electrons and the gas molecules creates positive ions that can balance the accumulation of negative charge on the sample surface. In addition, gas helps to prevent the evaporation of liquids from a sample. Water vapor is the most common gas used in ESEM due to its efficiency in signal amplification, because it results in images of superior quality and provides good sample stabilization [82].

ESEM can be used for the analysis of non-conducting samples without the need of coating, eliminates the drying step for liquid samples, and it is useful for the study of biological samples [77]. In addition, it allows studies of hydration/dehydration of structures and surfaces by changing the vapor pressure in the sample chamber [85].

Applications

SEM provides information about size, shape and other general aspects of nanoparticles through a direct image. Observation of particles smaller than 100 nm may be difficult depending on the type and configuration of the equipment available. Considering a conventional SEM analysis, the drying step of the material may lead to an underestimation of the particle diameter, since these may shrink during the process. Thus, for size determination, the combination of least two methods of analysis is highly recommended, one of which should be a microscopic method [86]. Most nanomaterials, such as carbon nanotubes and nanowires, can be observed directly by SEM, making sample preparation quite simple. However, non-conductive samples, such as bioorganic nanomaterials, require metallic coating, so sample preparation becomes a bit more complicated [76]. The following studies are examples of SEM application. They show that this is a technique with varied applications and used in different fields of study.

Bootz *et al.* [87] used SEM to determine the size and shape of poly (butylcyanocrylate) nanoparticles. The images showed a narrow size distribution of spherical particles with 167 nm. Two samples were analyzed by DLS and the average diameters obtained were 184 nm (with an additional stabilizer) and 199 nm. The difference in size observed between the two techniques is attributed to the shrinkage of the particles during the drying of the sample analyzed by SEM. Due to this reduction in size, the authors considered that the diameter determined by SEM should be used as the lower limit of particle size.

The influence of carbon nanoparticle morphology and filler content on the mechanical and electrical properties of epoxy matrix nanocomposites was investigated by Leopold *et al.* [88]. Spherical, tubular and layered particles were used and the modified epoxy carbon nanotubes exhibited a high potential for strength increase, but with critical dispersion and purity. The authors used the images obtained by SEM to discuss the toughness increase of the modified material depending on the morphology of the particles used, as well as to determine the mechanisms of damage in the different types of nanoparticles.

Carrillo *et al.* [89] developed cationic SLN capable of forming a complex with DNA plasmids and SEM was used as one of the characterization techniques for evaluation shape and size. In SEM images, SLN appeared as smooth spheres with a size similar to that obtained by LD (342 nm).

Silicon nanoparticles coated with poly (vinylpyrrolidone) were produced by Kwon *et al.* [90] and the field emission SEM was used for size determination and morphology analysis. The particles showed irregular surfaces with a diameter of 50 nm. The authors submitted the system to thermal sintering to remove the poly (vinylpyrrolidone) organic groups from the particle's surface. After this procedure, the particles had an estimated size of 10 nm, a result that agrees with the observation made by TEM.

Dalai *et al.* [91] evaluated the cytotoxicity of titanium dioxide nanoparticles and detoxification in a freshwater system. SEM analysis was performed to study the change in algal cell morphology (*Scenedesmus obliquus*) after the treatment with the nanoparticles, the fixation of the nanoparticles on cell surface and the tendency of the cells to agglomerate.

Transmission Electron Microscopy

TEM provides direct images and morphological, structural and chemical information of nanomaterials with resolution <1 nm. This technique, like SEM, allows the analysis of size, shape and heterogeneity of nanomaterials, besides allowing the observation of aggregation and dispersion [8]. Information on the crystal structure and granularity of a sample can also be obtained [92]. In addition, TEM is a valuable tool for studying intracellular localization and nanoparticle uptake and thereby developing and proposing a mechanism of action for these systems [93].

Equipment

A transmission electron microscope works similarly to an optical microscope. The light source is replaced by a source of electrons and magnetic lenses are used instead of glass lenses [94]. The equipment consists of a sequence of electromagnetic lenses arranged in a vertical column. The electron source is located at the top of the column and a fluorescent screen is present at the opposite end of the column, as shown in Fig. (**22**) [95].

As in SEM, the function of the electron gun is to produce electrons. The sources of electrons used are the same as those mentioned in item 4.2.2. The electrons produced are accelerated by the column with a high voltage, usually between 100 and 300 kV [95]. The electron beam has an energy between 60 and 300 keV; values much lower are not adequate, as the power of penetration in the sample would be insufficient [78, 83].

As in SEM, the function of the electron gun is to produce electrons. The sources of electrons used are the same as those mentioned in item 4.2.2. The electrons produced are accelerated by the column with a high voltage, usually between 100 and 300 kV [95]. The electron beam has an energy between 60 and 300 keV; values much lower are not adequate, as the power of penetration in the sample would be insufficient [78, 83].

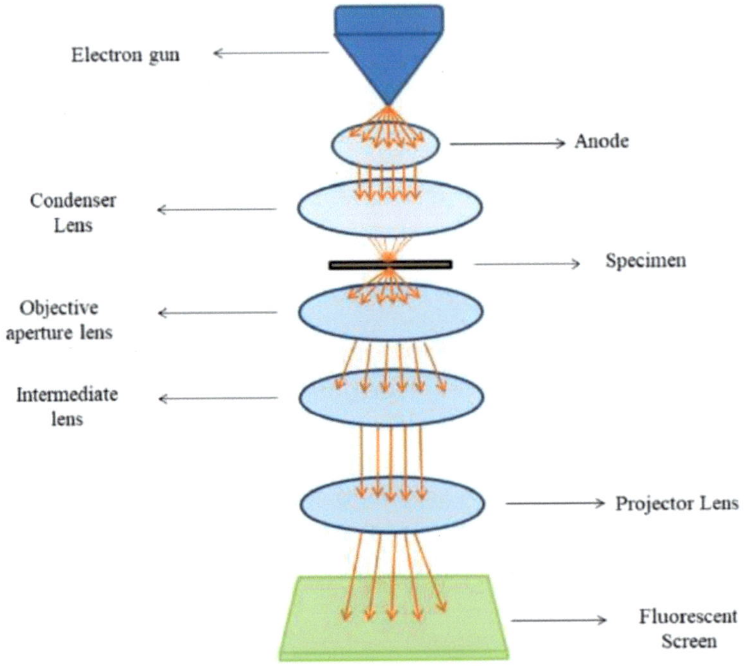

Fig. (22). Schematic representation of the basic composition of a transmission electron microscope.

The accelerated electrons are focused on a small, uniform beam by the condensing lenses. A small aperture of the condenser is used to exclude electrons that, during their trajectory, have large angles in relation to the optical axis. The beam then interacts with the sample and the diffraction pattern formed depends on the thickness of the material under analysis and its transmittance to the electrons [94]. The transmitted electrons are focused by the objective lens and the posterior lenses are responsible for the magnification of the image. By varying the current of these posterior lenses, a magnified diffraction pattern or a magnified image of the sample on the fluorescent screen can be obtained, with final magnification of 10^2 to 10^6 times [95]. This magnification is mainly determined by the ratio between the distance of the objective lens and the sample and the distance between the objective lens and the image plane [8]. A camera located below the screen is used to permanently record the diffraction pattern or the image [95].

Principles of the Technique

TEM, as well as SEM, involves the interaction of an electron beam with the internal structures of the sample. This interaction provides the generation of signals that will later form the image [96]. The technique relies on the transmission of electrons through a sufficiently thin specimen, whose thickness is smaller than 0.5 μm and, in particular cases, as in high-resolution TEM analysis, smaller than 50 nm [8, 83, 96]. This sample thickness is required to allow electron beam penetration and sufficient electron transmission to produce the image [8].

The incident electrons interact with the electrons and nuclei of the sample, being dispersed as a result of these interactions (Fig. **23**). The incident electrons that have not undergone any change in their direction after the interaction with the sample form the beam of transmitted electrons [94]. The sample modifies the phase and amplitude of the transmitted electrons and the resulting image is recorded by scanning the focused beam onto the sample [83].

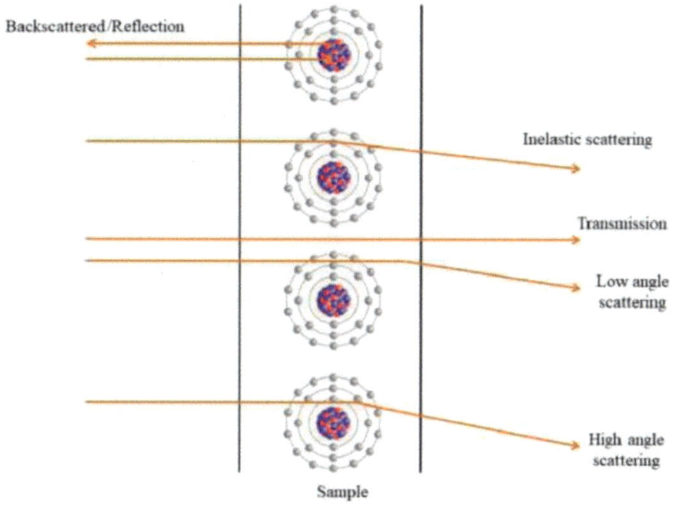

Fig. (23). Representative scheme of the interaction of the incident electrons with the sample atoms.

The major advantage of TEM is the high spatial resolution, which improves morphological and structural analyses. In comparison with optical microscopy, the obtained images have higher magnification and resolution. This is because the wavelength of the electron beam is smaller than that of light, which is associated with the Rayleigh criterion [96], the minimum resolvable detail, limited by diffraction. The equipment allows changing the current of the lenses and, thus, it is possible to obtain patterns of diffraction and image of electrons from the same sample area, only changing the strength of the lenses, which represents another advantage of TEM [94].

In comparison to SEM, TEM has the advantage of offering better resolution and capacity for additional analytical measurements since other analytical techniques can be coupled to TEM for different applications. For instance, energy dispersive X-ray spectroscopy can be used for quantitative analysis of the chemical composition of nanomaterials, while electron energy loss spectroscopy can be used to investigate their electronic structure [8]. For the current generation of equipment with a 300 keV electron beam, the resolution value can reach about 0.08 nm [97].

The need for high vacuum (as well as SEM) and the fact that the samples should be very thin represent disadvantages of TEM [8, 98]. The extensive sample preparation that may be required to obtain thin specimens increases the possibility of structural changes in the sample and makes the TEM analysis process very time-consuming [8]. Like any high-resolution imaging technique, the higher the resolution, the poorer the sampling skill, so only a small part of the sample is observed at the moment. Thus, techniques that offer a lower resolution, but a better sampling, such as SEM and optical microscopy, may be useful for an overview of the sample. Another problem is that TEM presents a 2D image of 3D specimens and, therefore, a single image has no depth sensitivity [99].

Another concern is: as the electron beam used presents high intensity and voltage, it promotes heating of the sample at the site of incidence and, thus, structural transformations, damage and even destruction of the specimen can occur during observation [8, 100]. The electron beam can also initiate nucleation of nanoparticles, so this effect must be carefully addressed in experiments using environmental TEM (see section 4.3.3) [100].

Ideally, sample preparation techniques should ensure that the specimen structure is preserved during the preparation process [98]. Conventional TEM requires the samples to be dry to be vacuum compatible (such as conventional SEM) and this drying process can significantly alter their structural characteristics [100]. The air drying of a drop containing nanoparticles is generally the preferred preparation protocol; however, the formation of artifacts may occur [101].

Domingos *et al.* [101] studied some protocols of preparation of nanoparticle samples for analysis by conventional TEM. The protocol that provided the most reproducible and representative results was the rapid freezing of nanoparticles drops placed on a grid covered with Formvar (commercial type of film used on the support grid), followed by sublimation of the water. Particles analyzed after the utilization of this sample preparation showed similar morphology and dimensions and a small proportion of artifacts compared to air drying of nanoparticle drops.

Like SEM, non-electron conducting samples must also be coated with heavy metals. Even for high electron density nanoparticles, such as metallic colloids, if they have multifunctional components on their surface, such as PEG or targeting antibodies, electron microscopy is not able to provide images of these surface groups without using cryogenic methods [92]. The addition of heavy-metals solutions produces high contrast. This technique is called ***negative staining*** since what is observed is the stain formed by the heavy metal and not the material itself. Negative staining provides useful structural information with a maximum resolution of 1.5 nm and is routinely used. As a disadvantage, the distortion of the material structure under analysis can occur during drying [98].

Normally, the samples are placed on a metal grid that contains a support film, which transmits the electrons and, therefore, does not interfere in the analysis [80, 102]. For determination of the mean particle diameter, it is recommended to record sufficient micrographs to create an image of 500 nanoparticles per sample, from a minimum of two regions widely separated in the grid [103].

The particle size results obtained by TEM may not coincide with other techniques, such as DLS. This is due both to differences in the weighted averages determined in each case (number of particles for TEM *versus* intensity for DLS) and differences in physical properties that are used to obtain the measurements (projected area for TEM *versus* hydrodynamic diameter for DLS). TEM is sensitive to the size of primary particles, while DLS is influenced by the presence of small amounts of large particles, small particle clusters, or adsorbed molecules on nanoparticles' surface [80].

Scanning Transmission Electron Microscopy and Environmental Transmission Electron Microscopy

Scanning transmission electron microscopy (STEM) is a rapid technique for the characterization of nanostructures. It has the same operating principle as conventional SEM, in which an electron beam is focused and scans the sample from point to point [104]. That differentiates STEM from conventional TEM, in which the image is generated by an electron beam interacting with a large area of the sample. However, as in conventional TEM, in STEM, the electron beam passes through a thin sample and the transmitted electrons are collected by a detector located after the sample. Therefore, it can be said that STEM combines TEM and SEM characteristics [105].

Due to electron-sample interactions, the trajectory of the electrons is scattered and different types of signals are recorded [106]. Thus, several detectors can be used simultaneously providing different, but complementary, information and this is one of the main advantages of STEM. The usual detectors include a BF detector that intercepts the transmitted beam and an annular dark field (ADF) detector that surrounds the transmitted beam to collect scattered electrons. Therefore, the BF detector includes the transmitted beam and the holes of the structures appear bright, while the ADF detector excludes the transmitted beam and the holes appear dark. The inner angle of the ADF detector can be altered, which provides higher efficiency for collecting the scattered electrons and improves the dependence of the atomic number on the contrast of the image. This latter configuration is referred to as Z-contrast or high-angle ADF (HAADF) [104].

The resolution in STEM is directly related to the size of the electron probe. A smaller probe corresponds to a higher spatial resolution. However, by decreasing the size of the probe, the beam current is reduced. Therefore, recent progress in STEM has focused on increasing the beam current in small probes. The improvement in field emission sources and the incorporation of spherical aberration correctors considerably increased the beam current density [107].

Moreover, there is also the environmental transmission electron microscopy, also known as *in situ* transmission electron microscopy. This technique differs from conventional TEM because the material is observed in its native environment, which may comprise pressurized gases or liquids under (ultra) high vacuum conditions [108]. The high vacuum in the sample region is obtained by the use of pump-limiting apertures, which separate the sample from the rest of the electron column [83].

For observations in liquids, it is necessary to use special environmental cells, with miniaturized versions that can be incorporated directly into the specimen holder. Nevertheless, it is complex and difficult to handle. The sample needs to be diluted and maintained within that gaseous or liquid environment [108]. TEM *in situ* technology allows the direct observation of nanomaterials in a fully hydrated environment with high spatial resolution [100]. It has been used to observe physical and chemical processes at the nanometric scale, such as phase transformation of nanoparticles, melting behavior, crystallization and coalescence [109].

Applications

Nanoparticles are detected by TEM when they have a sufficient combination of size and electron density. The observation of a large number of particles should provide data on mean diameter [101]. To report the results, it is recommended that the size and variance of the means are described. Other critical information to be reported includes: brand and model of the equipment used, operating voltage, beam current settings, objective lens excitation, and diffraction aperture size, if used [110].

As already mentioned, through TEM, it is possible to study the intracellular localization and the uptake of nanoparticles. Hondow *et al.* [111] obtained TEM images from thin sections of human osteosarcoma cells exposed to Qtracker® 705 quantum dots. The images indicated that nanoparticles were included in vesicles bounded to the cell membranes, which confirmed absorption by endocytosis. The images were also used to confirm the crystalline nature of the quantum dots.

Andrade-Mahecha *et al.* [112] used TEM to characterize achira nanofibers by morphology and size (diameter and length). Magnetite-carboxymethylcellulose nano-composites were prepared and characterized by Habibi [113] using TEM as one of the techniques for evaluating morphology, size and structural properties of the compound. This study used other techniques addressed in this chapter, such as X-ray diffraction, Fourier-transform IR spectroscopy (FTIR) and SEM using field emission source, showing that different techniques are used in a complementary way for a better characterization of nanosystems.

SLN containing sildenafil had their morphology evaluated by TEM using the negative staining technique in work by Paranjpe *et al.* [114]. In another study conducted by Rodrigues da Silva *et al.* [115], nanostructured lipid carriers (NLC) were analyzed by TEM before and after encapsulation of bupivacaine. The technique showed that the nanoparticles were spherical with regular and defined borders. The addition of the drug did not alter their morphology.

Atomic Force Microscopy

Atomic force microscopy (AFM) is based on the measure of force variation resulting from the interaction between the sample surface and a tip [116]. AFM has been used for the analysis of morphology and characterization of surface topography with high spatial resolution. In addition, inter and intramolecular forces can be measured with high sensitivity (in the range of 10 to 20 pN). Local surface interactions, nanomechanical and nano-adhesive properties of samples can be determined by using AFM to record force-distance curves [117, 118]. Its applications extend to several fields of science, including materials science and biology. It is an extremely versatile technique that can be practically applied to any sample, and it is possible to be operated in ambient air, in a vacuum or in a liquid medium [119].

Equipment

The atomic force microscope (Fig. **24**) consists of some parts common to any similar equipment: (1) a tip with a very small radius; (2) a support arm with a low spring constant (cantilever); (3) an instrument for measuring the cantilever deflections resulting from forces between the tip and the sample surface; (4) a three-dimensional scanning controller to move the tip; (5) a feedback system to obtain topography and other information, and (6) software capable of interpreting data as an image [120].

There is a great diversity of tips and cantilevers in use, the most common are manufactured from silicon and silicon nitride. The cantilevers can be rectangular or triangular with dimensions ranging from 60 to 200 μm in length, 10 to 40 μm in width and 0.3 to 2 μm in thickness. There are silicon cantilevers with a rectangular shape and spring constants from 0.1 N/m up to 40 N/m. The triangular cantilevers of silicon nitride are more suitable for the contact mode and have spring constants from 0.01 N/m to 0.58 N/m. Regarding the tips, the radium range from 10 to 20 nm for silicon nitride tips and about 10 nm for the silicon tips. There are still special probes with a radius as small as 1 nm. The sharper the tip, the better the image [119, 120].

The three techniques most used as deflection systems are the laser reflection method, the interferometric method and the piezoresistive method. In the laser reflection method, the most common in commercial systems, a laser beam is reflected from the surface of the cantilever towards the photodetector. The deflection of the cantilever is measured by the signal difference between the distinct parts of the photodetector [119, 120].

In the interferometric method, the deflection is determined by the difference between the optical path of a reflected beam from the fiber-air interface and the beam that is reflected by the cantilever. Piezoresistivity is the property of certain materials to change their resistivity through stress. Thus, in the method, the cantilever is coated with a piezoresistive material which the resistance is altered with deflection and, thus, this property is used to measure the deflections [120].

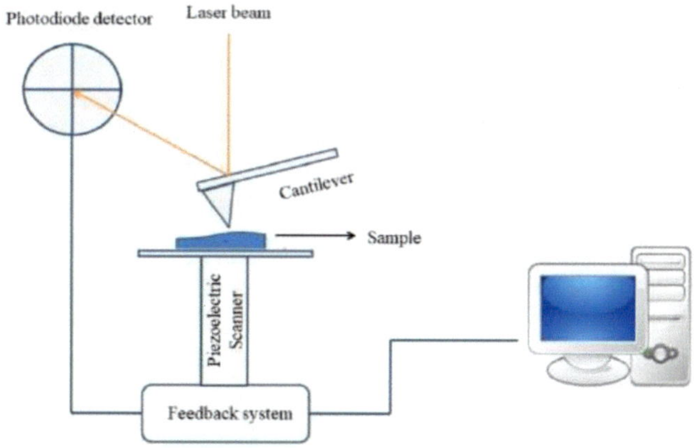

Fig. (24). Schematic representation of the basic composition of an atomic force microscope.

The piezoelectric scanner is the device responsible for moving the tip onto the surface of the sample. This device is manufactured with piezoelectric materials and scans the sample in a rectangular way, faster in the X-axis and slower in the Y-axis [120, 121]. The oscillating motion of the cantilever (Z-axis) and scanning on the surface (X- and Y-axis) are controlled by this device [122]. To obtain accurate measurements, the axial displacement (Z-axis) of the piezoelectric scanner needs to be calibrated with commercially available reference standards. If the height values obtained are markedly different from the certificate, the equipment manufacturer should be consulted on how to recalibrate the Z-displacement of the piezoelectric scanner [123]. The feedback system regulates the distance between the tip and the sample.

AFM software typically has the following features: ability to control sample scanning, feedback system, and data acquisition; ability to convert data points into pixels to form the image and the ability to analyze the image [120].

Regarding the resolution of the equipment, this depends on several factors such as tip radius, the spring constant of the cantilever, the sensitivity of the cantilever deflection detection, material resistance, mechanical and electrical noise [124]. Furthermore, it depends on the piezoelectric scanner's ability to apply sub-nanometric displacements in the cantilever [125].

Principles of the Technique

AFM is based on the interaction between a tip connected to a cantilever with a surface. This interaction can occur through various forces such as van der Waals, electric, magnetic, short-range, capillarity, or any other electromagnetic forces. This diversity of forces allows the technique to be used not only for obtaining surface images, but also for measuring physical and mechanical properties [120].

These inter and intramolecular forces can be measured by force-distance curves. The force-distance curve is a graph that correlates the deflection of the cantilever from the interaction tip-sample with the distance of the tip from the sample surface. To obtain this graph, two simultaneous events are measured: the cantilever movement towards the sample, which provides the distance information, and the cantilever deflection, through which the information about the AFM forces is obtained. The strength of the interaction between the tip and sample is described by Hooke's Law (Equation 10):

$$F = -k_c \Delta Z_c \quad \textbf{(10)}$$

where F = interaction force, k_c = cantilever spring constant, and ΔZc = deflection of the cantilever.

For analyzing these curves, several empirical and theoretical models have been proposed, such as the Hertz and Oliver-Pharr models [117]. The curve for the approximation of the tip towards the surface provides information on the repulsive or attractive forces and on the mechanical properties of the sample (such as Young's modulus and relaxation times). The retraction curve provides information on the adhesion, bonding, or unfolding forces between the tip and the sample. These experiments, commonly called force spectroscopy, are currently used to investigate non-specific interactions (hydrophobic and electrostatic forces); to quantify the receptor-binding forces; to study the mechanical unfolding of proteins; to investigate cell surface receptors, cell-cell interactions, cell adhesion and mechanics [124].

Regarding surface topography analyses, there are three main AFM modes of operation, which are contact mode, non-contact mode, and intermittent mode (Fig. **25**). In contact mode, there are strong repulsive forces. In the non-contact mode, there are weak attractive forces, and in the intermittent mode, there are intermediate forces between the tip and the surface. Therefore, the interaction between the tip and the surface lasts longer in the contact mode and shorter in the non-contact. The sample topography can be analyzed in the air or in a liquid medium by the contact or intermittent modes, or in a vacuum environment, for only contact or non-contact modes [120]. In these modes, the sample surface is scanned by a sharp tip mounted on a cantilever [118].

Contact mode is the simplest operating method and was the first to be developed. The tip is placed in contact with the surface and scans a square area to obtain a topographic map [119]. The cantilever deflections resulting from tip interactions with protrusions and/or cracks on the surface are detected by the photodiode through changes in the laser angle reflected by the cantilever [126].

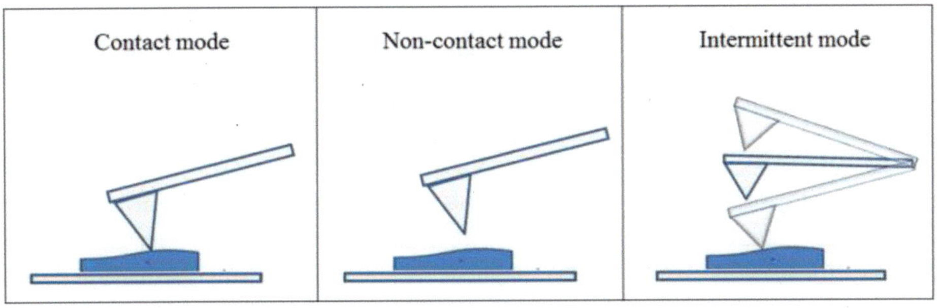

Fig. (25). Schematic of the main modes of operation used in atomic force microscopy.

There are two basic ways of obtaining topography images in this operation mode: constant and variable force mode [120]. In the constant force mode, the cantilever deflection is held constant so that a constant force is applied on the surface [118], while the variable force mode is considered a free deflection mode. In the free deflection mode, the feedback system is disabled and the tip scans the sample surface, leaving the cantilever free to deviate from the sample according to its topography, being the height equal to the cantilever deflection signal [120]. In the contact mode, the lateral forces applied to the surface are very high and can damage the sample. This is problematic mainly for biomolecules adsorbed on the surface since they are delicate samples [119]. In this way, the contact mode is used for the analysis of materials less susceptible to mechanical damages [124].

In non-contact mode, also known as static mode, van der Waals forces predominate and the cantilever oscillates near its resonance frequency, with a certain amplitude. When the tip is close to the surface, the resonance frequency of the cantilever is changed and it oscillates to lower amplitude because the driving frequency is kept constant. The main advantage of this mode in relation to contact mode is that since the tip is always located in the weak attractive region, the sample and tip are not easily damaged [120].

In the intermittent mode, also called dynamic or tapping mode, the cantilever oscillates near or at its resonance frequency and the oscillation amplitude is kept constant during the analysis [118, 119]. Depending on the topography of the sample, the oscillation amplitude changes. Then, the feedback system keeps the oscillation at a constant level by determining the height information of the sample [124]. This mode has a better resolution than the non-contact mode and has the same advantage, *i.e.* the sample is not easily damaged. It is currently more used than non-contact mode [120].

AFM has several advantages when compared to the microscopy techniques previously discussed. SEM requires a high vacuum to operate, and pretreatment of the sample that can lead to its destruction and generate artifacts. AFM, on the other hand, can work under environmental conditions, making sample preparation unnecessary and thus avoiding damage related to pre-treatment [116].

As AFM can operate on liquid surfaces, it allows the observation of nanoparticles in a hydrated state, closer to reality and less aggregated, when compared to SEM [127]. SEM resolution is limited by electrons wavelength, whereas the AFM resolution is not limited to any wavelength value, providing nanometric values precisely [116]. In addition, AFM provides three-dimensional images and both quantitative and qualitative information on physical properties such as size, morphology, texture and surface roughness [128].

A major disadvantage is that if the size of the cantilever tip is larger than the dimensions of the nanomaterials to be examined, this leads to an overestimation of the lateral dimensions of the sample. Moreover, unlike fluorescence techniques, AFM is not able to detect or localize specific molecules. This latter disadvantage can be eliminated by modifications at the cantilever tip with binders, cell adhesion molecules or chemical groups [8].

Sample Preparation and Data Analysis

Careful preparation of the sample and selection of the most suitable substrate for the analysis are essential for obtaining high-resolution images without artifacts. The substrate should be flat and the surface roughness should be much smaller than the diameter of the objects under analysis [129]. Mica, graphite and silicon are examples of substrates that are suitable for AFM [122]. The interaction between the sample and the substrate should be strong enough to avoid sample displacement during the analysis, but not so strong to alter sample morphology [129].

Different methods of sample preparation are described, like adsorption, droplet deposition, and ultracentrifugation [122]. The adsorption method depends on the particles' diffusion to the substrate and the interaction between the particles and sample substrate. Thus, it is possible to analyze particles with higher diffusion and those which bind more strongly to the AFM substrate. The droplet deposition method consists of depositing a drop of the sample on the substrate, followed by evaporation of the drop or blowing nitrogen into the sample [130]. This method has the disadvantage of inducing the formation of aggregation artifacts during the drying process [131].

The ultracentrifugation method forces the suspended particles toward the substrate, whose will be adhered to it, and therefore this preparation method provides a more representative average size and particle size distribution, and is considered more appropriate for sample preparation [131]. However, if the particles are not strongly bound to the substrate, too much loss can take place due to particle release from the substrate as a result of the centrifugation process itself or during the washing step, which is a part of the method. The washing process is necessary to decrease the occurrence of artifacts by removing excess salts from the sample and thus preventing the aggregation of particles due to increased salt concentration (ionic force) upon drying of the sample [132].

In liquid media, the particles are often attracted very weakly to the substrate and move around, resulting in image disturbances [122]. Furthermore, since probe-particle interactions cause a widening of the lateral dimensions, these can be overestimated while the height measurements are very accurate and reliable for diameter determination [122, 133]. This fact must be taken into account when interpreting the images obtained by AFM. For example, carbon nanotubes having the same height and width dimensions when analyzed by AFM may have a height of 1 nm but a width close to 50 nm. Thus, the tip utilized should be replaced with a smaller one for a more accurate analysis of small particles [122].

The particle size distribution can be performed by plotting a histogram containing the individual particle size results. The number of particles measured will influence the result: if it is too small, non-representative results will be obtained; however, if it is too high, it can be a waste of time and resources. The ideal number depends on the size of the sample, the degree of confidence required, and the resources available [133].

Sample preparation techniques used in microscopy have the drawback of having weak statistical power. Therefore, they require the analysis of a large number of nanoparticles to compensate the low and inconsistent recovery and non-uniform distribution of these nanoparticles [131].

Applications

DLS, SEM, TEM and AFM are techniques commonly employed for the determination of size, shape, and morphology of particles with nanoscale size. Although DLS is a technique capable of determining the mean diameter of a large number of particles, it may not reveal a real distribution, mainly due to the presence of aggregates or when non-spherical nanoparticles are analyzed. When dimensions of a single particle are desired, due to simple sample preparation and the possibility of liquid analysis, AFM generally provides faster results in comparison to SEM or TEM. Thus, size determination by AFM has become a standard technique for the characterization of drug delivery vehicles. AFM images can provide important information on surface modifications for drug targeting, on DL in the system, in addition to system degradation profile and drug release. Moreover, AFM has enough temporal and spatial resolution to monitor dynamic structures and reorganization of membrane models in nanosystems and drug complexes [134].

Eaton *et al.* [135] compared four techniques (DLS, SEM, TEM, and AFM) for the characterization of nanoparticles with different sizes (15 to 100 nm) and with different core compositions (gold, silica, and polystyrene). Regarding the microscopic techniques, there were no significant differences in the obtained sizes. High-quality images of silica and polystyrene nanoparticles were difficult to obtain by SEM, especially for smaller particles, due to the low contrast. AFM and TEM also provided accurate results even for smaller particles. As expected, DLS was considered unsuitable for polydisperse samples, while the microscopic techniques were able to distinguish particles with different sizes.

Moribe *et al.* [136] used AFM as one of the techniques for characterizing nanoparticles made of polyvinylpyrrolidone K17 and sodium dodecylsulfate, containing probucol. The average height of freshly prepared samples was 6 nm, while samples stored for 24 h had an average height of 15 nm. The analysis of the force-distance curve also showed the presence of probucol nanocrystals and agglomeration of the nanoparticles.

Polyurethane nanoparticles with and without PEG were prepared by Zanetti-Ramos *et al.* [137] and characterized by DLS and AFM. AFM analysis showed that most of the nanoparticles of both formulations were spherical. In AFM analysis, the formulation containing PEG showed two particle populations, one with a diameter of about 218 nm and the other around 130 nm. In the absence of PEG, the average diameter particle was 209 nm. In the meanwhile, DLS measurements showed a monodisperse size distribution for both formulations, with an average diameter of 246 nm and 262 nm for the particles without and with PEG, respectively.

Confocal Fluorescence Microscopy

Confocal microscopy excludes most of the light that is not in the focal plane of the microscope, making the specimen image less blurred, sharper and with better contrast when compared to the image of a conventional optical microscope. Confocal microscopy allows better observation of details and allows the construction of 3D images of a sample [138]. Samples can be observed in their hydrated state, and thus, processes such as phase separation, coalescence, aggregation and solubilization can be observed by confocal microscopy [139]. This technique generates significant improvements in resolution when compared to other light microscopies and is placed between conventional light microscopy and SEM/TEM [140].

Principles of Technique and Equipment

The confocal microscope can be used in transmission and reflection modes, but its most common use involves fluorescence imaging [141]. Fluorescence is a process in which a molecule absorbs an incident light and emits part of that light in a different color. When absorbing a photon of light, the electrons in that molecule increase their energy and change from the ground state to an excited one. Usually, the molecule quickly dissipates part of this absorbed energy, making the electron loses energy and returns to its ground state. This energy dissipation may be through collisions with surrounding molecules or through emission [138].

In fluorescence microscopy, a dichromatic mirror is used to reflect the light of a smaller wavelength while transmitting light of a longer wavelength. For example, when using a blue excitation light (shorter wavelength), the emitted light is green (longer wavelength). Thus, when the sample is struck by blue light, it is reflected and passes through the objective lens, but not through the dichromatic mirror. The longer wavelength (green) light passes through both the objective lens and the dichromatic mirror and will form the image (Fig. **26**) [138]. Confocal microscopy incorporates two main ideas: point-to-point illumination of the sample and the rejection of out-of-focus light [142]. The invention of the confocal microscope is attributed to Marvin Minsky. In Minsky's original microscope, a pinhole is placed in front of a light source. The light point is focused on the sample by an objective lens. The light reflected by the sample is focused by a second objective lens with a pinhole that has the same focus as the first one. The light passing through the second pinhole reaches a photomultiplier and produces a signal directly proportional to its brightness. The function of these pinholes is to reject the light that is out of focus and, as these two pinholes have the same focal plane conjugate, they are said to be confocal, so the technique is called confocal microscopy [143].

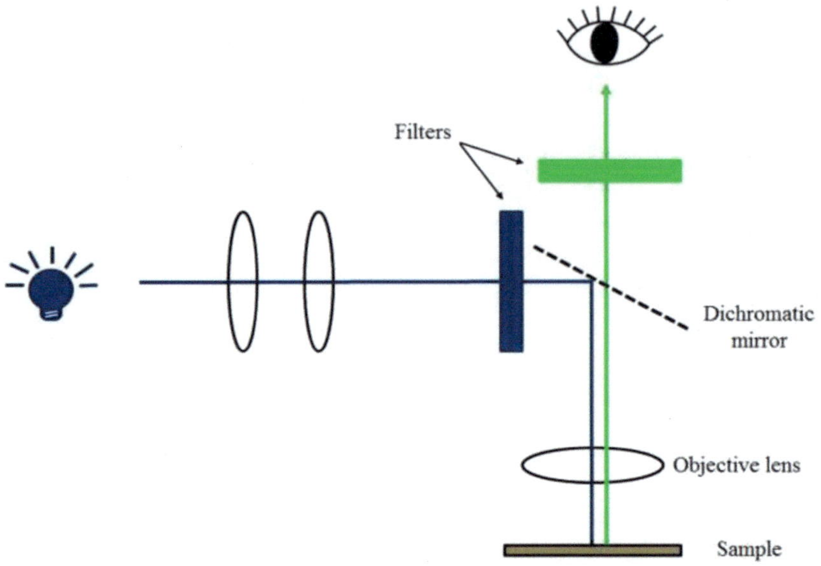

Fig. (26). Basic operation of a fluorescence microscope.

The main elements of Minsky's confocal microscope are maintained in current microscopes, which are the presence of pinhole and point-to-point illumination of the sample [138]. For the image to be constructed, the light spot must scan the sample somehow. In Minsky's original microscope, the sample was moved while the light beam remained stationary. Currently, the light beam is moved and scans the sample [143].

In the confocal microscope (Fig. **27**), the light source (usually a laser) is directed by a dichromatic mirror towards a pair of mirrors that digitize the light on the x and y axes. The light is focused at a point of the sample by the objective lens and excites the sample. The emitted fluorescence is focused by the same objective lens and passes through the pinhole blocking the light outside the focus plane, and thus only the light in the focus plane reaches a detector, as a photomultiplier tube [142, 144].

Only one point of the sample is observed at each moment [142], so when scanning successive planes of the sample, a three-dimensional image can be created [144].

This is one of the main advantages of confocal microscopy. In addition, fluorescence microscopy is more sensitive and specific than conventional microscopy. Samples can be labeled with one or more dyes capable of emitting fluorescence when stimulated by light (fluorophores), making them visible under the microscope [138]. Fluorophores can be used to mark specific parts of the sample allowing better and sharper detection of the target [145].

In comparison to conventional optical microscopy, a disadvantage would be the cost of the technique. Another limitation is the photobleaching phenomenon: the fluorophores tend to disappear irreversibly or react when exposed to the excitation light [145]. Several strategies have been developed to reduce the photodegradation rate, including the addition of free-radical scavengers during sample preparation to reduce the oxygen radicals that react with the fluorophore, and the use of high numerical aperture lenses to collect more fluorescence light and thereby using less excitation light [138].

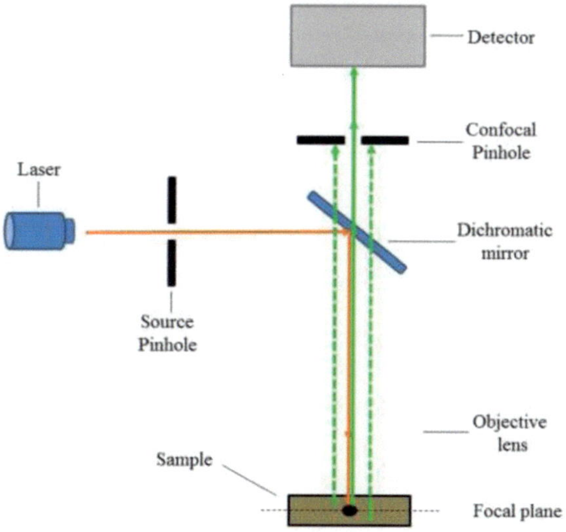

Fig. (27). Representative scheme of a confocal microscope.

Applications

Confocal fluorescence microscopy has been used to investigate drug delivery and characterization of pharmaceutical systems such as microspheres, granules, film coatings, and hydrophilic matrices [146]. Another application is viewing the penetration and localization of nanoparticles in the skin [147].

Sarmento *et al.* [148] investigated the pharmacological activity of insulin-containing alginate/chitosan nanoparticles after oral administration in diabetic rats. The drug was labeled with fluorescein and screened by confocal microscopy to investigate the absorption and localization of insulin after administration. The analysis by confocal microscopy showed that the mucoadhesive nanoparticles adhered to the mouse intestinal epithelium promoting the internalization of the insulin in the intestinal mucosa. Absorption was demonstrated by the presence of a labeled drug on the apical membrane of intestinal cells 3h after administration. Paracellular absorption and endocytosis were considered as synergistic mechanisms for the absorption of insulin.

The interaction of nanoparticles with living cells was investigated by Shang *et al.* [149] using confocal microscopy. The authors analyzed the uptake by living cells (HeLa cells and mesenchymal stem cells) of fluorescent nanoparticles with a size range from 3.3 to 100 nm. Gold nanoclusters with a diameter of approximately 3 nm, quantum semiconductor dots with a diameter of approximately 10 nm and polystyrene nanoparticles having a diameter of approximately 100 nm were labeled with a fluorophore. Particles smaller than 10 nm were first retained on the plasma membrane before being internalized by the cells, whereas those with larger diameters around 100 nm were directly internalized without prior accumulation in the plasma membrane.

CRYSTALLINE STATE, POLYMORPHISM AND DRUG NANOSYSTEM INTERACTION

Infrared Spectroscopy

When atoms or molecules absorb light, the energy excites a quantized structure to a higher energy level. The type of excitation depends on the wavelength of the light. The result of radiation interaction with the matter depends directly on the incident energy associated with the wavelength of electromagnetic radiation, according to Equation 11 [150]:

$$E = h\nu \tag{11}$$

where E = energy of the incident radiation, h = Planck constant and ν = radiation frequency (hertz).

Thus, E and ν are directly proportional. The wavelength (λ) and the frequency (ν) are related through the velocity of light propagation (c) in the medium and they are inversely proportional, as shown in Equation 12:

$$\nu = \frac{c}{\lambda} \tag{12}$$

Most scientists refer to the radiation in the IR region of the electromagnetic spectrum in terms of wavenumber rather than wavelength. The main reason for this is that wavenumber is directly proportional to the energy. Wavenumber is calculated as a reciprocal of the wavelength expressed in cm^{-1} [151].

The IR region in the electromagnetic spectrum includes radiations with a wavelength from 0.78 μm to 1,000 μm and wavenumber from 12,800 cm^{-1} to 10 cm^{-1}, which refers broadly to the part of the electromagnetic spectrum between the visible and microwave regions, as shown in Fig. (**28**). The spectrum in the IR region is conveniently divided into near-IR (12,800 to 4,000 cm^{-1}), mid-IR (4,000 to 200 cm^{-1}), and far-IR (200 to 10 cm^{-1}). The most used portion is the one between 4,000 and 400 cm^{-1}, corresponding to the wavelength from 2.5 to 25 μm. The IR spectrum is produced as a consequence of transitions between quantized vibrational energy states and is generally given as a graphic of transmittance (T, %) *versus* wavenumber (cm^{-1}) [150, 152, 153].

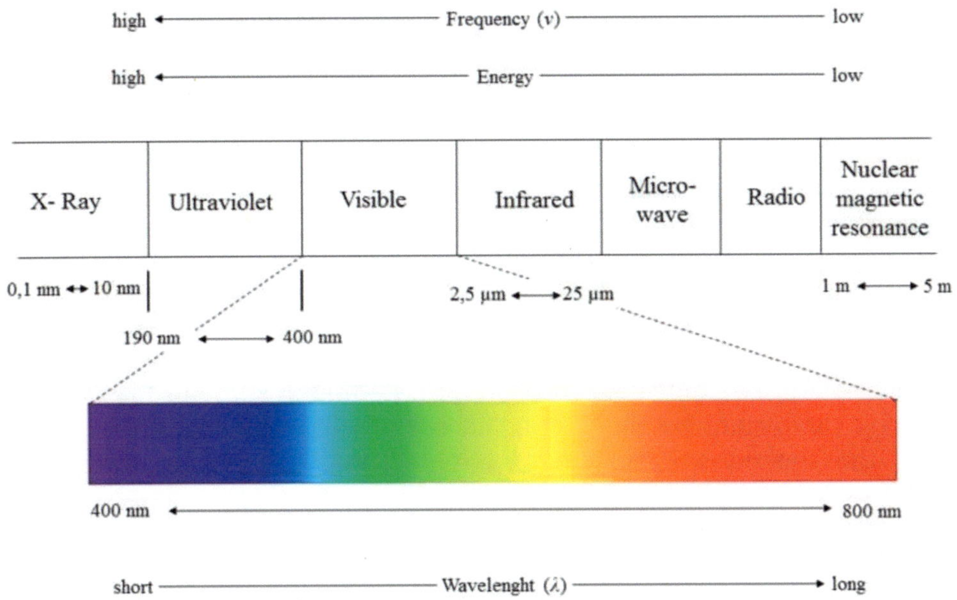

Fig. (28). Electromagnetic spectrum.

The interaction between radiation and matter can lead to redirection and/or transitions of the energy levels of atoms and molecules. Redirection of light due to its interaction with matter is called scattering, which may or may not occur with energy transfer. A transition from a higher energy level to a lower level is called emission. A transition from a lower level to a higher level with energy transference in the form of radiation is known as absorption. When atoms or molecules absorb electromagnetic radiation, the quantized energy excites the structure to a higher energy level, which depends on the incident wavelength [150, 151].

When molecules absorb energy in the X-ray region (higher energy and lower wavelength), bond breaking may occur. The energy correspondent to the ultraviolet or visible range is enough to promote electronic transitions, while the absorption in the IR and microwave regions results in vibrational and rotational transitions, respectively. At the other end of the electromagnetic spectrum, radiofrequencies have very low energy, only enough to cause nuclear or electronic spin transitions within molecules [151].

Typically, a molecule may absorb IR radiation if it possesses a time variant dipole moment and if its oscillating frequency is the same as the frequency of incident IR light. Only those vibrations that result in a rhythmical change in the dipole moment are capable of absorbing IR radiation. Symmetric bonds, such as diatomic molecules (*e.g.*, H_2 or Cl_2), do not absorb IR radiation because they do not present dipole moments. A bond must present an electrical dipole that is changing in the same frequency as the incoming radiation in order energy is transferred. The alternating electric field, produced by the change of charge distribution, couples the molecule vibration with the oscillating electric field of the electromagnetic radiation [151, 153, 154].

As every type of bond has a natural vibration frequency of its own, and since two bonds of the same type in different molecules are in different or slightly different environments, no molecule of the different structure has exactly the same IR absorption pattern, *i.e.*, the same IR spectrum. Although some of the frequencies absorbed by them may be the same, the set of all frequencies absorbed and, therefore, the spectra will not be identical. Thus, the IR spectrum offers a fingerprint of the molecular structure and the IR spectroscopy is a very useful tool for the identification of substances and structural elucidation [151].

The most common types of molecular vibrations are stretching and bending. The stretching vibration is a movement along the bond axis with the increasing or decreasing of the interatomic distance. The bending vibration consists of a change in the angle between two bonds with a common atom [151, 153].

To determine the IR spectrum of a compound, one must place the compound in a sample holder. Usually, solid samples are dispersed in ionic substances like potassium bromide – KBr, which not absorbs the commonly used IR range (4000 to 400 cm^{-1}). The sample preparation involves mixing the finely ground solid interest material with powdered KBr and pressing the mixture under high pressure. The result is a KBr pellet that can be inserted into a holder in the spectrometer. In the case of liquid samples, a drop of a liquid organic compound is placed between a pair of polished KBr or NaCl plates. When the plates are squeezed gently, a thin liquid film forms between them. The pair of plates is inserted into a holder that fits into the spectrometer. If no solvent is used for sample preparation, the spectrum is named neat spectrum [151].

Additionally, there is also a well-established sampling accessory in IR spectroscopy that is the attenuated total reflectance (ATR). In ATR mode, the total internal reflectance, occurring within the equipped internal reflection element (IRE) crystal, which has a high refractive index at certain angles, forms evanescent waves that extend from the IRE crystal-sample interface into the sample with a penetration depth of micrometers (0.5 to 5 μm). The intensity of the evanescent waves decays exponentially with distance from the interface of the crystal. The distance, in the order of micrometers, causes the ATR technique to be generally insensitive to the thickness of the sample, allowing analysis of thick samples or materials with high absorption. Therefore, the ATR spectrum carries information of thin sample layers that are closely placed on the crystal surface [8, 150, 154].

In ATR mode, the surface layer thickness reached by the beam depends on various factors, including the refractive index of both the sample and IRE, the angle of incidence, and the wavelength of the IR radiation (if it is in the wavenumber range from 4000 to 650 cm^{-1}). Due to the wavelength dependency on the surface sampling depth probed, in an ATR spectrum, the relative absorption band intensities will appear relatively more intense with decreasing wavenumber (increasing wavelength) than for a conventional transmission spectrum. ATR has the advantage of being a nondestructive method and does not require sample preparation [152, 155].

The instrument that determines the absorption spectrum for a compound is called an IR spectrometer or spectrophotometer. Basically, there are two types: dispersive and FTIR spectrometer. Both of these types of instruments provide spectra of compounds in the common range of 4000 to 400 cm^{-1}. The two instruments provide nearly identical spectra for a given sample; however, FTIR spectrometers provide the IR spectrum much more rapidly than dispersive instruments [151].

The operation of the dispersive spectrometer consists of the division of the beam and passing one portion through the sample cell and the other portion through the reference cell (double-beam instrument). The individual beams are then recombined into a single beam of alternating segments by means of a rotating sector mirror, and the absorption intensities are balanced by the attenuator in the reference beam. The detector senses the ratio between the intensity of the reference and sample beams and determines which frequencies have been absorbed by the sample and which frequencies are unaffected. After the signal from the detector is amplified, the recorder draws the resulting spectrum of the sample on a graphic. The spectrum is recorded as the frequency of IR radiation changes by rotation of the diffraction grating [151, 153].

The FTIR instrument *modus operandi* is usually described with reference to the basic Michelson interferometer, which replaces the prims and diffraction gratings. In these instruments, the radiation from the source is equally split by the beam splitter: one beam is focused and returned by a fixed position mirror and the other beam is focused and returned by a moving mirror. The varying distances between two pathlengths result in a sequence of constructive and destructive interferences, creating variations in intensities, thus generating an interferogram. The interferogram is a complex signal being essentially a plot of intensity *versus* time (a time-domain spectrum). A fast FT algorithm converts the interferogram, which is in the displacement domain, into a single-beam spectrum, in the frequency domain, identical to that obtained with a dispersive spectrometer [150, 152].

There are several advantages of FTIR, such as the acquirement of the interferogram in less than a second with high resolution (≤ 0.001 cm^{-1}), making it possible to collect various interferograms of the same sample. When FT is performed on the sum of the accumulated interferograms, a spectrum with a better signal-to-noise ratio can be plotted. Thus, this spectrometer is capable to generate spectra faster and with higher sensitivity than a dispersion instrument. Moreover, Fourier deconvolution is an excellent approach to resolve severely overlapped peaks [151, 153, 154, 156].

The technique called ATR-FTIR spectroscopy combines the advantages of the two techniques, ATR and FTIR, to probe the structure of adsorbed/deposited species in a solid/air or solid/liquid interface. Such advantages include avoiding the drawbacks of sample preparation complexity and spectral irreproducibility in conventional IR spectroscopy equipment [8, 154, 157].

Regarding characterization of nanosystems or nanomaterials, IR spectroscopy and its variations is a tool widely used and is capable of providing information mainly on interaction mechanisms between the nanosystems components and their molecular organization [42, 158].

As an example, in the study conducted by Karthivashan *et al.* [159], a flavanosome was developed as a promising nanodrug delivery system for loading multiflavonoids and the localization of flavonoids and the nature of the lipid membrane bonds were observed through FTIR spectroscopy. In another work reported by Ayan *et al.* [160], the analysis of the FTIR spectrum of radiolabeled curcumin-loaded SLN suggested an interaction between curcumin and the solid lipid matrix. Besides, alterations observed in specific regions of the formulation spectrum in relation to the SLN matrix spectrum could be interpreted as an indicator of curcumin encapsulation in SLN, according to the authors.

FTIR spectroscopy is also commonly employed to reveal nano material–biomolecule conjugation, as proteins bound to nanoparticle surfaces and to illustrate the conformational states of the bound proteins [8].

Ultraviolet-visible Spectroscopy

UV-vis spectroscopy is a versatile tool that can be applied both in the characterization and quantification of different types of materials as organic, inorganic, biological substances, and nanosystems. It is a technique accessible, easy and quick to execute [42, 150].

The UV-vis spectroscopy involves the absorption of electromagnetic radiation by atoms and molecules in the wavelength range from 190 to 400 nm (UV region) and from 400 to 800 nm (Vis region) (Fig. **28**), resulting in an electronic transition of a fundamental to an excited state of energy. When a molecule absorbs energy, an electron is promoted from an occupied orbital to an unoccupied orbital of greater potential energy. Generally, the most probable transition is from the highest occupied molecular orbital (HOMO) to the lowest unoccupied molecular orbital (LUMO). The electromagnetic radiation that is absorbed has the same energy of the difference between the excited and ground state energy [150, 151].

The UV or Vis absorption for molecules usually occurs over a wide range of wavelengths because molecules, unlike atoms, normally have many excited modes of vibration and rotation at room temperature. Consequently, a set of molecules generally has its members in many states of vibrational and rotational excitation. The energy levels of these states are quite closely spaced, corresponding to energy differences considerably smaller than those of electronic levels. The rotational and vibrational levels are then overlaid on the electronic levels. Because there are so many possible transitions, each differing from the others by only a slight amount, each electronic transition consists of a vast number of lines spaced so closely that the spectrophotometer cannot resolve them. Rather, the instrument traces an "envelope" over the entire pattern. The UV or Vis spectrum of a molecule usually consists of a broad band of absorption centered near the wavelength of the major transition, which is the result of the combined transitions [151].

The quantification of substances by UV-vis spectroscopy is governed by Beer-Lambert law or just Beer's law. Summarily, according to Beer's law, absorbance is directly proportional to the concentration of absorbing species and to the path length of the absorbing medium, as expressed by Equation 13:

$$A = \varepsilon bc \tag{13}$$

where ε = molar absorptivity (mol/L.cm); b = path length (cm) and c = concentration of absorbing specie (mol/L) [161].

The UV-vis spectrometer consists of a radiation source, a monochromator and a detector. The visible and UV light sources usually used are a tungsten and deuterium lamp, respectively. The monochromator is a diffraction grating, which spread the beam of light into its component wavelengths. A system of slits focuses on the desired wavelength on the sample cell. The light that passes through the sample cell reaches the detector, which records the intensity of the transmitted light. The detector is generally a photomultiplier tube or photodiodes in modern instruments. In a double-beam instrument, the light emanating from the light source is split into two beams, the sample beam and the reference beam. The samples are placed in a glass or plastic cuvettes for analysis in the region of the visible and in quartz for the UV region [150, 151].

UV-vis spectroscopy can be used for direct or indirect investigation of many properties of nanosystems. In the case of metal nanoparticles, UV-vis spectroscopy can be used in order to determine the presence of nanoparticles in solution and the position of the surface plasmon resonance band, which contains information about size, aggregation, and surface chemistry. The surface plasmon resonance is an important property of metal nanoparticles, also known as the resonance effect [6, 8, 162, 163].

UV–vis spectroscopy is also frequently applied to study the syntheses of new nanoparticles through the use of colorimetric sensors in catalysis studies. Moreover, the technique is widely used for monitoring the growth of nanostructured thin films produced by various techniques [42, 158].

X-ray Diffraction

In 1895, the German physicist Wilhelm Conrad Roentgen in experiments with cathode rays, discovered X-rays. As the nature of these rays was unknown at the time, Roentgen called them X-rays, until later, it was discovered that they were electromagnetic waves in the wavelength range from 0.1 nm to 10 nm (1 - 100 Å) [164].

X-rays used in diffraction have wavelengths approximately in the range 0.5-2.5 Å, since they are of the same order of magnitude as the shortest interatomic distances observed in both organic and inorganic materials. X-rays occupy the region between gamma and ultraviolet rays in the electromagnetic spectrum (Fig. **28**) and they are much more penetrating than light and can easily pass through solid materials of low density [164, 165].

Like all electromagnetic energy of wave nature, X-rays can undergo the phenomena of diffraction and interference. Diffraction occurs when a beam of radiation passes through a slit or obstacle present in the optical path. It corresponds to the change of direction of a wave by scattering centers whose spacing is approximately the same wavelength of the wave. This change of direction depends on the physical dimensions of the obstacle, the wavelength of the incident radiation, and the angle of incidence. This phenomenon promotes the formation of constructive and destructive interferences: if the wave peaks coincide, they are in phase and the amplitude of the resulting wave is equal to the sum of the amplitude of each one, in this case, constructive interference occurs; if the peak of one wave coincides with the valley of another wave, they are out of phase and the resulting wave amplitude is equal to the difference between the amplitudes of the waves, in this case, destructive interference occurs. A diffracted beam may be defined as a beam composed of a large number of scattered rays mutually reinforcing one another. Therefore, the diffraction is observed when constructive interference occurs and is essentially a scattering phenomenon [42, 150, 166].

For X-ray diffraction (XRD) analysis, an X-ray generator is required. These can be obtained in different ways, the most common being the bombardment of a metallic target with a high energy electron beam accelerated by the potential difference between the cathode and the anode. The electrons reaching the target excite the atoms at the anode and are abruptly decelerated, resulting in the emission of X-rays, which are collimated and directed to the analyzed material. Copper is usually used as a target in the production of X-rays. A nickel filter is used to select the most intense wavelength (K_α). It only allows the passage of X-rays with a wavelength of 1.5413 [150, 164].

Max von Laue (1879-1960), a German physicist, was the first to observe and explain the phenomenon of X-ray diffraction in 1912. Laue discovered that crystals are capable of diffracting X-rays and that the diffraction patterns reveal the structure of the crystal [164, 165]. In 1913, the English physicists William Henry Bragg and his son William Lawrence Bragg determined the first structures using X-ray diffraction as a new tool. They studied sodium chloride and other simple compounds, characterizing the atomic ordering of those giving rise to what we know today as crystallography [42].

W. H. Bragg and W. L. Bragg analyzed Laue's discovery and formulated the law most used in powder diffraction, the Bragg law. This law establishes certain relationships among the diffraction angle (Bragg angle), wavelength, and interplanar distance between atoms (Equation 14).

$$n\lambda = 2d \sin \theta \quad \textbf{(14)}$$

where n = order of reflection (n =1, 2, 3...), λ = wavelength of incident X-rays on the crystal at an angle θ = the Bragg angle, and θ is measured between the incident beam and the particular crystal planes under consideration, and d = atomic interplanar distance of the crystal (Fig. **29**).

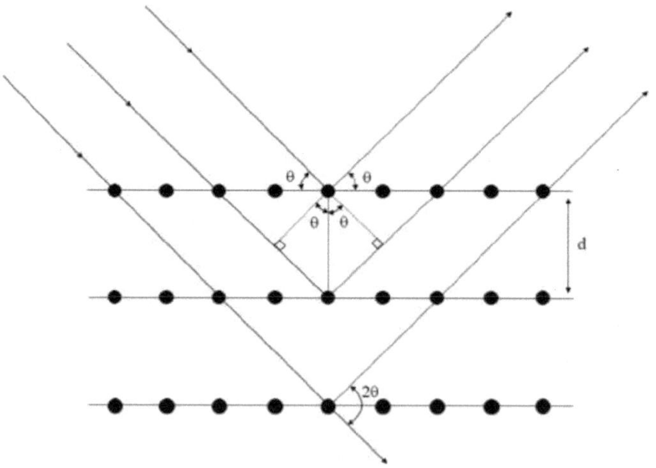

Fig. (29). Diffraction of X-rays by a crystal. The adjacent planes of the crystalline lattice spacing d. The beam strikes and is reflected at the same angle θ.

The Bragg law establishes the essential condition, which must be met for diffraction to occur: the path length difference between adjacent X-rays (Equation 14) is an integral number (n) of wavelength (λ), that is nλ [42, 150, 165].

The spectrums of XRD experiments are provided relating the diffraction angles (2θ) with their intensity (counts). Based on the wavelength values of the X-rays used and solving the Bragg equation for n=1, the corresponding d values can be obtained. In most equipment, there is software for this purpose. The diffractogram for a crystalline solid is almost zero, most of the time, except at certain angles where high sharp maxima occur – these are the diffracted beams. Both amorphous solids and liquids have structures characterized by an almost complete absence of periodicity, consequently, the X-ray spectrums show only one or two broad maxima [5, 150].

XRD is the most essential tool used to investigate and characterize crystal structures, since each crystalline solid has its unique X-ray diffraction pattern, which can be used as a "fingerprint". Currently, this technique covers a wide range of applications, solving diverse problems as chemical analysis and stress measurement, studying phase equilibria, measurement of particle size, determination of the preferential orientation of one crystal, the ensemble of orientations in a polycrystalline aggregate, and for polymorphism identification. Therefore, this technique can be used to characterize the most diverse materials from very simple structures to complex biological macromolecules, including proteins, nucleic acids, and ribosomes [5, 150].

The XRD technique is also important for the evaluation of crystal structure and phase identification of nanosystems, which are currently widely studied. The evaluation of nanomaterials by XRD can be performed with no significant difference in the case of microcrystalline materials. However, it should be noted that there is a considerable widening of the Bragg peaks, which is related to the fact that, in the case of nanocrystals, the theory of an ideal crystal, perfect and infinite, is no longer satisfied. In nanomaterials analysis, there is an inverse relationship between the mean crystallite size and the peak width [5, 42].

The enlargement of the Bragg peaks produced in nanocrystalline systems, with the consequent reduction in their height, may make it difficult to observe peaks that are very close to each other or that have low intensity. In these cases, it is necessary to use mathematical tools that allow to separate the contributions of each peak or to analyze using the Rietveld profile [5, 42].

The physical stability and the presence of polymorphism can also be studied by XRD. Due to the crystalline nature of many drugs and lipids used in the preparation of lipid nanoparticles for drug release, the heating/melting and cooling/recrystallization processes used during their production can lead to the occurrence of polymorphic transitions between their multiple states: unstable, metastable, and stable, affecting their physicochemical properties and stability of the final product. This should be considered for the production and long-term storage of SLN in order to predict if there are influences in their stability and capability for drug incorporation [98, 160, 167 - 169].

In a study described by Castro and colleagues [59], SLN formulations containing retinoic acid were analyzed by XRD and DSC to provide information regarding the formation of the ion pairing between the drug and stearylamine. XRD data revealed a slightly broader reflection for SLN formulation and the presence of X-ray peaks of retinoic acid crystals outside the lipid matrix of SLN without stearylamine. According to the authors, these results indicate the presence of an extensive association of the ion pairing components: retinoic acid and stearylamine. Moreover, the diffractogram of the formulations showed slightly broader reflections than those observed for the lipid matrix diffractogram. This result can be explained by the presence of a larger population of lattice defects in the formulations. It is conceivable that a larger population of lattice defects usually contributes to higher DL capacity [59, 170, 171].

Small-angle Scattering of X-rays

The development of nanomaterials for a proper application requires their detailed characterization as well as knowledge about the structural transformations occurring during materials processing. Another X-ray method of measuring the size of small particles is a small-angle scattering of X-rays (SAXS). It is a form of diffuse scattering very near the undeviated transmitted beam, at angles 2θ ranging from 0.1° to 10°, roughly. In SAXS, a portion of an incident X-ray beam elastically scattered from the sample forms a scattering pattern on a two-dimensional flat X-ray detector perpendicular to the incident X-ray beam direction [5, 26, 172, 173].

The basic process of X-ray scattering is photon-electron interaction. The scattering intensity produced by any material is a function of the scattering angle and of fluctuations in the electron density through FT (also refer to section 5.1). To obtain a good resolution in SAXS data, one must use a high-energy X-ray source such as a synchrotron [5, 172, 173].

An advantage of SAXS in relation to XRD is that the first can be applied not only to crystalline materials but also to amorphous materials, polymers, and biomolecules. The features of small-angle scattering in SAXS lead to the capability of studying non-repeating structures. Thus, perfect crystallized structures are not required, which simplifies sample preparation and makes SAXS a non-destructive method. The most basic application of SAXS is the determination of size, shape, morphology, size distribution, orientation and electron density of nanoparticles present as disperse dilutions. SAXS is also used to obtain information about the heterogeneity within amorphous materials and the pore sizes in porous materials, to evaluate long range ordering or interparticle distances in an assembly of nanocrystals or in polymeric materials, and the ratio of surface-to-volume of the particles (in the so-called Porod regime). Moreover, this has emerged as a powerful technique to study the structure and dynamics of biomacromolecular systems as proteins, enzymes, DNA in solution and near native environments [5, 172, 174 - 176].

SAXS is also described for the characterization of the surface hydrophobicity of nanoparticles. The surface chemistry of nanoparticles, including their hydrophobicity, is a key determinant of their fate, toxicity, and transport, and consequently of their bioavailability [169, 177, 178].

Additionally, interactions between components of the nanosystems for drug delivery can be investigate in association with various techniques, like in the study of Ferreira *et al.* [179]. Therein, interactions established between doxorubicin and the lipid bilayer in liposomes were verified both by SAXS and DSC.

Thermal Analyses

Thermal analyses are a useful tool to characterize nanoparticles and their bulk constituents. They are techniques that correlate temperature to the physical properties of pure substances, mixtures, and nanosystems. They can supply important data about the physical state, crystallinity, interactions between components, stability, among other information [180]. Herein, we will discuss the most important instruments and techniques used regarding thermal characterizations of nanosystems.

Differential Scanning Calorimetry

DSC is the most famous and used thermal analysis technique [181, 182]. Therein, a sample is heated (or cooled) to a certain temperature. Due to this temperature variation, the material can undergo physical state changes: it can melt (if it is heated) or crystallize (if it is cooled). When a material melts, it absorbs energy, making the process endothermic; in the case of crystallization, the opposite happens, *i.e.*, an exothermic process. The DSC detects the energy variation and converts it to electrical signals that are amplified and transferred to the control unit that is connected to the computer. The software builds DSC curves in which the endothermic and/or exothermic events are plotted *versus* the temperature and also controls equipment heat flow applied [167, 180, 183].

DSC equipment (Fig. **30**) is formed by an oven (furnace) connected to a computer. The samples first will be load in tiny pans (covered or not with lids), and these pans (crucibles) will be put in the oven. There will always be a pan without a sample, which is called a reference pan. There are two different types of DSC equipment: the power compensation DSC (patented by Perkin Elmer®), where there are two furnaces, one for the reference pan and the other for the sample pan, and the heat flux DSC where there is only one furnace where both reference and sample pans are in. The chamber where the samples are put is fulfilled with inert gas, nitrogen or helium. The environment therein is dynamic: the gas is constantly changed in order to avoid the accumulation of volatile gas that may be produced in the heating process. Usually, the temperature ranges between -60 °C to 350 °C. Regarding scan rates, they go from 1 to 10 °C/min, generally [167, 180].

Both reference and sample pans will be maintained at the same temperature. In older equipment, just one sample pan can be loaded each time. Newer equipment has carousels where many samples can be loaded.

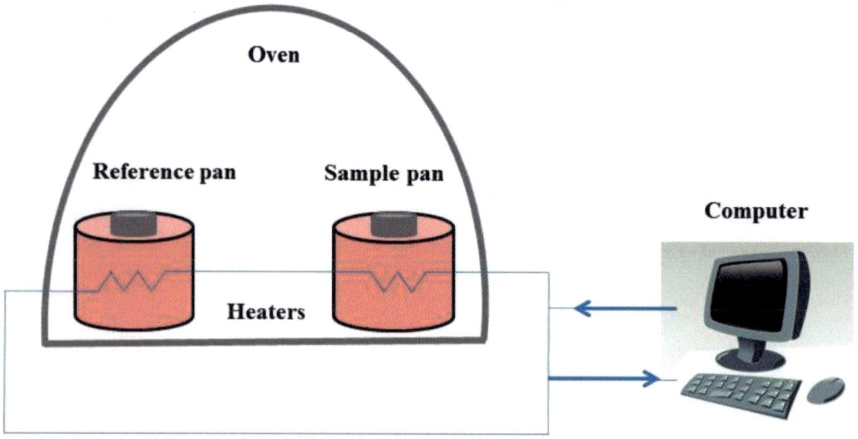

Fig. (30). Schematic representation of a DSC equipment.

Nevertheless, when a thermal event happens (melting, crystallization, *etc.*), the temperature stops changing in the sample pan and becomes constant. For example: when a material is melting, the energy applied to it will not be used to elevate the temperature but to the melting process, otherwise. Therefore, the DSC computer will have to supply more heat flow to the sample pan to maintain the same temperature as the reference pan. These heat flow values will be used to build the DSC curves [184]. It will be obtained by Equation 15:

$$q = \frac{\Delta T}{R} \qquad \textbf{(15)}$$

where q = heat flow; ΔT = difference between the temperature of the reference and sample pans and R = resistance of the thermoelectric disk (the place where the oven is settled in) [184].

Usually, an amount of 1-10 mg is put in each pan. There are many types of DSC pans and lids, depending on the temperature ramp and range used and if there is any volatile component. The experiments can be performed with the pan opened (no lid), partially closed or hermetically closed. Moreover, the lid can have a hole on it (to release gas produced in the heating process from the sample) or have a continuous surface. The operator will take these decisions based on sample characteristics as: physical state (solid or liquid), volatility, water content, degradation, among others. Pan and lids are made mainly of aluminum, but other materials can be used, *e.g.*, copper.

Moreover, the operator should choose some parameters when using the equipment. They are mainly: temperature range (from which temperature to what temperature the sample will be heated/cooled); temperature ramp (what is the degree of temperature variation, °C/min); temperature modulation (if the temperature will suffer minor variations during the ramp, ± °C/min every s) and the number of cycles (if just one heat/cool process will be performed or if there are several steps, *e.g.*, cool the sample after heated it).

Differential Thermal Analysis

DTA is a technique closely related to DSC, producing curves that, in theory, are not different from the ones obtained by the latter. The difference between DTA and DSC is basically one: in DSC, the temperature between the reference pan and the sample is maintained the same, being evaluated the difference in heat flow; in DTA, the heat flow is maintained the same between the reference pan and the sample and the temperature variation is compared between the samples. Nevertheless, DTA supplies only qualitative information, while DSC also gives quantitative ones. This is possible as the resistance of the thermoelectric disk in DSC is well defined, allowing the use of Equation 15 [180, 185]. The results of the thermal events between DSC and DTA are usually the same and will be discussed below.

Events Detected in DSC and DTA

Melting and recrystallization are the most easily detectable thermic events as they absorb or release high amounts of energy. However, other transitions that require a lower amount of energy to occur can have great importance. They are relaxations of the solid matter but not a melting event, *i.e.*, the material is still in its solid form, but now the molecules are less organized. Some examples are glass transition temperature (T_g), mainly used for amorphous drugs and polymers and Tp (pre-transition temperature), mainly used for liposomes. All of these events can tell us about the structuration of the nanocarriers, incompatibilities between the components, drug incorporation, among other information.

Melting Point

Melting point can be detected for crystalline solids mainly. It is characterized by an endothermic peak (first order transition). In a melting peak (Fig. **31**), we can analyze its peak high (PH), peak width (PW), enthalpy and melting peak temperature (PT). The lower the temperature scan rate, the more defined the peaks. Usually, all these parameters are related. The more enthalpy is required to melt something (more defined peaks), the values of PH and PT will be higher and the value of PW will be lower (less defined peaks).

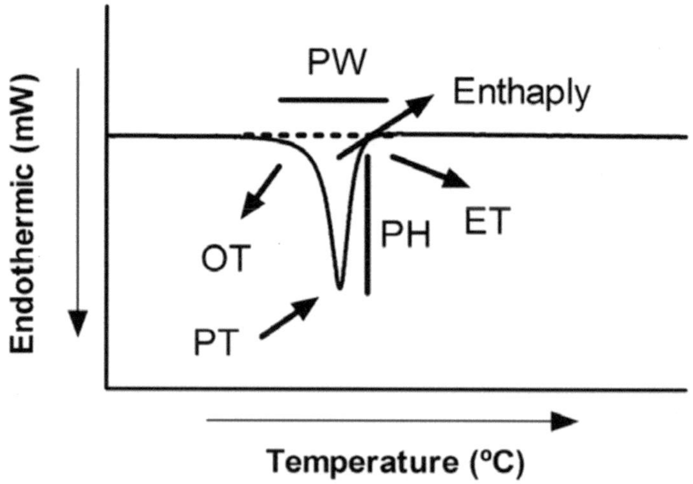

Fig. (31). Representation of a melting peak in DSC analysis.

Analyzing the melting of nanostructures and their bulk materials have the purpose to evaluate if the nanoparticles were formed (nanoparticles have less defines peaks than the correspondent bulk materials), to analyze whether there are interactions between the components (in the case of materials having some sort of interaction, there will be shifts in the melting point of each material), and to determine the extension of the impact of drug incorporation (differences in peak definition will give information about the differences that exist between a drug-loaded and a blank nanocarrier).

Glass Transition

Glass transitions are endothermic events that can be observed for amorphous materials. Therefore, they are usually studied by scientists who are dealing with polymers and polymeric nanoparticles. Below T_g, molecules are more static, conferring solid characteristics to the material. Above T_g, molecules gain more mobility and enter a liquid-like state. Cooling the material below T_g vitrifies the substance conferring to it the solid characteristics again [186]. Nevertheless, it should be stressed that a glassy state is not the same as a liquid. It is just a state where the molecules have high mobility, but they are still solid.

While melting points are represented as peaks in DSC and DTA curves, glass transitions are represented in the form of an "S" in the DSC curve (Fig. **32**). This happens due to a shift in the baseline due to a change in the heat capacity of the material (second order transition).

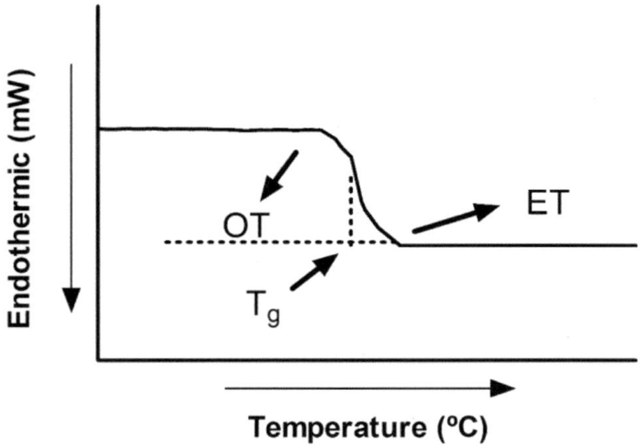

Fig. (32). Representation of a glass transition in DSC analysis.

T_g values of nanoparticles are usually lower than the ones observed for bulk materials. This is due to something called the confinement effect [187]. In sum, the small size and the big interface area reduce the interaction between the polymer chains and increase the interactions within the environment. A reduction in T_g is also an indication that the drug is entrapped in the nanoparticles. Whether the smaller drug molecules are located between polymer chains, they may contribute to decreasing the T_g values, by the plasticizing effect.

T_p and T_m

Phospholipids are lipids that, in an aqueous environment, organize themselves in double layer vesicles called liposomes. These lipids can be arranged in several forms in the space when in aqueous suspensions. From the most organized to the less, these forms are: planar (or interdigitated) gel phase, rippled gel phase, liquid crystalline phase and hexagonal phase. The changes between these phases can occur due to several factors, including temperature variation [181].

The temperature where the planar (or interdigitated) gel phase goes to the rippled gel phase is known as T_p. The temperature where the gel phase goes to the liquid crystalline phase is the phospholipid melting temperature, known as T_m. Some phospholipids derived from phosphatidylethanolamine groups also present the hexagonal phase. The temperature where the liquid crystalline phase goes to the hexagonal phase is known as T_h. Nevertheless, T_h detection is very difficult in calorimetric experiments [188].

T_p and T_m occur both as endothermic peaks (first order transitions) (Fig. **33**). T_m is the melting temperature of the lipid and the peak referent to it is very well defined. On the other hand, although T_p is represented by an endothermic peak, it has a much lower pH when comparing to T_m [188].

After liposomes are formed, it is important to know in which phase they are according to the temperature. This information helps the researcher to decide if certain temperatures are better for the preparation of the liposomes and if some behavior changes take place when liposomes are administered (usually, these formulations are stored at fridge temperatures and will be administrated at body temperatures, around 37°C).

Applications of DSC and DTA

In lipid nanocarriers, such as SLN and NLC, melting peaks have been used to compare the organization between different nanocarriers and to evaluate the impact of drug incorporation into them. A sharper peak (lower PW and higher pH) indicated better drug encapsulation [189]. Such data is important as it correlates directly with release and stability properties.

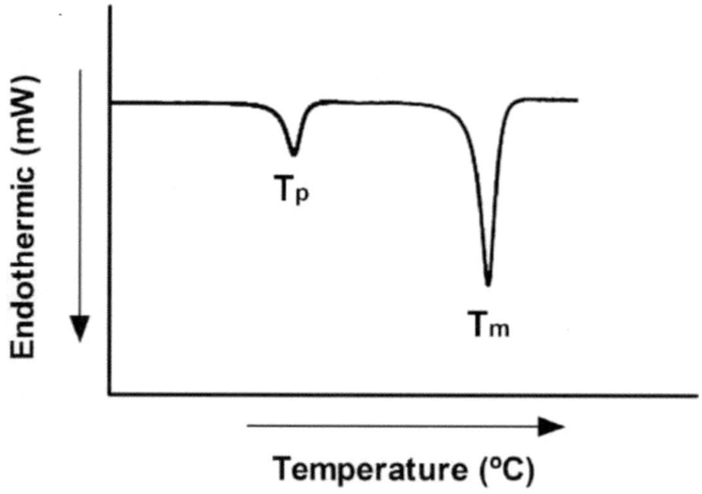

Fig. (33). Schematic representation of T_p and T_m in DSC analysis.

Regarding T_g, recent and interesting use of it is for thermosensitive polymeric nanocarriers. Heating the nanoparticles above the T_g can favor drug delivery as it will produce a restructuring of the nanoparticle. Therefore, it is possible to use such nanoparticles as vehicles for thermosensitive drug delivery. Lappe *et al.* [190] showed that when polymeric nanoparticles are in a temperature above Tg, a higher drug release is observed, showing that drug leakage is favored in this situation. Moreover, Zhu [191] observed that block copolymers with higher Tg formed more stable nanoparticles than ones with very low values.

Similarly, Preiss *et al.* [192] developed liposomes that showed higher drug releases when temperatures close to the transition phases (Tp and Tm) were applied. Moreover, knowing the T_m and the T_p of the liposomes prepared supplies important information to the scientist. One application regards the temperatures in the preparation of the liposomes. Above the T_p and T_m, incorporation of drugs in the liposome is facilitated as the double layers are more fluid [193, 194].

TGA

TGA is a technique that correlates the weigh of a sample to temperature variation. TGA equipment is composed of a precision balance and an oven (Fig. **34**). The chamber where the samples are put is fulfilled with inert gas.

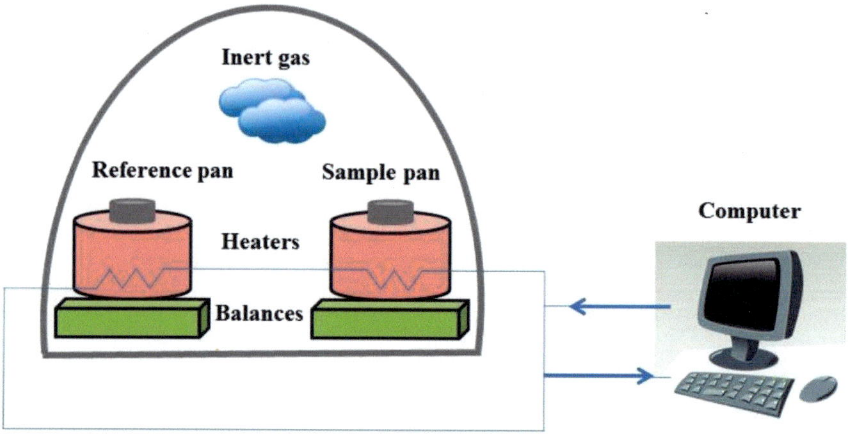

Fig. (34). Schematic representation of a TGA equipment.

There are three types of TGA analyses: isothermal, where the temperature in the equipment is maintained constant during all the analysis; almost isothermal, where the temperature increases until the point where a weight loss starts to be detected and during the fall in the weight, the temperature is maintained constant, and dynamic or conventional, where the temperature increases linearly throughout the analysis [180]. Therefore, it is possible to build TGA curves of weight *versus* temperature or weight *versus* time.

The sample is heated and when the temperature achieves a value where the sample starts to degrade or experience solvent evaporation, it is expected a diminishment in the sample weight and the balance will accuse so (Fig. **35**).

Although sometimes uniform, the weight loss in a sample can be due to several factors, like dehydration, vaporization, desorption, oxidation, reduction, among others. Therefore, the TGA curve can be formed just by a fall in the sample weight (Fig. **35a**) or can be in the format of a stair with different steps (Fig. **35b**). Using Fig. (**35b**) as an example, component 1 can be due to the elimination of the moisture, component 2 can be due to the water of crystallization and component 3 can be the elimination of solvent vestige in the sample. Looking at the format of the TGA curve and the temperatures where the steps appear, it is possible to take important information about the samples, regarding their contaminants, production process and, naturally, degradation temperature.

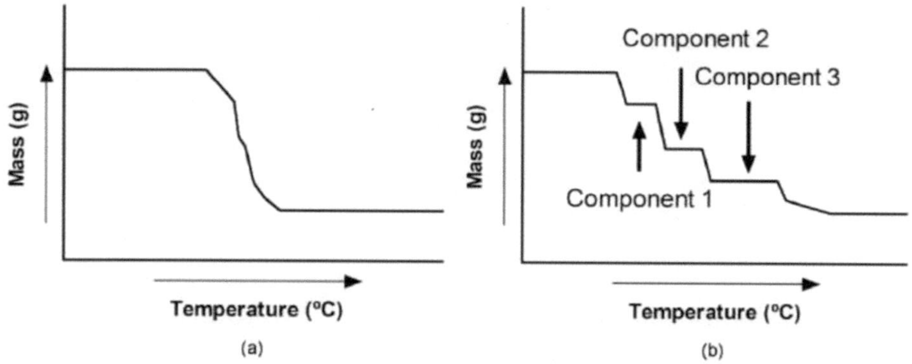

Fig. 35. Representative TGA curves with **(a)** a unique weight fall and **(b)** different weight falls (steps).

Applications of TGA

Some reports in the literature have shown that TGA has also been utilized in the characterization of nanosystems. Omwoyo *et al.* [195] used TGA analysis to determine moisture content in spray dried SLN in order to evaluate if the process was well conducted. Eloy *et al.* [196] used TGA to determine at which temperature the liposomes prepared were stable. Moreover, comparing the TGA curves of blank and curcumin-loaded mesoporous silica nanoparticles, it was possible to determine the curcumin content in the nanocarriers [197].

CONCLUSION

Throughout this chapter, many techniques were discussed, including microscopic, spectroscopic, spectrophotometric, and thermal analyses. Beyond the techniques herein discussed, there are many others, including NMR, HPLC, MS, Raman spectroscopy, and other microscopy techniques. It was not possible to discuss all of them, as we intended to focus on the most commonly used approaches to characterize nanosystems and their bulk constituents. Nevertheless, it is important to know that many other techniques that were not explained herein can be used efficiently in the process of nanoparticle characterization and provide further information.

Load a drug in a particle with nanometric size does not automatically enhance its properties. Depending on the intended use, the nanoparticles produced should have different characteristics. Therefore, it is essential to perform different tests to evaluate if the nanoparticles achieved the right characteristics after their production and during storage.

However, one problem that may occur is the inexistence of some equipment available for the scientist, which cannot allow the performance of one specific test. Therefore, the scientist should know the most important features that the nanoparticles produced should have in order to prioritize some analyses. Moreover, it is essential to know that combining two or more analysis techniques gives a complete view of the sample and helps to confirm whether a result obtained by an instrument can be obtained by others, attesting to their reliability.

Finally, a good physicochemical characterization will guarantee that the production and synthesis of nanoparticles were successful, assuring that the final goals will be obtained.

CONSENT FOR PUBLICATION

Not applicable.

CONFLICT OF INTEREST

The author declares no conflict of interest, financial or otherwise.

ACKNOWLEDGEMENTS

Declared none.

REFERENCES

[1] Fathi M, Barar J. Perspective highlights on biodegradable polymeric nanosystems for targeted therapy of solid tumors. Bioimpacts 2017; 7(1): 49-57. [http://dx.doi.org/10.15171/bi.2017.07] [PMID: 28546953]

[2] Baer DR, Engelhard MH, Johnson GE, *et al.* Surface characterization of nanomaterials and nanoparticles: Important needs and challenging opportunities. J Vac Sci Technol A 2013; 31(5): 50820. [http://dx.doi.org/10.1116/1.4818423] [PMID: 24482557]

[3] Bottaro E, Nastruzzi C. "Off-the-shelf" microfluidic devices for the production of liposomes for drug delivery. Mater Sci Eng C 2016; 64: 29-33. [http://dx.doi.org/10.1016/j.msec.2016.03.056] [PMID: 27127025]

[4] Wang Y, Wang S, Firempong CK, *et al.* Enhanced solubility and bioavailability of naringenin *via* liposomal nanoformulation: Preparation and *in vitro* and *in vivo* evaluations. AAPS PharmSciTech 2017; 18(3): 586-94. [http://dx.doi.org/10.1208/s12249-016-0537-8] [PMID: 27151135]

[5] Rao CNR, Biswas K. Characterization of nanomaterials by physical methods. Annu Rev Anal Chem (Palo Alto, Calif) 2009; 2(1): 435-62. [http://dx.doi.org/10.1146/annurev-anchem-060908-155236] [PMID: 20636070]

[6] Ju-Nam Y, Lead JR. Manufactured nanoparticles: an overview of their chemistry, interactions and potential environmental implications. Sci Total Environ 2008; 400(1-3): 396-414. [http://dx.doi.org/10.1016/j.scitotenv.2008.06.042] [PMID: 18715626]

[7] Li Y, Somorjai GA. Nanoscale advances in catalysis and energy applications. Nano Lett 2010; 10(7):

2289-95.
[http://dx.doi.org/10.1021/nl101807g] [PMID: 20524636]

[8] Lin P-C, Lin S, Wang PC, Sridhar R. Techniques for physicochemical characterization of nanomaterials. Biotechnol Adv 2014; 32(4): 711-26.
[http://dx.doi.org/10.1016/j.biotechadv.2013.11.006] [PMID: 24252561]

[9] Richman EK, Hutchison JE. The nanomaterial characterization bottleneck. ACS Nano 2009; 3(9): 2441-6.
[http://dx.doi.org/10.1021/nn901112p] [PMID: 19769400]

[10] Crist RM, Grossman JH, Patri AK, *et al.* Common pitfalls in nanotechnology: lessons learned from NCI's Nanotechnology Characterization Laboratory. Integr Biol 2013; 5(1): 66-73.
[http://dx.doi.org/10.1039/c2ib20117h] [PMID: 22772974]

[11] Fornaguera C, Solans C. Methods for the *in vitro* characterization of nanomedicines—Biological Component Interaction. J Pers Med 2017; 7(1): 2.
[http://dx.doi.org/10.3390/jpm7010002] [PMID: 28134833]

[12] Pettibone JM, Gigault J, Hackley VA. Discriminating the states of matter in metallic nanoparticle transformations: what are we missing? ACS Nano 2013; 7(3): 2491-9.
[http://dx.doi.org/10.1021/nn3058517] [PMID: 23425128]

[13] Panahi Y, Farshbaf M, Mohammadhosseini M, *et al.* Recent advances on liposomal nanoparticles: synthesis, characterization and biomedical applications. Artif Cells Nanomed Biotechnol 2017; 45(4): 788-99.
[http://dx.doi.org/10.1080/21691401.2017.1282496] [PMID: 28278586]

[14] Bhattacharjee S. DLS and zeta potential - What they are and what they are not? J Control Rel 2016; 235: 337-51.
[http://dx.doi.org/10.1016/j.jconrel.2016.06.017] [PMID: 27297779]

[15] Xia X, Liu H, Lv H, Zhang J, Zhou J, Zhao Z. Preparation, characterization, and *in vitro / vivo* studies of oleanolic acid-loaded lactoferrin nanoparticles. Drug Des Devel Ther 2017; 11: 1417-27.
[http://dx.doi.org/10.2147/DDDT.S133997] [PMID: 28533680]

[16] van Ommen JR, Kooijman D, Niet Md, Talebi M, Goulas A. Continuous production of nanostructured particles using spatial atomic layer deposition. J Vac Sci Technol A 2015; 33(2): 021513.
[http://dx.doi.org/10.1116/1.4905725]

[17] Dadkhah Tehrani A, Parsamanesh M. Preparation, characterization and drug delivery study of a novel nanobiopolymeric multidrug delivery system. Mater Sci Eng C 2017; 73: 516-24.
[http://dx.doi.org/10.1016/j.msec.2016.12.103] [PMID: 28183640]

[18] Cadogan SP, Hahn CJ, Rausch MH, Fröba AP. Study on the applicability of dynamic light scattering (DLS) to microemulsions including supercritical carbon dioxide-swollen micelles. J Colloid Interface Sci 2017; 499: 202-8.
[http://dx.doi.org/10.1016/j.jcis.2017.03.111] [PMID: 28384538]

[19] Fröhlich E. The role of surface charge in cellular uptake and cytotoxicity of medical nanoparticles. Int J Nanomed 2012; 5577.

[20] Bhattacharjee S, Rietjens IM, Singh MP, *et al.* Cytotoxicity of surface-functionalized silicon and germanium nanoparticles: the dominant role of surface charges. Nanoscale 2013; 5(11): 4870-83.
[http://dx.doi.org/10.1039/c3nr34266b] [PMID: 23619571]

[21] Axson JL, Stark DI, Bondy AL, *et al.* Rapid kinetics of size and pH-dependent dissolution and aggregation of silver nanoparticles in simulated gastric fluid. J Phys Chem C Nanomater Interfaces 2015; 119(35): 20632-41.
[http://dx.doi.org/10.1021/acs.jpcc.5b03634] [PMID: 28373899]

[22] Almeida A, Silva D, Goncalves V, Sarmento B. Synthesis and characterization of chitosan-grafted-polycaprolactone micelles for modulate intestinal paclitaxel delivery. Drug Deliv Transl Res 2017;

8(2): 387-97.
[http://dx.doi.org/10.1007/s13346-017-0357-8] [PMID: 28116655]

[23] Malvern-Instruments. Zetasizer Nano Series User Manual. 2004.

[24] Malvern-Instruments. Zetasizer 1000/2000/3000 - PCS theory. 1996.

[25] Einstein A. Investigations on the theory of the brownian movement. Dover Publications; 1956.

[26] Li Y, Barron AR. Dynamic Light Scattering. OpenStax-CNX. 2014.

[27] Haghiralsadat F, Amoabediny G, Sheikhha MH, Forouzanfar T, Helder MN, Zandieh-Doulabi B. A novel approach on drug delivery: Investigation of a new nano-formulation of liposomal doxorubicin and biological evaluation of entrapped Doxorubicin on various osteosarcoma cell lines. Cell J 2017; 19 (Suppl. 1): 55-65.
[PMID: 28580308]

[28] Wu S-Y, Debele TA, Kao Y-C, Tsai H-C. Synthesis and characterization of dual-sensitive fluorescent nanogels for enhancing drug delivery and tracking intracellular drug delivery. Int J Mol Sci 2017; 18(5): 1090.
[http://dx.doi.org/10.3390/ijms18051090] [PMID: 28534813]

[29] Carneiro G, Silva EL, Pacheco LA, *et al.* Formation of ion pairing as an alternative to improve encapsulation and anticancer activity of all-trans retinoic acid loaded in solid lipid nanoparticles. Int J Nanomedicine 2012; 7: 6011-20.
[PMID: 23251090]

[30] da Silva FLO, da Silva Tinoco LM, Ferreira LAM, Malagutti AR, Carneiro G. Determination of all-trans retinoic acid loaded in solid lipid nanoparticles by differential pulse voltammetry at glassy carbon electrode. Electrochim Acta 2015; 182: 929-34.
[http://dx.doi.org/10.1016/j.electacta.2015.10.018]

[31] Silva EL, Carneiro G, Caetano PA, *et al.* Nanostructured lipid carriers loaded with tributyrin as an alternative to improve anticancer activity of all-trans retinoic acid. Expert Rev Anticancer Ther 2015; 15(2): 247-56.
[http://dx.doi.org/10.1586/14737140.2015.1000868] [PMID: 25611812]

[32] Silva EL, Carneiro G, De Araujo LA, *et al.* Solid lipid nanoparticles loaded with retinoic acid and lauric acid as an alternative for topical treatment of acne vulgaris. J Nanosci Nanotechnol 2014; 14: 1-8.
[PMID: 26328443]

[33] Silva EL, Lima FA, Carneiro G, Gomes DA, de Souza-Fagundes EM, Ferreira LA. Improved *in vitro* Antileukemic activity of all-trans retinoic acid loaded in cholesteryl butyrate solid lipid nanoparticles. J Nanosci Nanotechnol 2016; 16(2): 1291-300.
[http://dx.doi.org/10.1166/jnn.2016.11677] [PMID: 27433579]

[34] Barbosa MV, Monteiro LO, Carneiro G, *et al.* Experimental design of a liposomal lipid system: A potential strategy for paclitaxel-based breast cancer treatment. Colloids Surf B Biointerfaces 2015; 136: 553-61.
[http://dx.doi.org/10.1016/j.colsurfb.2015.09.055] [PMID: 26454545]

[35] Almeida OP, Marcial SPS, Gouveia FPP, Carneiro G. Validation of a chromatographic analytical method for quantification of benznidazole incorporated in nanostructured lipid formulations. J Braz Chem Soc 2017; 28(2): 236-41.

[36] Li Y, Zhi X, Lin J, You X, Yuan J. Preparation and characterization of DOX loaded keratin nanoparticles for pH/GSH dual responsive release. Mater Sci Eng C 2017; 73: 189-97.
[http://dx.doi.org/10.1016/j.msec.2016.12.067] [PMID: 28183597]

[37] Ferreira M, Chaves LL, Lima SAC, Reis S. Optimization of nanostructured lipid carriers loaded with methotrexate: A tool for inflammatory and cancer therapy. Int J Pharm 2015; 492(1-2): 65-72.
[http://dx.doi.org/10.1016/j.ijpharm.2015.07.013] [PMID: 26169145]

[38] Fleischer CC, Payne CK. Nanoparticle-cell interactions: molecular structure of the protein corona and cellular outcomes. Acc Chem Res 2014; 47(8): 2651-9.
[http://dx.doi.org/10.1021/ar500190q] [PMID: 25014679]

[39] Liu W, Rose J, Plantevin S, Auffan M, Bottero JY, Vidaud C. Protein corona formation for nanomaterials and proteins of a similar size: hard or soft corona? Nanoscale 2013; 5(4): 1658-68.
[http://dx.doi.org/10.1039/c2nr33611a] [PMID: 23334428]

[40] Lesniak A, Fenaroli F, Monopoli MP, Åberg C, Dawson KA, Salvati A. Effects of the presence or absence of a protein corona on silica nanoparticle uptake and impact on cells. ACS Nano 2012; 6(7): 5845-57.
[http://dx.doi.org/10.1021/nn300223w] [PMID: 22721453]

[41] Sample dispersion and refractive index guide. 2007.

[42] Da Róz A, Leite F, Ferreira M, De Oliveira O Jr. Técnicas de Nanocaracterização. Rio de Janeiro: Elsevier 2015.

[43] Fernández JDS. Estabilidad coloidal de nanoestructuras liposómicas. PhD dissertation 2007.

[44] Hackley VA, Clogston JD. Measuring the size of nanoparticles in aqueous media using batch-mode dynamic light scattering. Technology NIoSa 2015. editor
[http://dx.doi.org/10.6028/NIST.SP.1200-6]

[45] Mastersizer 3000 - Smarter Particle Sizing

[46] de Boer GBJ, de Weerd C, Thoenes D, Goossens HWJ. Laser diffraction spectrometry: fraunhofer diffraction *versus* mie scattering. Part Part Syst Charact 1987; 4(1-4): 14-9.
[http://dx.doi.org/10.1002/ppsc.19870040104]

[47] Keck CM, Müller RH. Size analysis of submicron particles by laser diffractometry--90% of the published measurements are false. Int J Pharm 2008; 355(1-2): 150-63.
[http://dx.doi.org/10.1016/j.ijpharm.2007.12.004] [PMID: 18201848]

[48] Mitchell JP, Nagel MW, Nichols S, Nerbrink O. Laser diffractometry as a technique for the rapid assessment of aerosol particle size from inhalers. J Aerosol Med 2006; 19(4): 409-33.
[http://dx.doi.org/10.1089/jam.2006.19.409] [PMID: 17196072]

[49] Malvern-Instruments. Mastersizer 2000 User Manual. 2007:154.

[50] Agimelen OS, Mulholland AJ, Sefcik J. Modelling of artefacts in estimations of particle size of needle-like particles from laser diffration measurements. Chem Eng Sci 2017; 158(2): 445-52.
[http://dx.doi.org/10.1016/j.ces.2016.10.031]

[51] Manaia EB, Abuçafy MP, Chiari-Andréo BG, Silva BL, Oshiro Junior JA, Chiavacci LA. Physicochemical characterization of drug nanocarriers. Int J Nanomed 2017; 12: 4991-5011.
[http://dx.doi.org/10.2147/IJN.S133832] [PMID: 28761340]

[52] Zeta-Meter-Inc. Zeta Potential - A complete course in 5 minutes. 2011.

[53] Montes Ruiz-Cabello FJ, Maroni P, Borkovec M. Direct measurements of forces between different charged colloidal particles and their prediction by the theory of Derjaguin, Landau, Verwey, and Overbeek (DLVO). J Chem Phys 2013; 138(23): 234705.
[http://dx.doi.org/10.1063/1.4810901] [PMID: 23802974]

[54] Montes Ruiz-Cabello FJ, Trefalt G, Maroni P, Borkovec M. Electric double-layer potentials and surface regulation properties measured by colloidal-probe atomic force microscopy. Phys Rev E Stat Nonlin Soft Matter Phys 2014; 90(1): 012301.
[http://dx.doi.org/10.1103/PhysRevE.90.012301] [PMID: 25122297]

[55] Vincent B. Early (pre-DLVO) studies of particle aggregation. Adv Colloid Interface Sci 2012; 170(1-2): 56-67.
[http://dx.doi.org/10.1016/j.cis.2011.12.003] [PMID: 22257683]

[56] Hunter RJ. Zeta potential in colloid science: Principles and Applications. Sydney: Academic Press 1981.

[57] Clogston JD. Measuring zeta potential of nanoparticles. In: Laboratory NC, editor. 2009.

[58] Malvern-Instruments. Zeta Potential - An introduction in 30 minutes. 2011.

[59] Castro GA, Ferreira LAM, Oréfice RL, Buono VTL. Characterization of a new solid lipid nanoparticle formulation containing retinoic acid for topical treatment of acne. Powder Diffr 2008; 23(S1): 30-5. [http://dx.doi.org/10.1154/1.2903515]

[60] Castro GA, Oréfice RL, Vilela JM, Andrade MS, Ferreira LA. Development of a new solid lipid nanoparticle formulation containing retinoic acid for topical treatment of acne. J Microencapsul 2007; 24(5): 395-407. [http://dx.doi.org/10.1080/02652040701288519] [PMID: 17578730]

[61] Tran TH, Choi JY, Ramasamy T, *et al.* Hyaluronic acid-coated solid lipid nanoparticles for targeted delivery of vorinostat to CD44 overexpressing cancer cells. Carbohydr Polym 2014; 114: 407-15. [http://dx.doi.org/10.1016/j.carbpol.2014.08.026] [PMID: 25263908]

[62] Canevarolo S Jr. Técnicas de caracterização de polímeros. 1st ed., São Paulo: Artiliber 2003.

[63] Elliot MS, Poon WC. Conventional optical microscopy of colloidal suspensions. Adv Colloid Interface Sci 2001; 92(1-3): 133-94. [http://dx.doi.org/10.1016/S0001-8686(00)00070-1] [PMID: 11583297]

[64] Montgomery PC, Leong-Hoi A. Emerging optical nanoscopy techniques. Nanotechnol Sci Appl 2015; 8: 31-44. [http://dx.doi.org/10.2147/NSA.S50042] [PMID: 26491270]

[65] Egerton R. Physical Principles of Electron Microscopy: An Introduction to TEM, SEM, and AEM. 2nd ed., New York: Springer 2016. [http://dx.doi.org/10.1007/978-3-319-39877-8]

[66] Keller H, Watkins S. Contrast Enhancement in Light Microscopy. 2013.

[67] Koike-Tani M, Tani T, Mehta SB, Verma A, Oldenbourg R. Polarized light microscopy in reproductive and developmental biology. Mol Reprod Dev 2015; 82(7-8): 548-62. [http://dx.doi.org/10.1002/mrd.22221] [PMID: 23901032]

[68] Oldenbourg R. Polarized light microscopy: principles and practice. Cold Spring Harb Protoc 2013; 2013(11): 1023-36. [http://dx.doi.org/10.1101/pdb.top078600] [PMID: 24184765]

[69] Sibilia J. A Guide to Materials Characterization and Chemical analysis. 2nd ed., New York: Wiley-VCH 1996.

[70] Gulsuner HU, Gengec NA, Kilinc M, Erbil HY, Tekinay AB. Osteoselection supported by phase separated polymer blend films. J Biomed Mater Res A 2015; 103(1): 154-61. [http://dx.doi.org/10.1002/jbm.a.35164] [PMID: 24619979]

[71] Joksimovic R, Watanabe S, Riemer S, Gradzielski M, Yoshikawa K. Self-organized patterning through the dynamic segregation of DNA and silica nanoparticles. Sci Rep 2014; 4(1): 3660. [http://dx.doi.org/10.1038/srep03660] [PMID: 24413900]

[72] Verma P, Ahuja M. Optimization, characterization and evaluation of chitosan-tailored cubic nanoparticles of clotrimazole. Int J Biol Macromol 2015; 73: 138-45. [http://dx.doi.org/10.1016/j.ijbiomac.2014.10.065] [PMID: 25463320]

[73] Xiao L, Qiao Y, He Y, Yeung ES. Three dimensional orientational imaging of nanoparticles with darkfield microscopy. Anal Chem 2010; 82(12): 5268-74. [http://dx.doi.org/10.1021/ac1006848] [PMID: 20469837]

[74] Wang S-H, Lee C-W, Tseng F-G, Liang K-K, Wei P-K. Evolution of gold nanoparticle clusters in

living cells studied by sectional dark-field optical microscopy and chromatic analysis. J Biophotonics 2016; 9(7): 738-49.
[http://dx.doi.org/10.1002/jbio.201500182] [PMID: 29943945]

[75] Bose S, Keller SS, Alstrøm TS, Boisen A, Almdal K. Process optimization of ultrasonic spray coating of polymer films. Langmuir 2013; 29(23): 6911-9.
[http://dx.doi.org/10.1021/la4010246] [PMID: 23631433]

[76] Zhou W, Wang Z. Scanning Microscopy for Nanotechnology. 1st ed., New York: Springer 2006.

[77] Goldestein J, Newbury D, Joy D, *et al.* Scanning Electron Microscopy and X-ray Microanalysis. 3rd ed., New York: Springer 2003.
[http://dx.doi.org/10.1007/978-1-4615-0215-9]

[78] Collett B. Scanning electron microscopy: a review and report of research in wood science. Wood Fiber Sci 1970; 2: 113-33.

[79] Zhu F-Y, Wang Q-Q, Zhang X-S, Hu W, Zhao X, Zhang HX. 3D nanostructure reconstruction based on the SEM imaging principle, and applications. Nanotechnology 2014; 25(18): 185705.
[http://dx.doi.org/10.1088/0957-4484/25/18/185705] [PMID: 24739789]

[80] Vladár A, Ming B. Measuring the size of colloidal gold nano-particles using high-resolution scanning electron microscopy. Tecnology NIoSa 2011. editor

[81] Suga M, Asahina S, Sakuda Y, *et al.* Recent progress in scanning electron microscopy for the characterization of fine structural details of nano materials. Prog Solid State Chem 2014; 42(1-2): 1-21.
[http://dx.doi.org/10.1016/j.progsolidstchem.2014.02.001]

[82] Bogner A, Jouneau PH, Thollet G, Basset D, Gauthier C. A history of scanning electron microscopy developments: towards "wet-STEM" imaging. Micron 2007; 38(4): 390-401.
[http://dx.doi.org/10.1016/j.micron.2006.06.008] [PMID: 16990007]

[83] de Jonge N, Ross FM. Electron microscopy of specimens in liquid. Nat Nanotechnol 2011; 6(11): 695-704.
[http://dx.doi.org/10.1038/nnano.2011.161] [PMID: 22020120]

[84] Stabentheiner E, Zankel A, Pölt P. Environmental scanning electron microscopy (ESEM)--a versatile tool in studying plants. Protoplasma 2010; 246(1-4): 89-99.
[http://dx.doi.org/10.1007/s00709-010-0155-3] [PMID: 20446004]

[85] Méndez-Vilas A, Jódar-Reyes AB, González-Martín ML. Ultrasmall liquid droplets on solid surfaces: production, imaging, and relevance for current wetting research. Small 2009; 5(12): 1366-90.
[http://dx.doi.org/10.1002/smll.200800819] [PMID: 19507183]

[86] Gaumet M, Vargas A, Gurny R, Delie F. Nanoparticles for drug delivery: the need for precision in reporting particle size parameters. Eur J Pharm Biopharm 2008; 69(1): 1-9.
[http://dx.doi.org/10.1016/j.ejpb.2007.08.001] [PMID: 17826969]

[87] Bootz A, Vogel V, Schubert D, Kreuter J. Comparison of scanning electron microscopy, dynamic light scattering and analytical ultracentrifugation for the sizing of poly(butyl cyanoacrylate) nanoparticles. Eur J Pharm Biopharm 2004; 57(2): 369-75.
[http://dx.doi.org/10.1016/S0939-6411(03)00193-0] [PMID: 15018998]

[88] Leopold C, Augustin T, Schwebler T, Lehmann J, Liebig WV, Fiedler B. Influence of carbon nanoparticle modification on the mechanical and electrical properties of epoxy in small volumes. J Colloid Interface Sci 2017; 506: 620-32.
[http://dx.doi.org/10.1016/j.jcis.2017.07.085] [PMID: 28763766]

[89] Carrillo C, Sánchez-Hernández N, García-Montoya E, *et al.* DNA delivery *via* cationic solid lipid nanoparticles (SLNs). Eur J Pharm Sci 2013; 49(2): 157-65.
[http://dx.doi.org/10.1016/j.ejps.2013.02.011] [PMID: 23454134]

[90] Kwon H, Lim E, Lee S, Lee KK. Synthesis and characterization of poly(vinylpyrrolidone)-capped silicon-nanoparticles. J Nanosci Nanotechnol 2014; 14(8): 5983-7. [http://dx.doi.org/10.1166/jnn.2014.8434] [PMID: 25936041]

[91] Dalai S, Pakrashi S, Joyce Nirmala M, *et al.* Cytotoxicity of TiO_2 nanoparticles and their detoxification in a freshwater system. Aquat Toxicol 2013; 138-139: 1-11. [http://dx.doi.org/10.1016/j.aquatox.2013.04.005] [PMID: 23680676]

[92] Hall JB, Dobrovolskaia MA, Patri AK, McNeil SE. Characterization of nanoparticles for therapeutics. Nanomedicine (Lond) 2007; 2(6): 789-803. [http://dx.doi.org/10.2217/17435889.2.6.789] [PMID: 18095846]

[93] Hondow N, Wang P, Brydson R, *et al.* TEM analysis of nanoparticle dispersions with application towards the quantification of *in vitro* cellular uptake. J Phys Conf Ser 2012; 371: 012020. [http://dx.doi.org/10.1088/1742-6596/371/1/012020]

[94] Zuo J, Spence J. Advanced Transmission Electron Microscopy. 1st ed., New York: Springer 2017. [http://dx.doi.org/10.1007/978-1-4939-6607-3]

[95] Sharif A. Review on advances in nanoscale microscopy in cement research. Micron 2016; 80: 45-58. [http://dx.doi.org/10.1016/j.micron.2015.09.010] [PMID: 26447783]

[96] Mammadov R, Tekinay AB, Dana A, Guler MO. Microscopic characterization of peptide nanostructures. Micron 2012; 43(2-3): 69-84. [http://dx.doi.org/10.1016/j.micron.2011.07.006] [PMID: 21821422]

[97] Urban KW. Studying atomic structures by aberration-corrected transmission electron microscopy. Science 2008; 321(5888): 506-10. [http://dx.doi.org/10.1126/science.1152800] [PMID: 18653874]

[98] Müller SA, Aebi U, Engel A. What transmission electron microscopes can visualize now and in the future. J Struct Biol 2008; 163(3): 235-45. [http://dx.doi.org/10.1016/j.jsb.2008.05.008] [PMID: 18614377]

[99] Williams DB, Carter CB. Transmission Electron Microscopy. 2nd ed., New York: Springer 2009. [http://dx.doi.org/10.1007/978-0-387-76501-3]

[100] Chen X, Wen J. *In situ* wet-cell TEM observation of gold nanoparticle motion in an aqueous solution. Nanoscale Res Lett 2012; 7(1): 598. [http://dx.doi.org/10.1186/1556-276X-7-598] [PMID: 23107519]

[101] Domingos RF, Baalousha MA, Ju-Nam Y, *et al.* Characterizing manufactured nanoparticles in the environment: multimethod determination of particle sizes. Environ Sci Technol 2009; 43(19): 7277-84. [http://dx.doi.org/10.1021/es900249m] [PMID: 19848134]

[102] Thompson RF, Walker M, Siebert CA, Muench SP, Ranson NA. An introduction to sample preparation and imaging by cryo-electron microscopy for structural biology. Methods 2016; 100: 3-15. [http://dx.doi.org/10.1016/j.ymeth.2016.02.017] [PMID: 26931652]

[103] Rice SB, Chan C, Brown SC, *et al.* Particle size distributions by transmission electron microscopy: an interlaboratory comparison case study. Metrologia 2013; 50(6): 663-78. [http://dx.doi.org/10.1088/0026-1394/50/6/663] [PMID: 26361398]

[104] Pennycook S, Lupini A, Varela M, *et al.* Scanning transmission electron microscopy for nanostructure characterization.Scanning Microscopy for Nanotechnology. 1st ed., New York: Springer 2006. [http://dx.doi.org/10.1007/978-0-387-39620-0_6]

[105] Deepak F, Mayoral A, Arenal R. Advanced Transmission Electron Microscopy. 1st ed., New York: Springer 2015. [http://dx.doi.org/10.1007/978-3-319-15177-9]

[106] Ponce A, Mejía-Rosales S, José-Yacamán M. Scanning transmission electron microscopy methods for

the analysis of nanoparticles.Nanoparticles in Biology and Medicine. 1st ed., New York: Springer 2012.
[http://dx.doi.org/10.1007/978-1-61779-953-2_37]

[107] Buban JP, Ramasse Q, Gipson B, Browning ND, Stahlberg H. High-resolution low-dose scanning transmission electron microscopy. J Electron Microsc (Tokyo) 2010; 59(2): 103-12.
[http://dx.doi.org/10.1093/jmicro/dfp052] [PMID: 19915208]

[108] Walther T. What environmental transmission electron microscopy measures and how this links to diffusivity: thermodynamics *versus* kinetics. J Microsc 2015; 257(2): 87-91.
[http://dx.doi.org/10.1111/jmi.12193] [PMID: 25401262]

[109] Sharma R. An environmental transmission electron microscope for *in situ* synthesis and characterization of Nanomaterials. J Mater Res 2005; 20(07): 1695-707.
[http://dx.doi.org/10.1557/JMR.2005.0241]

[110] Bonevich J, Haller W. Measuring the size of nanoparticles using transmission electron microscopy (TEM). Technology NIoSa 2010. editor

[111] Hondow N, Brydson R, Brown A. The use of transmission electron microscopy in the quantification of nanoparticle dose. J Phys Conf Ser 2014; 522: 012055.
[http://dx.doi.org/10.1088/1742-6596/522/1/012055]

[112] Andrade-Mahecha MM, Pelissari FM, Tapia-Blácido DR, Menegalli FC. Achira as a source of biodegradable materials: Isolation and characterization of nanofibers. Carbohydr Polym 2015; 123: 406-15.
[http://dx.doi.org/10.1016/j.carbpol.2015.01.027] [PMID: 25843874]

[113] Habibi N. Preparation of biocompatible magnetite-carboxymethyl cellulose nanocomposite: characterization of nanocomposite by FTIR, XRD, FESEM and TEM. Spectrochim Acta A Mol Biomol Spectrosc 2014; 131: 55-8.
[http://dx.doi.org/10.1016/j.saa.2014.04.039] [PMID: 24820322]

[114] Paranjpe M, Finke JH, Richter C, *et al.* Physicochemical characterization of sildenafil-loaded solid lipid nanoparticle dispersions (SLN) for pulmonary application. Int J Pharm 2014; 476(1-2): 41-9.
[http://dx.doi.org/10.1016/j.ijpharm.2014.09.031] [PMID: 25261709]

[115] Rodrigues da Silva GH, Ribeiro LNM, Mitsutake H, *et al.* Optimised NLC: a nanotechnological approach to improve the anaesthetic effect of bupivacaine. Int J Pharm 2017; 529(1-2): 253-63.
[http://dx.doi.org/10.1016/j.ijpharm.2017.06.066] [PMID: 28655546]

[116] Chatterjee S, Biswas N, Datta A, Dey R, Maiti P. Atomic force microscopy in biofilm study. Microscopy (Oxf) 2014; 63(4): 269-78.
[http://dx.doi.org/10.1093/jmicro/dfu013] [PMID: 24793174]

[117] Kurland NE, Drira Z, Yadavalli VK. Measurement of nanomechanical properties of biomolecules using atomic force microscopy. Micron 2012; 43(2-3): 116-28.
[http://dx.doi.org/10.1016/j.micron.2011.07.017] [PMID: 21890365]

[118] Pillet F, Chopinet L, Formosa C, Dague E. Atomic Force Microscopy and pharmacology: from microbiology to cancerology. Biochim Biophys Acta 2014; 1840(3): 1028-50.
[http://dx.doi.org/10.1016/j.bbagen.2013.11.019] [PMID: 24291690]

[119] de Pablo PJ, Carrión-Vázquez M. Imaging biological samples with atomic force microscopy. Cold Spring Harb Protoc 2014; (2): pdb.top080473.
[http://dx.doi.org/10.1101/pdb.top080473]

[120] Zavala G. Atomic force microscopy, a tool for characterization, synthesis and chemical processes. Colloid Polym Sci 2007; 286(1): 85-95.
[http://dx.doi.org/10.1007/s00396-007-1791-9]

[121] Fleming A, Leang K. Design, Modeling and Control of Nanopositioning Systems. 1st ed., New York: Springer 2014.

[http://dx.doi.org/10.1007/978-3-319-06617-2]

[122] Hassellöv M, Readman JW, Ranville JF, Tiede K. Nanoparticle analysis and characterization methodologies in environmental risk assessment of engineered nanoparticles. Ecotoxicology 2008; 17(5): 344-61.
[http://dx.doi.org/10.1007/s10646-008-0225-x] [PMID: 18483764]

[123] Grobelny J, DelRio FW, Pradeep N, *et al.* Size measurement of nanoparticles using atomic force microscopy. Technology NIoSa 2009. editor

[124] Kainz B, Oprzeska-Zingrebe EA, Herrera JL. Biomaterial and cellular properties as examined through atomic force microscopy, fluorescence optical microscopies and spectroscopic techniques. Biotechnol J 2014; 9(1): 51-60.
[http://dx.doi.org/10.1002/biot.201300087] [PMID: 24265117]

[125] Churnside AB, Perkins TT. Ultrastable atomic force microscopy: improved force and positional stability. FEBS Lett 2014; 588(19): 3621-30.
[http://dx.doi.org/10.1016/j.febslet.2014.04.033] [PMID: 24801176]

[126] Eghiaian F, Rico F, Colom A, Casuso I, Scheuring S. High-speed atomic force microscopy: imaging and force spectroscopy. FEBS Lett 2014; 588(19): 3631-8.
[http://dx.doi.org/10.1016/j.febslet.2014.06.028] [PMID: 24937145]

[127] Dubes A, Parrot-Lopez H, Abdelwahed W, *et al.* Scanning electron microscopy and atomic force microscopy imaging of solid lipid nanoparticles derived from amphiphilic cyclodextrins. Eur J Pharm Biopharm 2003; 55(3): 279-82.
[http://dx.doi.org/10.1016/S0939-6411(03)00020-1] [PMID: 12754001]

[128] Tan YH, Schallom JR, Ganesh NV, Fujikawa K, Demchenko AV, Stine KJ. Characterization of protein immobilization on nanoporous gold using atomic force microscopy and scanning electron microscopy. Nanoscale 2011; 3(8): 3395-407.
[http://dx.doi.org/10.1039/c1nr10427f] [PMID: 21750834]

[129] Sitterberg J, Özcetin A, Ehrhardt C, Bakowsky U. Utilising atomic force microscopy for the characterisation of nanoscale drug delivery systems. Eur J Pharm Biopharm 2010; 74(1): 2-13.
[http://dx.doi.org/10.1016/j.ejpb.2009.09.005] [PMID: 19755155]

[130] El Kirat K, Burton I, Dupres V, Dufrene YF. Sample preparation procedures for biological atomic force microscopy. J Microsc 2005; 218(Pt 3): 199-207.
[http://dx.doi.org/10.1111/j.1365-2818.2005.01480.x] [PMID: 15958012]

[131] Baalousha M, Prasad A, Lead JR. Quantitative measurement of the nanoparticle size and number concentration from liquid suspensions by atomic force microscopy. Environ Sci Process Impacts 2014; 16(6): 1338-47.
[http://dx.doi.org/10.1039/C3EM00712J] [PMID: 24668140]

[132] Baalousha M, Lead JR. Characterization of natural and manufactured nanoparticles by atomic force microscopy: Effect of analysis mode, environment and sample preparation. Colloids Surf A Physicochem Eng Asp 2013; 419: 238-47.
[http://dx.doi.org/10.1016/j.colsurfa.2012.12.004]

[133] Boyd RD, Pichaimuthu SK, Cuenat A. New approach to inter-technique comparisons for nanoparticle size measurements; using atomic force microscopy, nanoparticle tracking analysis and dynamic light scattering. Colloids Surf A Physicochem Eng Asp 2011; 387(1-3): 35-42.
[http://dx.doi.org/10.1016/j.colsurfa.2011.07.020]

[134] Lamprecht C, Hinterdorfer P, Ebner A. Applications of biosensing atomic force microscopy in monitoring drug and nanoparticle delivery. Expert Opin Drug Deliv 2014; 11(8): 1237-53.
[http://dx.doi.org/10.1517/17425247.2014.917078] [PMID: 24809228]

[135] Eaton P, Quaresma P, Soares C, *et al.* A direct comparison of experimental methods to measure dimensions of synthetic nanoparticles. Ultramicroscopy 2017; 182: 179-90.

[http://dx.doi.org/10.1016/j.ultramic.2017.07.001] [PMID: 28692935]

[136] Moribe K, Limwikrant W, Higashi K, Yamamoto K. Structural evaluation of probucol nanoparticles in water by atomic force microscopy. Int J Pharm 2012; 427(2): 365-71. [http://dx.doi.org/10.1016/j.ijpharm.2012.02.020] [PMID: 22370474]

[137] Zanetti-Ramos BG, Fritzen-Garcia MB, de Oliveira CS, *et al.* Dynamic light scattering and atomic force microscopy techniques for size determination of polyurethane nanoparticles. Mater Sci Eng C 2009; 29(2): 638-40. [http://dx.doi.org/10.1016/j.msec.2008.10.040]

[138] Semwogerere D, Weeks E. Confocal Microscopy. Encyclopedia of Biomaterials and Biomedical Engineering. 2nd ed., Washington: Taylor & Francis 2005.

[139] Dürrenberger MB, Handschin S, Conde-Petit B, Escher F. Visualization of food structure by confocal laser scanning microscopy (CLSM). Lebensm Wiss Technol 2001; 34(1): 11-7. [http://dx.doi.org/10.1006/fstl.2000.0739]

[140] Watson TF. Fact and artefact in confocal microscopy. Adv Dent Res 1997; 11(4): 433-41. [http://dx.doi.org/10.1177/08959374970110040901] [PMID: 9470501]

[141] Hepler P, Gunning B. Confocal fluorescence microscopy of plant cells. Protoplasma 1998; 201: 121-7. [http://dx.doi.org/10.1007/BF01287411]

[142] Prasad V, Semwogerere D, Weeks ER. Confocal microscopy of colloids. J Phys Condens Matter 2007; 19(11): 113102. [http://dx.doi.org/10.1088/0953-8984/19/11/113102]

[143] Paddock SW. Principles and practices of laser scanning confocal microscopy. Mol Biotechnol 2000; 16(2): 127-49. [http://dx.doi.org/10.1385/MB:16:2:127] [PMID: 11131973]

[144] Cox G. Biological confocal microscopy. Mater Today (Kidlington) 2002; 5(3): 34-41. [http://dx.doi.org/10.1016/S1369-7021(02)05329-4]

[145] Nwaneshiudu A, Kuschal C, Sakamoto FH, Anderson RR, Schwarzenberger K, Young RC. Introduction to confocal microscopy. J Invest Dermatol 2012; 132(12): e3. [http://dx.doi.org/10.1038/jid.2012.429] [PMID: 23187113]

[146] Furrer P, Gurny R. Recent advances in confocal microscopy for studying drug delivery to the eye: concepts and pharmaceutical applications. Eur J Pharm Biopharm 2010; 74(1): 33-40. [http://dx.doi.org/10.1016/j.ejpb.2009.09.002] [PMID: 19755157]

[147] Zhang LW, Monteiro-Riviere NA. Use of confocal microscopy for nanoparticle drug delivery through skin. J Biomed Opt 2013; 18(6): 061214. [http://dx.doi.org/10.1117/1.JBO.18.6.061214] [PMID: 23224242]

[148] Sarmento B, Ribeiro A, Veiga F, Sampaio P, Neufeld R, Ferreira D. Alginate/chitosan nanoparticles are effective for oral insulin delivery. Pharm Res 2007; 24(12): 2198-206. [http://dx.doi.org/10.1007/s11095-007-9367-4] [PMID: 17577641]

[149] Shang L, Nienhaus K, Jiang X, *et al.* Nanoparticle interactions with live cells: Quantitative fluorescence microscopy of nanoparticle size effects. Beilstein J Nanotechnol 2014; 5: 2388-97. [http://dx.doi.org/10.3762/bjnano.5.248] [PMID: 25551067]

[150] Oréfice RL, Pereira MM, Mansur HS. Biomateriais: Fundamentos & aplicações. Rio de Janeiro: Cultura Médica 2012.

[151] Pavia DL, Lampman GM, Kriz GS, Vyvyan JR. Introduction to Spectroscopy. 5th ed., Stamford: Cengage Learning 2015.

[152] Hilfiker R. Polymorphism: in the Pharmaceutical Industry. Weinheim: WILEY-VCH 2006. [http://dx.doi.org/10.1002/3527607889]

[153] Silverstein RM, Webster FX, Kiemle DJ, Bryce DL. Spectrometric Identification of Organic Compounds. 8th ed., New Jersey: John Wiley & Sons 2015.

[154] Johal MS. Understanding Nanomaterials. Boca Raton: CRC Press 2011.

[155] Sharma S, Uttam KN. Investigation of the manganese stress on wheat plant by attenuated total reflectance Fourier transform infrared spectroscopy. Spectrosc Lett 2016; 49(8) [http://dx.doi.org/10.1080/00387010.2016.1212897]

[156] Li Y, Pan C, Xue Y, Meng X, Ding Y. A novel signal enhancement method for overlapped peaks with noise immunity. Spectrosc Lett 2016; 49(4) [http://dx.doi.org/10.1080/00387010.2016.1144614]

[157] Hind AR, Bhargava SK, McKinnon A. At the solid/liquid interface: FTIR/ATR--the tool of choice. Adv Colloid Interface Sci 2001; 93(1-3): 91-114. [http://dx.doi.org/10.1016/S0001-8686(00)00079-8] [PMID: 11591110]

[158] Alessio P, Aoki PHB, Furini LN, Aliaga AE, Leopoldo Constantino CJ. Spectroscopic techniques for characterization of Nanomaterials. 2017; 65-98. [http://dx.doi.org/10.1016/B978-0-323-49778-7.00003-5]

[159] Karthivashan G, Masarudin MJ, Kura AU, Abas F, Fakurazi S. Optimization, formulation, and characterization of multiflavonoids-loaded flavanosome by bulk or sequential technique. Int J Nanomedicine 2016; 11: 3417-34. [http://dx.doi.org/10.2147/IJN.S112045] [PMID: 27555765]

[160] Ayan AK, Yenilmez A, Eroglu H. Evaluation of radiolabeled curcumin-loaded solid lipid nanoparticles usage as an imaging agent in liver-spleen scintigraphy. Mater Sci Eng C 2017; 75: 663-70. [http://dx.doi.org/10.1016/j.msec.2017.02.114] [PMID: 28415513]

[161] Skoog DA, West DM, Holler FJ, Crouch SR. Fundamentals of Analytical of Chemistry. 9th ed., Belmont: Brooks/Cole 2014.

[162] Biju V, Mundayoor S, Omkumar RV, Anas A, Ishikawa M. Bioconjugated quantum dots for cancer research: present status, prospects and remaining issues. Biotechnol Adv 2010; 28(2): 199-213. [http://dx.doi.org/10.1016/j.biotechadv.2009.11.007] [PMID: 19969062]

[163] Sapsford KE, Tyner KM, Dair BJ, Deschamps JR, Medintz IL. Analyzing nanomaterial bioconjugates: a review of current and emerging purification and characterization techniques. Anal Chem 2011; 83(12): 4453-88. [http://dx.doi.org/10.1021/ac200853a] [PMID: 21545140]

[164] Cullity BD, Stock SR. Elements of X-Ray Diffraction. 3rd ed., New Jersey: Prentice-Hall 2001.

[165] Pecharsky VK, Zavalij PY. Fundamentals of Powder Diffraction and Structural Characterization of Materials. New York: Springer 2005.

[166] Wang H, Yuan XZ, Li H. PEM Full Cell Diagnostic Tools. New York: Taylor & Francis Group 2012.

[167] Bunjes H, Unruh T. Characterization of lipid nanoparticles by differential scanning calorimetry, X-ray and neutron scattering. Adv Drug Deliv Rev 2007; 59(6): 379-402. [http://dx.doi.org/10.1016/j.addr.2007.04.013] [PMID: 17658653]

[168] Mehnert W, Mäder K. Solid lipid nanoparticles: production, characterization and applications. Adv Drug Deliv Rev 2001; 47(2-3): 165-96. [http://dx.doi.org/10.1016/S0169-409X(01)00105-3] [PMID: 11311991]

[169] Silva AC, González-Mira E, García ML, *et al.* Preparation, characterization and biocompatibility studies on risperidone-loaded solid lipid nanoparticles (SLN): high pressure homogenization *versus* ultrasound. Colloids Surf B Biointerfaces 2011; 86(1): 158-65. [http://dx.doi.org/10.1016/j.colsurfb.2011.03.035] [PMID: 21530187]

[170] Jenning V, Gohla SH. Encapsulation of retinoids in solid lipid nanoparticles (SLN). J Microencapsul 2001; 18(2): 149-58.
[http://dx.doi.org/10.1080/02652040010000361] [PMID: 11253932]

[171] Jenning V, Schäfer-Korting M, Gohla S. Vitamin A-loaded solid lipid nanoparticles for topical use: drug release properties. J Control Rel 2000; 66(2-3): 115-26.
[http://dx.doi.org/10.1016/S0168-3659(99)00223-0] [PMID: 10742573]

[172] Craievich A. Synchrotron SAXS Studies of Nanostructured materials and colloidal Solutions: A review. Mater Res 2002; 5(1)
[http://dx.doi.org/10.1590/S1516-14392002000100002]

[173] Craievich AF. Small-angle X-ray Scattering by Nanostructured Materials Handbok of sol-gel science and technology. New York: Kluwer Academic Publishers 2005.

[174] Dencheva N, Stribeck A, Denchev Z. Nanostructure development in multicomponent polymer systems characterized by synchrotron X-ray scattering. Eur Polym J 2016; 81: 447-69.
[http://dx.doi.org/10.1016/j.eurpolymj.2016.02.004]

[175] Mertens HDT, Svergun DI. Combining NMR and small angle X-ray scattering for the study of biomolecular structure and dynamics. Arch Biochem Biophys 2017; 628: 33-41.
[http://dx.doi.org/10.1016/j.abb.2017.05.005] [PMID: 28501583]

[176] Zienkiewicz-Strzałka M, Skibińska M, Pikus S. Small-angle X-ray scattering (SAXS) studies of the structure of mesoporous silicas. Nucl Instrum Methods Phys Res B 2017; 411(15): 72-7.
[http://dx.doi.org/10.1016/j.nimb.2017.03.028]

[177] Mishra DK, Dhote V, Bhatnagar P, Mishra PK. Engineering solid lipid nanoparticles for improved drug delivery: promises and challenges of translational research. Drug Deliv Transl Res 2012; 2(4): 238-53.
[http://dx.doi.org/10.1007/s13346-012-0088-9] [PMID: 25787030]

[178] Xiao Y, Wiesner MR. Characterization of surface hydrophobicity of engineered nanoparticles. J Hazard Mater 2012; 215-216: 146-51.
[http://dx.doi.org/10.1016/j.jhazmat.2012.02.043] [PMID: 22417396]

[179] Ferreira DdosS, Faria SD, Lopes SCA, *et al.* Development of a bone-targeted pH-sensitive liposomal formulation containing doxorubicin: physicochemical characterization, cytotoxicity, and biodistribution evaluation in a mouse model of bone metastasis. Int J Nanomed 2016; 11: 3737-51.
[http://dx.doi.org/10.2147/IJN.S109966] [PMID: 27563241]

[180] Matos J, Mercuri L, Barros G. Análise térmica aplicada a fármacos e medicamentos.Biofarmacotécnica. 1st ed., Rio de Janeiro: Guanabara Koogan 2009.

[181] Demetzos C. Differential Scanning Calorimetry (DSC): a tool to study the thermal behavior of lipid bilayers and liposomal stability. J Liposome Res 2008; 18(3): 159-73.
[http://dx.doi.org/10.1080/08982100802310261] [PMID: 18770070]

[182] Gill P, Sauerbrunn S, Reading M. Modulated differential scanning calorimetry. J Therm Anal Calorim 1993; 40: 931-9.
[http://dx.doi.org/10.1007/BF02546852]

[183] Lampman S, Ed. Characterization and failure analysis of plastics. Novelty: ASM International. 2003.
[http://dx.doi.org/10.31399/asm.tb.cfap.9781627082815]

[184] Gill P, Moghadam TT, Ranjbar B. Differential scanning calorimetry techniques: applications in biology and nanoscience. J Biomol Tech 2010; 21(4): 167-93.
[PMID: 21119929]

[185] Yang J, Roy C. Using DTA to quantitatively determine enthalpy change over a wide temperature range by the "Mass Difference Baseline Method. Thermochim Acta 1999; 333(2): 131-40.
[http://dx.doi.org/10.1016/S0040-6031(99)00106-9]

[186] Roos YH. Glass transition temperature and its relevance in food processing. Annu Rev Food Sci Technol 2010; 1(1): 469-96.
[http://dx.doi.org/10.1146/annurev.food.102308.124139] [PMID: 22129345]

[187] Chen R, Huang D. Preparation of polymer nanoparticles, and the effect of nanoconfinement on glass transition, structural relaxation and crystallization. Front Chem China 2011; 6(4): 332-40.
[http://dx.doi.org/10.1007/s11458-011-0262-5]

[188] Smith E, Dea P. Differential scanning calorimetry studies of phospholipid membranes: The interdigitated gel phase.Applications of Calorimetry in a Wide Context - Differential Scanning Calorimetry. Isothermal Titration Calorimetry and Microcalorimetry: InTech 2013.

[189] Oliveira MS, Mussi SV, Gomes DA, *et al.* α-Tocopherol succinate improves encapsulation and anticancer activity of doxorubicin loaded in solid lipid nanoparticles. Colloids Surf B Biointerfaces 2016; 140: 246-53.
[http://dx.doi.org/10.1016/j.colsurfb.2015.12.019] [PMID: 26764108]

[190] Lappe S, Mulac D, Langer K. Polymeric nanoparticles - Influence of the glass transition temperature on drug release. Int J Pharm 2017; 517(1-2): 338-47.
[http://dx.doi.org/10.1016/j.ijpharm.2016.12.025] [PMID: 27986475]

[191] Zhu Z. Effects of amphiphilic diblock copolymer on drug nanoparticle formation and stability. Biomaterials 2013; 34(38): 10238-48.
[http://dx.doi.org/10.1016/j.biomaterials.2013.09.015] [PMID: 24070569]

[192] Preiss MR, Hart A, Kitchens C, Bothun GD. Hydrophobic nanoparticles modify the thermal release behavior of liposomes. J Phys Chem B 2017; 121(19): 5040-7.
[http://dx.doi.org/10.1021/acs.jpcb.7b01702] [PMID: 28441023]

[193] Kan P, Tsao C-W, Wang A-J, Su W-C, Liang H-F. A liposomal formulation able to incorporate a high content of Paclitaxel and exert promising anticancer effect. J Drug Deliv 2011; 2011: 629234.
[http://dx.doi.org/10.1155/2011/629234] [PMID: 21490755]

[194] Maia ALC, Fernandes C, de Oliveira CNP, *et al.* Liposomes containing Gadodiamide: preparation, physicochemical characterization, and *in vitro* cytotoxic evaluation. Curr Drug Deliv 2017; 14(4): 566-74.
[http://dx.doi.org/10.2174/1567201813666160907095404] [PMID: 27605018]

[195] Omwoyo WN, Ogutu B, Oloo F, *et al.* Preparation, characterization, and optimization of primaquine-loaded solid lipid nanoparticles. Int J Nanomed 2014; 9: 3865-74.
[PMID: 25143734]

[196] Eloy JO, Petrilli R, Topan JF, *et al.* Co-loaded paclitaxel/rapamycin liposomes: Development, characterization and *in vitro* and *in vivo* evaluation for breast cancer therapy. Colloids Surf B Biointerfaces 2016; 141: 74-82.
[http://dx.doi.org/10.1016/j.colsurfb.2016.01.032] [PMID: 26836480]

[197] Kotcherlakota R, Barui AK, Prashar S, *et al.* Curcumin loaded mesoporous silica: an effective drug delivery system for cancer treatment. Biomater Sci 2016; 4(3): 448-59.
[http://dx.doi.org/10.1039/C5BM00552C] [PMID: 26674254]

CHAPTER 3

Liposomes: The New Developments in Topical Drug Delivery

Gajanand Sharma[1], Kanika Thakur[1], Nitika Dehal[1], Akanksha Mahajan[1], Kaisar Raza[2] and **O.P. Katare[1,*]**

[1] *University Institute of Pharmaceutical Sciences, UGC Centre of Advanced Studies, Panjab University, Chandigarh-60014, India*

[2] *UGC-Centre of Excellence in Applications of Nanomaterials, Nanoparticles & Nanocomposites (Biomedical Sciences), Panjab University, Chandigarh-160014, India*

Abstract: Liposomes are spherical vesicular systems composed of lipid bilayers. Among the various novel drug delivery systems, it has been found that the liposomes follow an advanced technology for the delivery of drugs to the target site. In the present era, there are various liposomal formulations which are in clinical use. Due to the extensive efforts of the researchers, there has been tremendous progress in liposomal technology, *i.e.,* from, conventional vesicles to the 2nd generation liposomes. The current review article deals with the materials involved in the preparation of liposomes, techniques for preparation, characterization studies and application of liposomes for drug delivery.

Keywords: Characterization, Chemotherapy, Phospholipids, Preparative techniques, Vaccine delivery, Vesicular systems.

INTRODUCTION

Liposomes represent the first generation of nanocarriers that have successfully evolved from laboratory set up to clinical applications. Liposomes as drug delivery systems are well-established technology platform with great impact on patient healthcare [1, 2]. It has been more than 5 decades since liposomes were described as closed bilayer phospholipid systems by late Sir Alec Bangham and his colleagues [3, 4]. The pioneering work done by number of other scientists in coming years established the concept that liposomes could entrap drug molecules

* **Corresponding author O.P. Katare:** University Institute of Pharmaceutical Sciences, UGC Centre of Advanced Studies, Panjab University, Chandigarh - 60014, India; E-mail: drkatare@yahoo.com

Surendra Nimesh, Nidhi Gupta and Ramesh Chandra (Eds.)

and can be employed as drug delivery systems by changing pharmacokinetics and pharmacodynamics of entrapped drug molecules [5 - 8].

Liposomes are spherically shaped micro-vesicles composed of a bilayer of natural or synthetic non-toxic phospholipids, which spontaneously form closed structures when they are hydrated. These vesicles have one or more bilayer membranes, which are known as lamellae [9 - 11]. The unique structure of liposomes allows encapsulation of drugs with different lipophilicities,such as strongly hydrophilic or lipophilic drugs, and drugs with intermediate log P, which makes them an ideal drug delivery system [12]. The hydrophobic molecules are entrapped within the lipid membrane, whereas hydrophilic molecules can be enclosed in the aqueous core, enabling the encapsulation of a number of low and high molecular weight molecules to be released at the targeted site [1, 13]. The major advantages of liposomes include biocompatibility, biodegradability and non-toxicity; hence they are an ideal candidate for delivering enzymes, antibacterial, anti-inflammatory, antiviral, antiparasitic drugs, fungicides, diagnostic tools and adjuvants for vaccines [14 - 18].

The properties of liposomes differ considerably with lipid composition, surface charge, size and method of preparation. A number of characteristic features of liposomes have been modified over the years to achieve better entrapment efficiency and drug targeting profile. These include changes in lamellarity, shape and fluidity of the bilayer [19, 20] along with surface modification for active or passive targeting [21]. The conventional liposomes have been modified to increase the circulation time (stealth liposomes), improve their elasticity to enhance permeability (transferosomes) and formation of drug-lipid complexes by covalent bonding [7, 22]. The surface of liposomes can also be easily modified with a suitable ligand for targeted delivery to improve stability by reduced opsonization and clearance by the mononuclear phagocytic system [9, 23]. A number of research studies have focused on improving the encapsulation efficiency of liposomes. These strategies developed controlled delivery systems using microfluidic technology, which has emerged as an alternative for the assembly of vesicles. Additionally, microfluidics has been recognized as an emerging new technology providing a controlled environment to address issues of size and structure heterogeneity [22, 24].

A number of liposomal products have been approved for clinical use in several countries, such as doxorubicin (Doxil®, Sequus Pharmaceutical Inc., USA), daunorubicin (Daunoxome®, Nexstar Pharmaceutical Inc., USA), and amphotericin B (Ambisome®, Nexstar Pharmaceutical Inc., USA) [25]. Doxil, a long-circulating liposomal formulation, was the first liposomal product approved for clinical use in the United States. The liposomal formulation of amphotericin B

(Ambisome®) available as lyophilized reconstitute powder for injection has been shown to possess superior therapeutic index over the deoxycholate-based formulation (Fungizone®, Bristol-Myers Squibb, USA). The overall aim of this manuscript is to provide a comprehensive overview of the liposomal technology for various drug delivery applications.

COMPOSITION OF LIPOSOMES

Liposomes are potential drug carriers for various pharmaceutical drug molecules, dietary and nutritional supplements. Liposomes form spontaneously when phospholipids are dissolved in an aqueous solution leading to the formation of closed bilayered structures called vesicles [26, 27]. The amphiphilic nature of phospholipids results in the formation of polar shells in an aqueous solution due to the hydrophobic effect of the acyl chains with the aqueous medium. This thermodynamically favored transition is further improved by hydrogen bonding, van der Waals forces and other electrostatic interactions [28]. The resultant vesicles having a phospholipid bilayer membrane possess the ability to transport aqueous or lipid drugs, depending on their nature. Under certain conditions, stabilizers and electrolytes can be added to enhance the lipid bilayer formation depending on the application of liposomes [29]. The components of liposomes are discussed here below.

Phospholipid

A phospholipid is an amphipathic molecule containing a phosphate head group on one side, which is hydrophilic in nature and the acyl hydrocarbon chains on the other side being hydrophobic in nature, as represented in Fig. (**1**). When phospholipids are dispersed in an aqueous medium, hydrophilic interaction between the lipid head group and water results in the formation of vesicles. All lipids possess a specific temperature at which their fluidity changes, known as the phase transition temperature (T_M) [30]. T_M is directly proportional to the length of the acyl hydrocarbon chain; a longer chain leads to higher T_M and greater rigidity of the membrane. The lipids exist in either a fluid state ($T>T_M$) or a gel state ($T<T_M$) depending on their temperature [9, 31]. At body temperature ($T \approx 37°C$), the fluid state makes the liposomes leaky and the encapsulated drug tends to leach out before reaching the desired site of action. Thus, the choice of phospholipids having gel state at physiological conditions is often desirable to stabilize liposomes [9]. A variety of natural and synthetic phospholipids are employed for the preparation of liposomes, as discussed below:

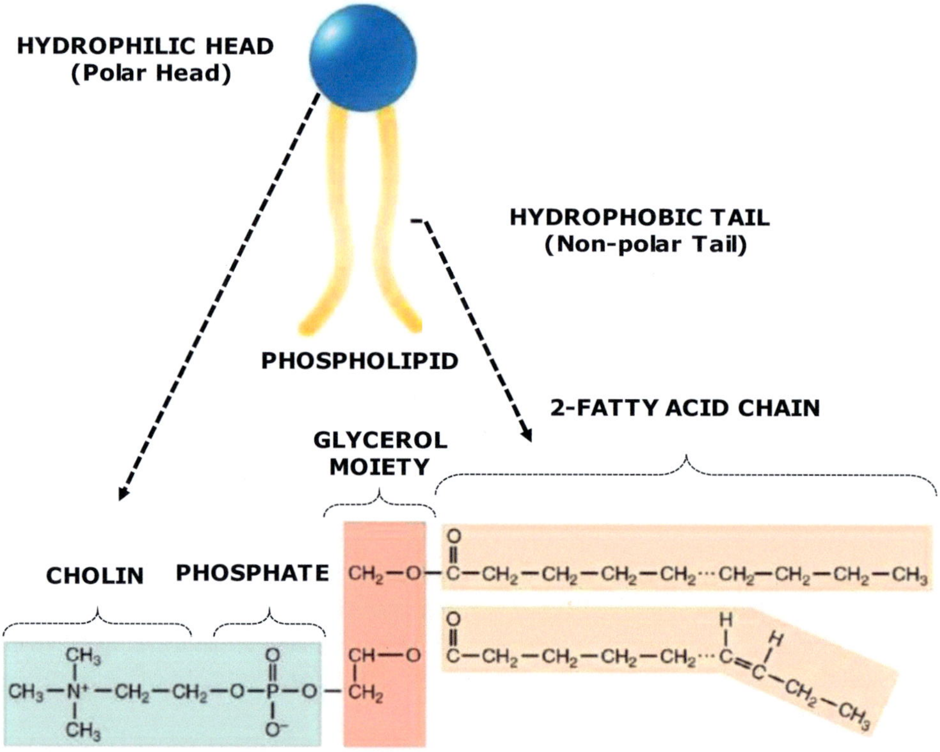

Fig. 1. Basic structure of phospholipid.

Natural Phospholipids

The most abundant natural phospholipid used for the preparation of liposomes is Phosphatidylcholine (PC), which is often referred to as lecithin. It is the major component of biological membranes, abundantly found in plants and animals, and plays an important role in imparting rigidity to the liposomal membrane [32]. Phosphatidylethanolamine (PE) is another naturally obtained phospholipid found in biological membranes.

Synthetic Phospholipids

Some of the commonly used synthetic phospholipids are phosphatidylglycerol, Dioleoyl phosphatidylcholine (DOC), Dioleoylphosphatidylethanolamine (DOPE), Distearoylphosphatidylcholine (DSPC) and Distearoyl phosphati dylethanolamine (DSPE) [33].

Cholesterol

Cholesterol is a component of the cell membrane and has been comprehensively studied in varied liposomal systems as a membrane stabilizer [34, 35]. Cholesterol acts as "fluidity buffer" and suppresses the tilts within the membrane. Cholesterol, being an amphipathic molecule, orients in such a manner that hydroxyl groups are towards the aqueous surface and aliphatic chains are parallel to the acyl chains in the centre of the bilayer. This arrangement of cholesterol within the phospholipid molecule makes the bilayered membrane more ordered above the transition temperature, thus decreasing its fluidity along with decreasing the permeability of lipid bilayer [36, 37]. The concentration of cholesterol is an important consideration to formulate liposomes of desired characteristics [38].

Collado *et al.*, in 2016, demonstrated that cholesterol favors surface localization of polyphenols in model membranes and enhances the stability of liposomes against lysozyme but does not offer much protection from reactive-oxygen species [39]. Kaur *et al.*, in 2014, included cholesterol within the dimethyl dioctadecylammonium (DDA) and trehalose 6,6′-dibehenate (TDB) (DDA:TDB) liposome-adjuvant delivery system in order to alter bilayer fluidity and potentially alter the pharmacokinetic profile of the liposomal–adjuvant system [34].

CLASSIFICATION OF LIPOSOMES

Liposomes can be classified on the basis of structural parameters, methods of preparation, composition and applications, as represented in Fig. (**2**).

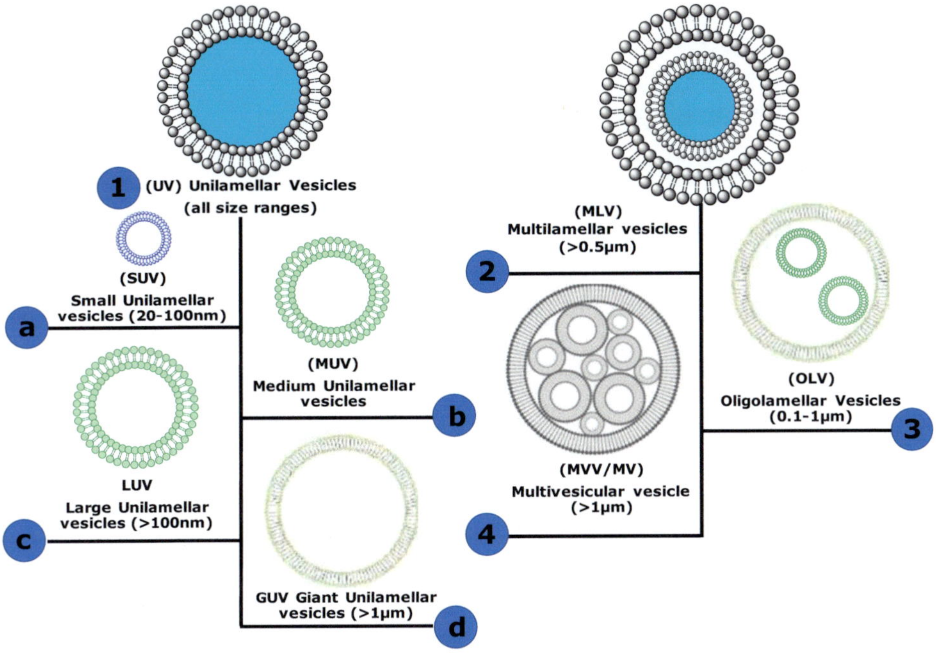

Fig. 2. Classification based on the structural size.

Based on Structural Parameters

According to structural parameters, liposomes can be classified as:

1. Unilamellar vesicles (UV)
    a. Small unilamellar vesicles (SUV) (20-100nm)
    b. Medium unilamellar vesicles (MUV)
    c. Large unilamellar vesicles (LUV) (>100nm)
    d. Giant unilamellar vesicles (GUV) (>1μm)
2. Multilamellar vesicles (MLV) (>0.5μm)
3. Oligolamellar vesicles (OLV) (0.1-1μm)
4. Multivesicular vesicles (MVV) (>1μm)

Based on the Method of Preparation

Different production methods can have different effects on liposomal parameters [13, 40, 41]. Hence, on the basis of methods of preparation, liposomes can be classified as:

1. DRV (vesicles prepared by dehydration rehydration method)
2. REV (SUVs/OLVs prepared by reverse-phase evaporation method)
3. MLV-REV (MLVs prepared by reverse-phase evaporation method)
4. VET (vesicles prepared by extrusion technique)
5. FAT-MLV (frozen and thawed MLVs)
6. SPLV (stable plurilamellar vesicles)

Based on Composition and Applications

Normally, the membrane of liposomes is composed of natural components, but these constituents can be varied and synthetic materials can be included according to the specific requirements. On the basis of composition and applications, liposomes can be classified as:

1. Conventional liposomes (neutral/negatively charged phospholipids and cholesterol)
2. Cationic liposomes (cationic lipids)
3. Fusogenic liposomes (RSVE-reconstituted sendai virus)
4. pH-sensitive liposomes (pH-sensitive phospholipids)
5. Long circulatory/stealth liposomes (Polyethylene Glycol, PEG)
6. Immunoliposomes (with attached monoclonal antibodies)

PREPARATION OF LIPOSOMES

All the methods for preparing liposomes consist of four basic steps [13].

a. Drying lipids from organic solvent
b. Hydration of lipid film with aqueous media
c. Purification of resultant liposomes
d. Analysis of the final product

Liposomes are manufactured in the majority using various procedures. Some approaches are easy at a laboratory scale, while others are useful for scale-up using special equipment [42, 43]. The various methods of liposome preparation are categorized as:

Conventional Methods

The first method described for preparing liposomes was the Bangham method (after the name of the scientist A.D. Bangham who discovered liposomes) or commonly called thin lipid film hydration method [9, 44] (Fig. **3**). The method involves the formation of a thin film of lipid on the inner surface of the round-bottom flask (RBF) by evaporation of organic solvents in which lipids are dissolved. Further, it is freeze-dried to ensure complete removal of solvents followed by the rehydration of lipid film with aqueous solvents. Upon rehydration, the lipids swell and peel off from the wall of RBF, forming the vesicles. The mechanical energy required for swelling and dispersion of lipids is imparted by handshaking [45]. Sonication is a widely used method for size reduction and the MLVs are exposed to ultrasonic radiation.

The size can also be reduced by extrusion of preformed liposomes in a French Press under very high pressure or by multiple extrusions through a polycarbonate membrane [46, 47]. Other methods for preparing liposomes involve mixing of organic and aqueous phases (Fig. **4**). The phospholipids dissolved in an organic solvent (ethanol or ether) are injected through a narrow needle at the vaporization temperature of organic solvent into the aqueous solution. The organic solvent is diluted instantaneously in water resulting in formation of liposomes. The major shortcoming of this method is limited solubility of lipids in ethanol or ether and incomplete removal of organic solvent especially ethanol [40].

Another commonly used method includes reverse phase evaporation method which is also termed as modified emulsification method (Fig. **5**). This method consists of emulsion formation by mixing of two phases and further conversion by slow evaporation to semisolid gel under reduced pressure which results in an aqueous suspension of liposomes [48 - 50]. The dispersion is dialyzed further to remove the remaining traces of organic solvent. Detergent removal methods are also used for preparation of liposomes. These methods involve intimate contact of phospholipids with the aqueous phase *via* detergents. The phospholipids dissolved in detergent micelles are added to aqueous media and further, the detergent is removed by dialysis/column chromatography/detergent adsorption by bio-beads [51, 52].

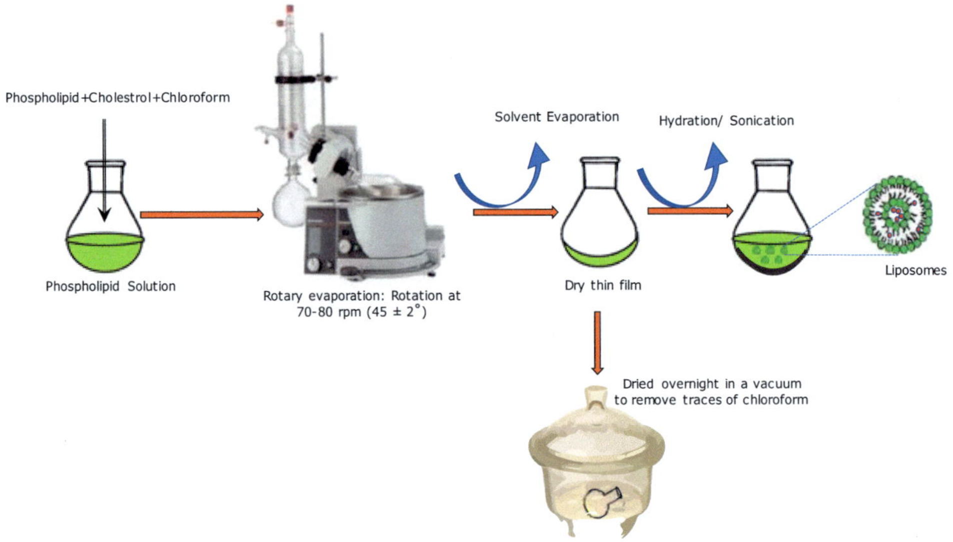

Fig. 3. Preparation of liposomes by thin film hydration technique.

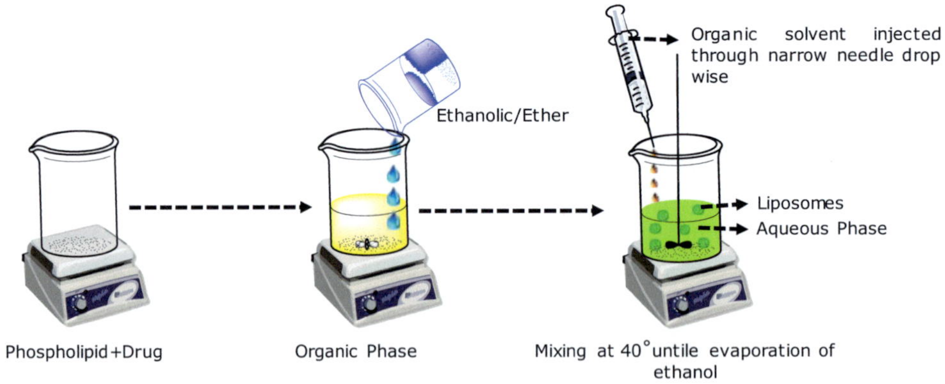

Fig. 4. Preparation of Liposomes by Injection method.

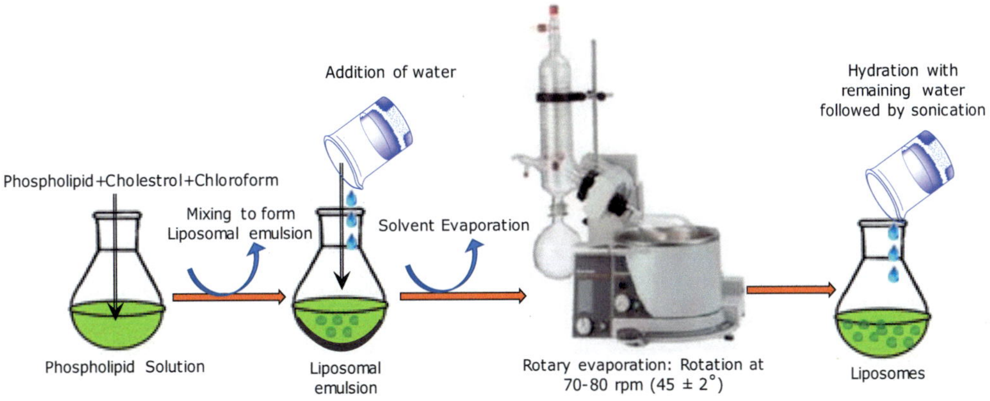

Fig. 5. Preparation of liposomes by reverse phase evaporation method.

All these conventional techniques are simple and easy at laboratory scale but cannot be employed at an industrial scale. The limitations associated with conventional method of liposomal preparation include broad size distribution, variable encapsulation efficiency and requirement of post processing granulation. The traces of organic solvent remain in the final product which can adversely affect the stability of encapsulated drug molecules. A number of novel techniques have been studied for efficient production of liposomes that can also be scaled up as discussed in the following section:

Novel Methods

The techniques employed nowadays include supercritical fluid technology, dual asymmetric centrifugation, freeze drying technology, cross-flow filtration technology and membrane contactor technology [13].

Supercritical Fluid Technology

Supercritical carbon dioxide ($scCO_2$) is an excellent solvent for phospholipids [53, 54]. It is non-toxic, non-inflammable and has a low cost. There are two liposome preparation methods based on supercritical fluid technology, including supercritical reverse phase evaporation (SRPE) method. This method was first used in 2001 by Otake and his colleagues for the preparation of liposomes [55]. The apparatus for this method consists of three basic parts including a variable volume view cell, an HPLC pump for feeding the aqueous solution of drug into the view cell and a high-pressure gas pump for CO_2.

The ethanolic solution of phospholipids is fed into the cell, followed by the introduction of gaseous CO_2 into the cell. The temperature is increased to a value that could reach both the phase transition temperature of phospholipids and supercritical temperature of CO_2 [56].The second method is supercritical anti-solvent (SAS), in which $scCO_2$ is used as an antisolvent. An organic solution of lipids (phospholipid and cholesterol) and a source of carbon dioxide are connected to pumps which are further linked to a precipitation vessel.

The pressure in the precipitation vessel is high and CO_2 is pumped into the vessel by spraying through capillary tubes. The temperature and pressure are changed to reach the supercritical value and organic solution is evaporated leading to supersaturation of solute as the $scCO_2$ and the lipid materials precipitate. Further on hydration with the aqueous phase, liposomes are obtained [57]. Zhang *et al.*, in 2012 prepared sirolimus-loaded liposomes by the expansion of supercritical solution (RESS) technology and studied the effect of temperature, pressure, and equilibrium time on the average particle size and encapsulation rate of liposomes [58]. Yang *et al.* in 2015 attempted to improve the bioavailability of silymarin using bile salt-containing liposomes prepared employing supercritical fluid technology. The liposomes produced exhibited highest entrapment efficiency, drug loading, smallest particle size and best stability compared to liposomes produced by the conventional methods [59].

Dual Asymmetric Centrifugation (DAC)

It is a special kind of centrifugation technique in which the vials turn around the main rotation axis at a defined distance and speed as well as turn around their own centre [60, 61]. This method is particularly useful for viscous materials and since lipids are viscous materials, so DAC can be used for liposome preparation [62]. This new technique combines rapid and sterile liposome preparation with very high entrapping efficiency.

Freeze Drying Technology

In this method, liposomal dispersion is lyophilized in the presence of suitable lyoprotectants [63]. The lyophilized liposome powder can be stored and can generate liposomal colloidal suspension on rehydration with aqueous buffer [64]. The method suffers from difficulties in terms of residual water content and drug leakage due to ice crystals as well as phase transition of bilayers during rehydration [65].

Vincourt *et al.* in 2010 examined the key parameters that influence freeze-drying process *viz.* colloidal characteristics and drug leakage, in order to stabilize liposomal preparations [66]. Yang *et al.* in 2013 developed freeze-dried liposomes encapsulating vitamin C and medium-chain fatty acids (MCFAs) by combining double emulsion method with dynamic high pressure micro fluidization [67].

Benjakul *et al.* in 2011 developed one-step method of liposome preparation by freeze-drying at room temperature as well as investigated the physicochemical properties of dry reconstituted liposomal powder [68].

Cross-flow Filtration Technology

The micellar solution is filtered under pressure through a suitable membrane and the solution flows tangentially to the membrane surface [69]. Detergent monomers pass through the membrane and are collected as waste while the micellar solution is pumped as retentate until liposome formation is completed [70]. This technique combines the advantage of conventional detergent removal method and accelerated removal of detergent leading to production of homogeneously sized liposomes [71, 72].

Membrane Contactor Technology

This technique is based on the conventional ethanol-injection technique and uses a membrane contactor module. Shirasu porous glass membranes have been extensively used for the preparation of microparticles. The phospholipids and cholesterol dissolved in ethanol are placed in a pressurised vessel and passed through a membrane having defined pore size in which aqueous phase flows tangentially. Ethanol gets diluted locally as the aqueous phase touches the organic phase and leads to liposome formation. The liposomes are collected in a collector and subjected to rotary evaporation to remove the ethanol [73, 74].

CHARACTERIZATION OF LIPOSOMES

The characterization parameters can be classified into three broad categories:

a. Physical parameters
b. Chemical parameters
c. Biological parameters

Physical Characterization

Vesicle Shape and Lamellarity

The morphological shape of liposomes has recently been considered as an important factor. Some morphological changes of spherical liposomes have been shown to activate the complement system and determination of liposomal morphology is essentially recommended in draft guidelines [75]. Freeze-fracture and freeze-etch electron microscopy is a technique which can be used to assess shape, lamellarity and surface morphology of the liposomes [74 - 76]. ^{31}P Nuclear magnetic resonance analysis is another accurate technique to determine the lamellarity of liposomes by monitoring the phospholipid phosphorous signal intensity [77].

Vesicle Size and Size Distribution

Size can be measured by several techniques including dynamic light scattering (DLS), field flow fractionation, nuclear magnetic resonance (NMR), size exclusion chromatography (SEC) and microscope technologies including optical microscopy, transmission electron microscopy (TEM), cryogenic-TEM (cryo-TEM) and atomic force microscopy (AFM) [78].

Optical Microscopy

The optical microscopy method uses a bright field, phase contrast microscope and fluorescent microscope (if liposomes are loaded with fluorescent probes). The dispersion is appropriately diluted and wet mounted on a hemocytometer and is photographed with the phase contrast microscope. The negatives are then projected on a calibrated paper and diameter of approximately 500 vesicles is measured. The major limitations of this method include tediousness and limited resolution [16, 79].

Negative Stain Transmission Electron Microscopy (TEM)

The negative stains used are ammonium molybdate or phosphotungstic acid. These stains are anionic in nature; hence do not bind to liposomes containing either neutral or negatively charged phospholipids. But they may bind to or precipitate the liposomes composed of positively charged lipids. So, in such cases cationic uranyl acetate should be used. A complete structure of liposomes is observed by this method unlike freeze-fracture technique [80, 81].

Cryo-TEM

Cryo-TEM is a two-dimensional (2-D) technique employed to evaluate liposomal morphology. Cryo-TEM easily allows direct observation of morphology and interior structure of liposomes in amorphous ice, which models its native condition [82]. The films are then vitrified by quick freezing in liquid nitrogen and are then transferred for TEM analysis [83, 84]. Wibroe *et al.* in 2016 compared shape and size distribution characteristics of a number of liposomal doxorubicin products and assessed vesicular-mediated complement activation in sera of human subjects with the help of Cryo-TEM [85, 86].

Small Angle X-ray Scattering (SAXS)

A recent study compared the membrane electron density, thickness and density of PEG layers and the structure of entrapped doxorubicin between certain batches of Doxil®, Caelyx® and Lipodox® by SAXS and implied that these products were structurally similar [87, 88].

Atomic Force Microscopy (AFM)

AFM enables the direct evaluation of 3-D morphology by scanning samples on a supportive substrate [82, 89]. The analysis of liposomes by AFM is a challenging task because these tend to collapse easily and form either a lipid bilayer or fused vesicles on solid substrates [90, 91]. PEG layers on liposomal surface can hinder the binding of liposomes to mica surface, and it is not easy to observe PEG liposomes [92].

In addition, the effect of membrane rigidity of positively charged liposomes on the efficiency of their cellular uptake was examined [93]. Takahashi *et al.* in 2017 characterized 3-D morphological changes of doxorubicin-loaded liposomes in an aqueous environment employing AFM and cryo-TEM. Finally, the effects of osmotic stress and storage on liposomal morphology were evaluated by AFM to elucidate the underlying mechanisms of doxorubicin-loading induced structural changes [94]. Takechi *et al.* in 2017 determined the permeability coefficient of a model hydrophilic drug, calcein, encapsulated within saturated lipid-based nano-sized liposomes. The increase in the permeability of the liposomal membrane to calcein was determined by means of AFM [94].

Dynamic Light Scattering (DLS)

When a particle is suspended in a solution and illuminated by light, it scatters the light if its refractive index is different from that of the suspending solvent. Due to collision between suspended particles and solvent molecules, particles undergo Brownian motion [95]. DLS measures these time-dependent fluctuations in light scattered by particles under Brownian motion. The major drawback of this technique is that it does not yield information about particle shape [96].

Field Flow Fractionation (FFF) Technique

Sedimentation FFF and flow FFF measure different vesicle properties because of difference in their driving forces. The most commonly used detector is UV/VIS spectrophotometer which provides direct measurement of vesicle size and size distribution. The retention times in this technique are proportional to the hydrodynamic diameter of particles. Sedimentation FFF measures the effective mass and mass distribution.

Size Exclusion Chromatography (SEC)

SEC can be used in combination with HPLC to separate and quantify the liposomes according to their size [52, 97]. The liposomes having larger size exclude first from the column leading to their shorter retention on the column. Though separation is based on particle size but chemistry between liposomes and column packing can also play a role in the assay. The technique is reliable and reproducible but sometimes, liposomes may become attached to packing in the columns, which can lead to problems in recovery of liposomes [98]. Xu *et al.* in 2016 reported a facile method based on SEC with fluorescence detection (SEC-FD) for quantification of small vesicles [99].

Nanoparticle Tracking Analysis (NTA)

It is a recently developed technique which detects the liposomes by light scattering. The liposomal suspension is injected into the view cell which is illuminated by laser light. The movement of liposomes in the media is captured by a digital camera and the rate of particle movement is related to sphere equivalent hydrodynamic radius which is calculated by Stokes-Einstein equation [100].

Entrapment Efficiency (EE)

The percent of aqueous phase and hence the percent of hydrophilic drug entrapped in liposomes is termed as the entrapment efficiency. It is usually expressed as % entrapment/mg lipid. EE can be determined by two methods:

Mini-column Centrifugation Method

This method is based on the difference of size between drug loaded liposomes and un-entrapped drug. A sephadex or sepharose gel column is pre-saturated with dispersion media and undiluted liposomal dispersion (200 μL) is run through the column and then centrifuged at 2000 rpm for 3 min. Then water/saline (250 μL) is added and centrifugation is repeated. The unentrapped drug remains bound to the column and vesicles are collected.

Ultracentrifugation Technique

It can also be used to separate drug-loaded liposomes from their medium. The sample is centrifuged at 50,000 rpm for 50 min at 4°C. Protamine aggregation is another method to determine encapsulation efficiency in which the liposomal dispersion is precipitated with a protamine solution and then centrifuged at 2000 rpm. Further, EE can be calculated by analyzing the material in the supernatant and in the pellet [101].

Surface Charge

The surface charge on the vesicles can be calculated from the mobility of liposomal dispersion in a suitable buffer. Zeta potential measurement involves the calculation of overall charge on the surface of liposomes in dispersion [102]. An electric field is applied to the cell and fluctuation in the detected light proportional to the particle speed is detected and zeta potential is calculated [103, 104].

Chemical Characterization

Phospholipid Concentration

The phospholipid concentration can be determined employing two methods *viz.* Barlet assay and Stewart assay [105]. In case of barlet assay, the phosphorous in phospholipid is hydrolyzed to inorganic phosphate and then converted to phosphomolybdic acid by adding ammonium molybdate. The chemical entity formed is then quantitatively reduced to a blue colored compound by aminonaphthyl sulfonic acid. The intensity of the blue color is measured spectrophotometrically and is compared with the standard curve to give phosphorous and hence the phospholipid content [106, 107]. In stewart assay, the phospholipid forms a complex with ammonium ferrocyanate in organic solution. The standard curve is prepared by adding ammonium ferrocyanate (0.1M) solution with different known concentrations of phospholipids in chloroform. The samples are treated in the same way and optical density of solution is measured and compared with standard curve to get the phospholipid concentration.

Lipid Hydrolysis

The phospholipids convert into lyso-phospholipid upon hydrolysis which leads to enhanced permeability of liposomal content. The content of lyso-phospholipid can be estimated using densitometry. The hydrolysis of ester bond at the C-position of the glycerol moiety leads to lipid hydrolysis and hydrolysis can also be quantified using HPLC [108].

Lipid Oxidation and Peroxidation

The methods to determine lipid peroxidation are either based on the disappearance of unsaturated fatty acids or the appearance of conjugated dienes. The various techniques like UV absorbance method, iodometric method (for hydroperoxidase), TBA reagent (for endoperoxidase) and gas liquid chromatography can be used to estimate lipid oxidation [109].

Cholesterol Analysis

The cholesterol in liposomal formulation is qualitatively analyzed using capillary column of flexible fused silica, whereas it is quantitatively estimated by measuring the absorbance of purple complex produced with iron upon reaction with a reagent containing ferric perchlorate, ethyl acetate and sulfuric acid at 610 nm [110].

Biological Characterization

The biological characterization is done to establish the safety and suitability of the formulation for *in vivo* use. This includes sterility testing performed with aerobic and anaerobic cultures.

LIPOSOMAL DRUG LOADING METHODS

The major achievement in liposomal clinical application is the ability to load sufficient amount of drug required to achieve therapeutic efficacy. Hence, around 15 liposomal based drug products are in clinical use. The physical and chemical properties of drugs vary considerably requiring different methods of drug loading. Nichols and Deamer were the first scientists to express remote loading of amphipathic amines by pH gradient method [111]. This method was also used by Cullis and his co-workers to obtain the remote loading of drugs such as doxorubicin [112]. The pharmaceutically acceptable drug loading methods should be fast, efficient and should result in formation of stable vesicles. The method should be applicable to various drugs having similar physicochemical properties and must not be affected by liposomal lipid composition [113, 114]. There are mainly two methods by which drugs can be incorporated into liposomes namely, passive loading method and active (remote) loading method.

Passive Loading Method

In this method, the drug is loaded during the manufacturing of liposomes. The drugs are co-solubilized with lipids in an organic solvent and then dispersed in an aqueous buffer, thus achieving the entrapment during liposome formation. The passive loading techniques include mechanical dispersion, solvent dispersion and detergent removal method. For lipophilic and positively charged drugs, the trapping efficiency can be enhanced by using acidic (negatively charged) lipids into the liposome [115, 116].

Active (remote) Loading Method

This method includes all those techniques where the drug is loaded after the liposomes have been formed. The drugs are loaded into the preformed liposomes by using a certain driving force [117]. Uptake of lipophilic and positively charged drugs can be attained after short incubation by adding drug to the external media [114, 118]. Still, loading mechanism in response to transmembrane potential gradient is not so clear yet. This kind of uptake is independent of the lipid composition and high drug/lipid ratio and major advantages include high entrapment efficiency and enhanced retention properties [115]. Several methods of active drug release have being employed, including pH sensitive liposomes, thermo sensitive liposomes and many more to release the content of drug loaded liposomes at the desired site [113, 117]. The emerging active loading method is the EDTA ion gradient in which low solubility drugs precipitate with EDTA molecules at low pH. This method was successfully used to encapsulate hydrophobic anthracycline idarubicin [119]. The recently approved liposomal bupivacaine injection (Exparel) contains bupivacaine loaded liposomes. This preparation increases the duration of action of bupivacaine as a local anesthetic by slow release from the liposomes. It has been approved by US-FDA for local infiltration for pain relief after bunionectomy and hemorrhoidectomy [120].

TARGETING STRATEGIES FOR LIPOSOMES

Passive Targeting

It involves natural anatomical structures and physiological features like recognition by reticuloendothelial system (RES), enhanced permeation and retention effect (EPR-effect) leading to retention of liposomes in extracellular fluid [121]. Upon injection of liposomes, they are progressively cleared from circulation or sequestered in tissues for long period of time.

To generate long circulating liposomes, the properties of liposomes like composition and size can be varied. It was observed that circulation time of small vesicles was higher than those of larger and MLVs [122]. Passive delivery of liposomes is also useful where the structure and function of diseased area is altered.

Active Targeting

Active targeting of liposomes is aimed to deliver the drug exclusively to and retain it in a specific tissue or cell. It can be achieved by surface modification of liposomes using ligands and antibodies [123, 124]. The attachment of specific ligands to the liposome surface assists in the fusion of liposomes with the targeted cells, facilitating the release of entrapped material at the targeted site.

The conventional approach involves addition of lipid-polymer-ligand conjugate in the lipid composition at the time of liposome preparation. A recent approach involves incubation of micellar solution of the lipid-polymer-ligand conjugation with a suspension of liposomes at high temperature and the lipid-polymer-ligand conjugate is inserted into the pre-formed liposomes. This method is known as Insertion method [125, 126].

Triggered Drug Targeting

Ligand mediated targeting faces several problems like ligand-target affinity, PEGylation acting as barrier against interaction of ligand with receptors and the number of receptors present on the cell surface [127]. Hence, other targeting strategies that use the disease microenvironment or external stimuli can be used to increase the efficiency of liposomal drug delivery.

pH Sensitive Liposomes

These liposomes can be designed by the use of fusogenic peptides. GALA is such a peptide having repeated sequences of Glu-Ala-Leu-Ala [128]. At neutral and high pH, these peptides are monomeric and water soluble, but at low pH these peptides assume monomeric transmembrane helix structure that inserts into the membrane leading to the fusion of liposomes with the cell membrane. So, at low pH areas, the release of drug is enhanced by fusion of liposomes to the membrane [129]. The pH sensitivity can also be increased by adding small pH-sensitive lipids, such as oleic acid, to liposome preparations.

Redox potential can also be used as a driving force to trigger the drug release. At the tumor site, there is an imbalance in the production of reactive oxygen species and cell's antioxidant defense leading to deregulation of redox homeostasis [130]. Redox responsive systems lose their structural integrity in response to the redox potential and deliver the entrapped drug at required site of action [131]. Fan *et al.* in 2017 studied the pH-response mechanism of various molecules that compose liposomes like oleic acid (OA), linoleic acid (LA) and cholesteryl hemisuccinate (CHEMS) [132]. Ferraria *et al.* in 2017 developed pH-sensitive, doxorubicin containing liposomes as a promising approach to selectively deliver doxorubicin to bone tumors while minimizing cardiac toxicity [133].

External Stimuli-dependent Release

These systems are governed by a stimulus given from outside the body. The light-sensitive liposomes make use of photosensitizing agents which can either be light sensitive lipids or other light sensitive molecules like porphyrin derivatives, chlorins, azobenzene, coumarins, and pyrene derivatives [134]. On laser irradiation, these molecules cause disruption of the lipid bilayers to release the drug and also generate the reactive oxygen species to kill the cancer cells [135, 136].

Thermosensitive liposomes are triggered by changes in heat. The phospholipids have a transition temperature below which they are in gel state and above which in fluid state. Hence, all the liposomes are inherently thermosensitive. On applying heat locally at the desired site, the drug release from liposomes occurs due to disruption of liposomal membrane [137]. The low transition temperature makes thermosensitive liposomes suitable for targeted and triggered drug delivery to cells, tissues, organs in combination with local hyperthermia [138]. Local heating can be done by using water bath, heat applicators or focused ultrasound [139].

Fite *et al.* in 2016 developed strategies to improve the *in vivo* stability of doxorubicin-loaded liposomes, including remote loading strategy utilizing copper gluconate and triethanolamine (TEA) to load thermosensitive liposomes and long-circulating liposomes. Doxorubicin has also been encapsulated in temperature-sensitive liposomes using ammonium sulfate remote loading and these particles are in clinical trials for liver and breast cancer treatment [140]. A combination of pH sensitive and temperature sensitive liposomes were prepared by Sugimoto *et al.* in 2017 by synthesizing dual-signal-sensitive copolymers composed of methoxy diethylene glycol methacrylate, methacrylic acid and lauroxy tetraethylene glycol methacrylate [141]. Affram *et al.* in 2017 studied the ability of thermosensitive liposomal nanoparticles as a delivery system to deliver poorly membrane-permeable anticancer drug, gemcitabine to solid pancreatic tumor with the aid of local mild hyperthermia [142].

Ultrasound stimulus transfer energy to the encapsulated air pockets in liposomes leading to expansion and collapsing of bubble resulting in liposomal bilayer disruption and release of drug [143]. Echogenic liposomes have been considered for the treatment of various diseases like atherosclerosis, thrombosis and cancer [144].

Magnetically Sensitive Liposomes

These liposomes contain stabilized iron oxide particles. On magnetic stimulation, these particles allow triggered release of entrapped drug. Lee *et al.* formulated liposomal gel with iron oxide magnetic nanoparticles (IMNs) and showed that a small amount of IMNs can trigger the drug release from liposome gel [9, 145]. Zhao *et al.* in 2014 developed amphotericin B magnetic liposomes (AmB-MLPs) to improve drug concentration in brain and reduce drug toxicity in the presence of magnetic field. The drug concentration in brain tissue in the presence of magnetic field and AmB-MLPs group in absence of magnetic field was obviously increased, whereas the drug none or scarcely reached to brain tissue by tail vein injection [146].

Electro Responsive Liposomes

This class of liposomes employs the conductive properties of certain organic or inorganic materials to deliver drugs in response to an external electric field. Nahire *et al.* in 2014 prepared folate receptor targeted liposomes encapsulating ammonium bicarbonate to grant pH sensitivity. The CO_2 gas bubbles are generated in acidic pH leading to release of encapsulated drug. The drug release is further enhanced by simultaneous application of ultrasound [147].

APPLICATIONS OF LIPOSOMES

Liposomes are nontoxic, non-immunogenic and biodegradable in nature. Liposomal formulations reduce serious side-effects, increase drug accumulation at the target site, encapsulate both hydrophilic and lipophilic drugs and can be used as sustained release drug delivery systems for drugs like hormones, corticosteroids and antineoplastic drugs [7, 148]. On systemic administration, liposomes are easily recognized by reticulo-endothelial system (RES) so vaccines and antimalarials can be targeted to RES by encapsulating them in liposomes. Liposomes as a carrier for cosmeceuticals have various advantages as lipids are well hydrated, hence reduce the dryness of the skin. Also, liposomes help to fix the active ingredients to the outermost skin layers and improve the dermal drug delivery [149]. Liposomes have also been used to treat acne, melasma, hyperpigmentation, vitiligo, alopecia, psoriasis, dermatitis and hirsutism [150]. The various applications of liposomes have been discussed under.

Liposomes for Topical Delivery

The benefit of topical liposomal formulations includes reduction in adverse effects, significantly enhance the accumulation of drug at the site of action and can easily incorporate a wide variety of hydrophilic and hydrophobic drugs [14]. Liposomal entrapment of drug increases its deposition at the skin target site and reduces its absorption into the central blood supply. Mezei and Gulasekharam reported that topical application of triamcinolone acetonide for 5 days resulted in drug concentration four times higher in the epidermis and dermis than that obtained using a control ointment [151]. Table **1** details the various drug molecules that have been incorporated into liposomes for better targeting efficacy.

Liposomes for Pulmonary Delivery

Liposomes are established drug carriers for inhalation because they are safe and can provide controlled drug release in the lungs [174]. The development of successful liposomal carriers for inhalation depends on formulation composition and nebulizer design. A number of nebulizable liposome formulations like Arikace® (liposomal amikacin) and Pulmaquin® (liposomal ciprofloxacin) are currently under development. Liposomes are potential carriers for anticancer drugs. Various inhalation devices like pressurized metered dose inhalers, dry powder inhalers, soft mist inhalers and medical nebulizers are used for the pulmonary delivery of liposomes [177].

Liposomes for Ophthalmic Delivery

Liposomes have been widely investigated for treatment of various eye disorders. Liposomes are biodegradable and biocompatible and they can enhance the permeation of poorly absorbed drug molecules by binding to the corneal surface and improving its residence time. Various other drug molecules like levofloxacin, acetazolamide and tacrolimus are also delivered to eye as liposomal formulations with improved activity and reduced toxicity by controlled and targeted delivery [191].

Liposomes for Parenteral Delivery

Parenteral route of administration is the most effective route for delivery of active pharmaceutical substances especially those prescribed to unconscious patients. Frequent injections are required to maintain effective therapeutic concentration which leads to patient discomfort. So, novel drug delivery systems like liposomes can be used to provide targeted and sustained release of drugs. Various anticancer drugs and anti-infective drugs are injected as liposomal formulations. For example, liposomal amphotericin B (AmBisome) is used for treatment of systemic

Table 1. List of drug molecules encapsulated into liposomes along with their route of administration.

Route of Administration	Drug Molecule	Composition	Remarks	Reference
Topical	Amphotericin B	SPC, sodium cholate	Ultradeformable amphotericin B demonstrated increased skin deposition for topical fungicidal and leishmanicidal treatment.	[152]
	Psoralen	Cholesterol, L-α-Egg Lecithin, cardiolipin, cholesterol	Penetration and skin deposition of psoralen is enhanced when used with UV-A for treatment of psoriasis.	[153]
	Resveratrol	Lipoid S100, chitosan	Liposomal formulation solubilized the drug and enhanced its anti-oxidative and anti-inflammatory properties.	[154]
	Meglumine antimoniate (MA)	SPC, cholesterol, vitamin E, propylene glycol, methyl and propyl paraben	Topical liposomes containing MA might be an appropriate choice for clinical trials for the treatment of cutaneous leishmaniasis .	[155]
	5-aminolevulinic acid	SPC, cholesterol	Liposomal formulation has few side effects with the same effectiveness and can be effectively used for wrinkle reduction.	[156]
	Aceclofenac	Phospholipon® 90H	The drug permeation and retention offered by the vesicular system was superior to that of conventional cream.	[17]
		Phospholipon® 90G, cholesterol	The studies indicated superiority of lysine conjugated liposome-entrapped nanocarriers for improved management of conditions like arthritis over the marketed product.	[15]
	Acyclovir	Phospholipon® 90G, Phospholipon® 90H	The pharmacodynamic studies on murine model for HSV-1 infection demonstrated immense potential and safety of topically applied formulation.	[16]
	Tamoxifen	Phospholipon® 90H, cholesterol, sorbitan monooleate	Tamoxifen-loaded liposomal systems demonstrated immense potential in the management of skin cancer.	[11]
	Minoxidil	Phospholipon® 90H, Cholesterol, Dicetylphosphosphate, Stearylamine	The neutral liposomes delivered the drug molecules into pilosebaceous units more effectively than the other studied formulations.	[157]
	Fusidic acid (FA)	Phosphatidylcholine, cholesterol	FA-liposomal system demonstrated enhanced safety and efficacy	[158]
	Cyclosporine	Phosphatidylcholine, cholesterol	Topical liposomal formulation of cyclosporine may prove to be an effective treatment for limited chronic plaque psoriasis.	[159]
	Capsicin	Phospholipon® 90H, cholesterol	The vesicles showed higher skin retention, higher analgesic potential and better flexibility in regaining their size.	[18]
	Interferon alpha-2β	Phosphatidylcholine, cholesterol, polyethylene glycol	The penetration of interferon alpha-2β entrapped in PEGylated liposomes is increased and improved performance in human papilloma virus (HPV) vaginal infections.	[160]
	Coenzyme Q10	SPC	Liposomal formulation promotes the penetration of CoQ10 to reduce photo aging.	[161]
	Usnic acid	Gelatin, phosphatidylcholine, propylene glycol	The gelatin membranes containing usnic acid/liposomes promoted the healing and macroscopic shrinkage of the wound due to controlled release of drug.	[162]
	Azithromycin	α-phosphatidylcholine, Cholesterol	Topical liposomal azithromycin has the same efficacy as intralesional glucantime in the treatment of leshminasis	[163]
	Hyaluronic acid (HA)	L–α phosphatidylcholine, tween 80, transcutol® P	*In vitro* and *ex vivo* permeation assays showed that penetration enhancers enhanced HA-release from liposomes.	[164]

(Table 1) cont.....

Route of Administration	Drug Molecule	Composition	Remarks	Reference
Oral	Itraconazole	Lecithin, sodium deoxycholate, cholesterol	Treatment of fungal infections of the toenails and fingernails.	[165]
	Myrcludex B	Lecithin, cholesterol, GCTE	Improved oral uptake of Myrcludex B for hepatitis B treatment.	[166]
	Vancomycin	Lecithin, GCTE, cholesterol	Oral availability of peptide drug is increased using the liposomal formulation.	[167]
	Insulin	SPC, DPPG, cholesterol	Liposomes containing ergosterol protect insulin against degradation in gastro-intestinal fluids.	[168]
	Amphotericin B	EPC, L-α-Distearoyl phosphatidylcholine, cholesterol	Ceramide-based liposomes enhanced the stability of drug on oral administration.	[169]
	Chlorogenic acid	Soybean lecithin, cholesterol	Relative bioavailability of liposomal formulation is 1.29-fold higher than chlorogenic acid.	[170]
	Docosahexaenoic acid (DHA)	DPGP, DDGP	Liposomal formulation protects DHA from oxidation and pH fluctuations.	[171]
	Adefovir dipivoxil	EPC or soya bean, cholesterol, spray dried mannitol, maltodextrin	The optimized proliposomal formulation offered good stability, improved drug dissolution and enhanced bioavailability.	[172]
	Quercetin and Resveratrol	Lipoid S75, cholesterol	The vesicular formulations may be of value in the treatment of pre-cancerous/cancerous intestinal conditions associated with inflammation and oxidative stress.	[173]
Pulmonary	Calcifediol	DPPC	The antibacterial action of calcifediol was significantly improved on incorporation into liposomes.	[174]
	Fasudil	DPPC, DSPE, PEG 2000, Cholesterol	Magnetic liposomes extended the half life and area under curve by 27-fold and 14-fold respectively as compared to plain fasudil.	[175]
	Procaterol hydrochloride	EPC, HSPC	Liposomal formulation achieved sustained release and long term effect.	[176]
	Gemcitabine hydrochloride	HSPC, DSPG, mPEG2000-DSPE, cholesterol	Liposomal dry powder inhaler (DPI) showed better uptake of drug than plain DPI.	[177]
Ocular	Distamycin	Phospholipon® 80, cholesterol	The bioavailability of drug in the tear fluid uptake into the cornea was increased after instillation of liposomes in rabbits.	[178]
	Timolol maleate	SPC, cholesterol	Chitosan loaded liposomes is a potentially useful carrier for ocular drug delivery, which could improve the efficacy of Timolol.	[179]
	Diltiazem HCl	SPC, cholesterol	The *in vivo* studies proved that the liposomal formulations have enhanced the intraocular pressure reducing activity when compared with the solution form of the same concentration.	[180]
	Voriconazole	SPC, DOTAP, cholesterol	The formulation was classified as 'non-irritant' after HET-CAM test with good permeation across corneal barrier.	[181]
	Fluconazole	Lipoid S100, Hylauronic acid	The novel hyalugel-integrated liposomes significantly enhanced corneal permeability compared to conventional liposomes and drug suspension.	[182]
	Infliximab	DOPC, cholesterol, DSPE-PE-NH_2	The use of biodegradable particles for therapeutic antibody delivery may provide a promising approach for the treatment of ocular diseases.	[183]
	Minocycline	EPC, dihexadecyl phosphate, cholesterol	The nontoxic nanoliposomes delivered 40% of encapsulated minocycline to the retina after a subconjunctival injection.	[184]
Vaccine	Gp2, a HER2/neu-derived peptide	DMPC, DMPG, DOPE, cholesterol, MPL	GP2 peptide conjugation on the surface of liposomes increased splenocytes INF-γ production, which can be applied prophylactically as well as therapeutically to reduce tumor growth in HER2/neu-overexpressing breast cancer model.	[185]
Gene	pDNA	DDCTMA, cholesterol	The results demonstrated FA coated DDCTMA/Chol (3:1) cationic liposome could serve as a promising candidate for the gene delivery.	[186]

(Table 1) cont.....

Route of Administration	Drug Molecule	Composition	Remarks	Reference
Nasal	Opiorphin	Sodium glycerophosphate hydrate, chitosan, EPC	The liposomal hydrogel formulation enabled a sustained and controlled delivery of opiorphin over more than 5 h and highlighted the role of the liposomal carrier in enhancing up to 6 times permeability coefficient.	[187]
	Ovalbumin	TPGS, SPC, DDA, ovalbumin	Novel liposomes have potential for development as a mucosal vaccine delivery system.	[188]
	Quetiapine Fumarate	EPC, cholesterol	The diffusion study proved the liposomal dispersion as a virtuous delivery system for brain drug delivery through nasal route.	[189]
Parenteral	Doxorubicin	Lecithin S100, DOTAP, Cholesterol	The liposomes exhibited sustained drug release and have synergic anticancer effect which may be promising as drug delivery vehicle.	[190]
	Docetaxel	PE, DMPA, DCC	The liposomes significantly and safely accumulated in tumor tissue in xenografted nude mice.	[131]

DMPC: Dimyristoylphosphatidylcoline; DMPG: Dimyristoylphosphoglycerol; DOPE: Dioleoylphosphatidylethanolamine; MPL: Monophosphoryl Lipid A; SPC: Soyabean Phosphatidylcholine; GCTE: Glycerylcaldityltetraether; EPC: Egg yolk phosphatidylcholine; DPPC: Dipalmitoyl phosphatidylcholine; DSPE: Distearoyl phosphoethanolamine; DOPC: 2-dioleoyl-*sn*-glycero-3-phosphocholine; FA: Folic acid; TPGS: Tocopheryl Polyethylene Glycol Succinate; DDA: Dimethyldioctadecylammonium; DOTAP: 1,2-dioleoyl-3-trimethylammonium-propane chloride; DMAP: 4-Dimethylaminopyridine; DCC: Dimethylcyclohexylcarbodiimide; DPGP: Diphytanyl glycerol phosphoethanolamine; DDGP: dihexadecyl glycerol phosphocholine; DPPG: Dipalmitoyl phosphatidylglycerol,

fungal infections. Liposomal Cytarabine, Daunorubicin and doxorubicin are employed as antineoplastic agents. Liposomal vaccines can also be prepared by associating microbes, soluble antigens, cytokines or DNA with liposomes [190].

Liposomes for Oral Delivery

Oral delivery is the dominant route for drug administration. Orally delivered liposomal carriers can enhance the drug solubility and can protect the encapsulated therapeutic agents from the extreme environment found in the GI tract to improve their bioavailability [192, 193]. Liposomes can encapsulate both unstable hydrophilic and lipophilic drugs to protect and release them in a controlled manner [166].

Liposomes as Vaccine Delivery

Liposomes have emerged as an important carrier system in vaccine development. Liposome composition and preparation can be varied to achieve desired features such as the selection of lipid, charge, size, size distribution, entrapment and location of antigens or adjuvants. Depending on the chemical behavior, water-soluble compounds (proteins, peptides, nucleic acids, carbohydrates, haptens) are entrapped within the aqueous inner space, whereas lipophilic compounds (lipopeptides, antigens, adjuvants, linker molecules) are incorporated into the lipid bilayer and antigens can be attached to the liposome surface either by adsorption or stable chemical linking [158, 194, 195].

CLINICAL USES OF LIPOSOMAL DRUGS

A number of liposome-based drug formulations are available in the market for human use and various products are under clinical trials. Encapsulation of drugs in liposomes enhances the therapeutic indices of drugs and also reduces their toxicity. The first successfully produced liposome-based product in 1995 was Doxil for the treatment of AIDS related Kaposi's sarcoma and breast/ovarian cancer. Later, various other liposomal products came into market like DaunoXome in 1996, Ambisome in 1997, Depocyt in 1999 and various others. The examples of liposomal based drug products are listed in Table **2**. The liposome-based formulations in clinical trials are listed in Table **3**. Table **4** gives a detailed account of the patents related to liposomes in recent years.

Table 2. Clinically approved liposomal drug products in the market for various therapeutic indications.

Product	Drug	Lipid Composition	Indication
Ambisome	Amphotericin B	HSPC, DSPG, Cholesterol	Serious fungal infections and leishmaniasis
Abelcet	Amphotericin B	DMPC, DMPG	Serious fungal infections and leishmaniasis
Amphotec	Amphotericin B	Cholesteryl sulfate	Serious fungal infections and leishmaniasis
Caelyx	Doxorubicin	HSPC, Cholesterol	Kaposi's sarcoma in AIDS, ovarian/breast cancer
DaunoXome	Daunorubicin	DSPC, Cholesterol	Blood tumors
Depocyt	Cytarabine	DOPC, DPPG, cholesterol and triolein	Lymphomatous and neoplastic meningitis
DepoDur	Morphine sulfate	DOPC, DPPG, cholesterol and triolein	Pain
Doxil	Doxorubicin	HSPC, cholesterol, PEG 2000-DSPE	Kaposi's sarcoma in AIDS, ovarian/breast cancer
Epaxal	Inactivated hepatitis A virus	DOPC, DOPE	Hepatitis A
Inflexal	Inactivated hemaglutinine of influenza virus strains A and B	DOPC, DOPE	Influenza
Lipodox	Doxorubicin	HSPC, cholesterol and PEG 2000-DSPE	Kaposi's sarcoma, ovarian/breast cancer

(Table 2) cont.....

Myocet	Doxorubicin	EPC, Cholesterol	Metastatic breast cancer with cyclophosphamide
Marqibo	Vincristine	Sphingomyelin, cholesterol	Acute lymphoblastic leukemia
Visudyne	Verteporphin	EPG, DMPC	Macular degeneration

HSPC: Hydrogenated soybean phosphatidylcholine; DSPG: 1, 2-Distearoyl-sn-glycero-3-phosphoglycerol; DMPC: Dimyristoylphosphatidylcoline, DMPG: Dimyristoylphosphoglycerol, DSPC: 1,2-Distearoyl--n-glycero-3-phosphocholine; DPPG: 1,2-dipalmitoyl-sn-glycero-3-phosphoglycerol, PEG 2000-DSPE: 1,2-Distearoyl-sn-glycero-3-phosphoethanolamine; DOPE: 1,2-Dioleoyl-sn-glycero-3-phosphoethanolamine.

Table 3. List of liposomal based drug products in clinical trials.

Product	Drug	Indication	Clinical Trial Status
Onco-TCS	Vincristine	Non-Hodgkin Lymphoma	Clinical phase II
LEP-ETU	Paclitaxel	Ovarian/breast/lung cancer	Clinical phase II
Aroplatin	Cisplatin	Colorectal cancer	Clinical phase I/II
OSI-211	Lurtotecan	Lung cancer/recurrent ovarian cancer	Clinical phase II
SPI-77	Cisplatin	Head and neck cancer/lung cancer	Clinical phase III
ThermoDox	Doxorubicin	Hepatocellular carcinoma	Clinical phase III
Atragen	trans-retinoic acid	Acute promyelocytic leukemia	Clinical phase II
Lipoplatin	Cisplatin	Pancreatic and breast cancer	Clinical phase III
EndoTAG-1	Paclitaxel	Breast neoplasms	Clinical phase II

Table 4. Patents related to the liposomes for the treatment of various therapeutic indications.

Title of Patent	Route of Administration	Therapeutic Indication	Publication Number	Reference
Concentrated, inhalable antibiotic formulation	Inhalation	Respiratory tract infections	US9028864 B2	[196]
Liposome system for ocular administration	Ocular	Ophthalmic indications	WO2011098578 A2	[197]
Cationic liposomal drug delivery system for specific targeting of human CD14+ monocytes in whole blood	Parenteral	Cancer treatment, inflammatory diseases	WO2013135800 A1	[198]
Pain reducing parenteral liposome formulation	Parenteral	Pain	EP0975330 A1	[199]
Amphotericin B liposome preparation	Parentral	Fungal infections	US5965156 A	[200]
Sustained-release liposomal anesthetic compositions	Parenteral	Anesthesia	US8182835 B2	[201]
Liposome based topical vitamin D formulation	Topical	Psoriasis	US5834016 A	[202]
Liposome formulations of immunomodulating drugs for the topical and aerosol administrations	Topical	Immunomodulation	EP0450352 A1	[203]

(Table 4) cont.....

Title of Patent	Route of Administration	Therapeutic Indication	Publication Number	Reference
Stable liposomal formulations for ocular drug delivery	Ocular	Glaucoma	US20150190359 A1	[204]
Liposome eye drops	Ocular	Dry eye syndrome	WO1996019211 A1	[205]
Synergistic liposomal tamoxifen composition for topical application and method of preparing thereof	Topical	Cutaneous melanoma	WO2006001035 A2	[206]
Novel Inter and Intra Multilamellar Vesicular Composition	Topical	Psoriasis	US20080171795 A1, WO2006100692 A2, EP1871342 A2, CN101180037 A	[207 - 210]
Method of treatment using liposomally formulated reduced glutathione to cooperate with IL-10 to modulate inflammatory response triggered in HIV+ and TB immune-compromised diabetic patients	Oral	Immune compromised diabetes	WO1996019211 A1	[211]
Therapy for human cancers using Cisplatin and other drugs or genes encapsulated into liposomes	Intravenous	Cancer	EP1156789 B9	[211]
Sustained timolol maleate delivery from liposomes for glaucoma therapy and ocular hypertension	Ocular	Glaucoma and ocular hypertension	WO2016048242 A1	[213]
Method for regressing pancreatic tumor by a liposomal formulation along with DNA vaccines	Intravenous	Cancer chemotherapy	US 20170267719	[214]
Kit for the formulation of liposomal doxorubicin modified with bioactive peptides for selective targeting of receptors	Intravenous	Cancer treatment	EP3202424 A1	[215]

CONCLUSION

The journey of liposomes from the time of Sir Bangham till the 21st century has been long and innovative. The applications of liposomes have been diverse and ranged from imaging tumours and infection sites to vaccine delivery, cancer treatment, lung and tropical diseases. The ease of chemical modification, the ability to load drugs/genes of varied types and the potential to be administered by different routes have made them attractive. Further, liposomes have demonstrated the ability to "passively" accumulate at sites of increased vasculature permeability and reduce the side effects of the incorporated drug relative to a free drug molecule. Liposomal drug delivery has become an established technology platform and has gained considerable clinical acceptance. However, still, the full potential of liposomes has not been realized. Joint collaborative efforts are desired between regulatory bodies as well as industry and academia to explore the full medical potential of liposomal drug delivery systems.

CONSENT FOR PUBLICATION

Not applicable.

CONFLICT OF INTEREST

The author declares no conflict of interest, financial or otherwise.

ACKNOWLEDGEMENTS

Declared none.

REFERENCES

[1] Sercombe L, Veerati T, Moheimani F, Wu SY, Sood AK, Hua S. Advances and challenges of liposome assisted drug delivery. Front Pharmacol 2015; 6: 286. [http://dx.doi.org/10.3389/fphar.2015.00286] [PMID: 26648870]

[2] Huang X, Li M, Bruni R, Messa P, Cellesi F. The effect of thermosensitive liposomal formulations on loading and release of high molecular weight biomolecules. Int J Pharm 2017; 524(1-2): 279-89. [http://dx.doi.org/10.1016/j.ijpharm.2017.03.090] [PMID: 28377318]

[3] Bangham AD, Standish MM, Watkins JC, Weissmann G. The diffusion of ions from a phospholipid model membrane system. Protoplasma 1967; 63(1): 183-7. [http://dx.doi.org/10.1007/BF01248030] [PMID: 6037197]

[4] Bangham AD, Standish MM, Watkins JC. Diffusion of univalent ions across the lamellae of swollen phospholipids. J Mol Biol 1965; 13(1): 238-52. [http://dx.doi.org/10.1016/S0022-2836(65)80093-6] [PMID: 5859039]

[5] Gregoriadis G. The carrier potential of liposomes in biology and medicine (second of two parts). N Engl J Med 1976; 295(14): 765-70. [http://dx.doi.org/10.1056/NEJM197609302951406] [PMID: 785256]

[6] Gregoriadis G. The carrier potential of liposomes in biology and medicine (second of two parts). Q Rev Biophys 1980; 13(2): 121-200. [PMID: 7015403]

[7] Allen TM, Cullis PR. Liposomal drug delivery systems: from concept to clinical applications. Adv Drug Deliv Rev 2013; 65(1): 36-48. [http://dx.doi.org/10.1016/j.addr.2012.09.037] [PMID: 23036225]

[8] Kimelberg HK, Tracy TFJ Jr, Biddlecome SM, Bourke RS. The effect of entrapment in liposomes on the *in vivo* distribution of [3H]methotrexate in a primate. Cancer Res 1976; 36(8): 2949-57. [PMID: 819137]

[9] Pattni BS, Chupin VV, Torchilin VP. New developments in liposomal drug delivery. Chem Rev 2015; 115(19): 10938-66. [http://dx.doi.org/10.1021/acs.chemrev.5b00046] [PMID: 26010257]

[10] Bhatia A, Singh B, Bhushan S, Katare OP. Tamoxifen-encapsulated vesicular systems: cytotoxicity evaluation in human epidermal keratinocyte cell line. Drug Dev Ind Pharm 2010; 36(3): 350-4. [http://dx.doi.org/10.3109/03639040903173549] [PMID: 19719395]

[11] Bhatia A, Singh B, Raza K, Shukla A, Amarji B, Katare OP. Tamoxifen-loaded novel liposomal formulations: evaluation of anticancer activity on DMBA-TPA induced mouse skin carcinogenesis. J Drug Target 2012; 20(6): 544-50.
[http://dx.doi.org/10.3109/1061186X.2012.694887] [PMID: 22643315]

[12] Rahman M, Kumar V, Beg S, Sharma G, Katare OP, Anwar F. Emergence of liposome as targeted magic bullet for inflammatory disorders: current state of the art. Artif Cells Nanomed Biotechnol 2016; 44(7): 1597-608.
[http://dx.doi.org/10.3109/21691401.2015.1129617] [PMID: 26758815]

[13] Akbarzadeh A, Rezaei-Sadabady R, Davaran S, *et al.* Liposome: classification, preparation, and applications. Nanoscale Res Lett 2013; 8(1): 102.
[http://dx.doi.org/10.1186/1556-276X-8-102] [PMID: 23432972]

[14] Sharma G, Goyal H, Thakur K, Raza K, Katare OP. Novel elastic membrane vesicles (EMVs) and ethosomes-mediated effective topical delivery of aceclofenac: a new therapeutic approach for pain and inflammation. Drug Deliv 2016; 23(8): 3135-45.
[http://dx.doi.org/10.3109/10717544.2016.1155244] [PMID: 26960815]

[15] Sharma G, Saini MK, Thakur K, *et al.* Aceclofenac cocrystal nanoliposomes for rheumatoid arthritis with better dermatokinetic attributes: a preclinical study. Nanomedicine (Lond) 2017; 12(6): 615-38.
[http://dx.doi.org/10.2217/nnm-2016-0405] [PMID: 28186461]

[16] Sharma G, Thakur K, Setia A, *et al.* Fabrication of acyclovir-loaded flexible membrane vesicles (FMVs): evidence of preclinical efficacy of antiviral activity in murine model of cutaneous HSV-1 infection. Drug Deliv Transl Res 2017; 7(5): 683-94.
[http://dx.doi.org/10.1007/s13346-017-0417-0] [PMID: 28801835]

[17] Sharma G, Kaur M, Raza K, Thakur K, Katare OP. Aceclofenac-β-cyclodextrin-vesicles: a dual carrier approach for skin with enhanced stability, efficacy and dermatokinetic profile. RSC Advances 2016; 6(25): 20713-27.
[http://dx.doi.org/10.1039/C5RA24516H]

[18] Raza K, Singh B, Mahajan A, Negi P, Bhatia A, Katare OP. Design and evaluation of flexible membrane vesicles (FMVs) for enhanced topical delivery of capsaicin. J Drug Target 2011; 19(4): 293-302.
[http://dx.doi.org/10.3109/1061186X.2010.499464] [PMID: 20615093]

[19] Mouritsen OG. Lipids, curvature, and nano-medicine. Eur J Lipid Sci Technol 2011; 113(10): 1174-87.
[http://dx.doi.org/10.1002/ejlt.201100050] [PMID: 22164124]

[20] Sharma VK, Sarwa KK, Mazumder B. Fluidity enhancement: a critical factor for performance of liposomal transdermal drug delivery system. J Liposome Res 2014; 24(2): 83-9.
[http://dx.doi.org/10.3109/08982104.2013.847956] [PMID: 24160895]

[21] Hua S, Wu SY. The use of lipid-based nanocarriers for targeted pain therapies. Front Pharmacol 2013; 4: 143.
[http://dx.doi.org/10.3389/fphar.2013.00143] [PMID: 24319430]

[22] Zylberberg C, Matosevic S. Pharmaceutical liposomal drug delivery: a review of new delivery systems and a look at the regulatory landscape. Drug Deliv 2016; 23(9): 3319-29.
[http://dx.doi.org/10.1080/10717544.2016.1177136] [PMID: 27145899]

[23] Hwang SY, Kim HK, Choo J, Seong GH, Hien TB, Lee EK. Effects of operating parameters on the efficiency of liposomal encapsulation of enzymes. Colloids Surf B Biointerfaces 2012; 94: 296-303.
[http://dx.doi.org/10.1016/j.colsurfb.2012.02.008] [PMID: 22398367]

[24] Matosevic S. Synthesizing artificial cells from giant unilamellar vesicles: state-of-the art in the development of microfluidic technology. BioEssays 2012; 34(11): 992-1001.
[http://dx.doi.org/10.1002/bies.201200105] [PMID: 22926929]

[25] Torchilin VP. Multifunctional, stimuli-sensitive nanoparticulate systems for drug delivery. Nat Rev Drug Discov 2014; 13(11): 813-27.
[http://dx.doi.org/10.1038/nrd4333] [PMID: 25287120]

[26] Eloy JO, Claro de Souza M, Petrilli R, Barcellos JPA, Lee RJ, Marchetti JM. Liposomes as carriers of hydrophilic small molecule drugs: strategies to enhance encapsulation and delivery. Colloids Surf B Biointerfaces 2014; 123: 345-63.
[http://dx.doi.org/10.1016/j.colsurfb.2014.09.029] [PMID: 25280609]

[27] Israelachvili JN, Marcelja S, Horn RG. Physical principles of membrane organization. Q Rev Biophys 1980; 13(2): 121-200.
[http://dx.doi.org/10.1017/S0033583500001645] [PMID: 7015403]

[28] Lasic DD. Novel applications of liposomes. Trends Biotechnol 1998; 16(7): 307-21.
[http://dx.doi.org/10.1016/S0167-7799(98)01220-7] [PMID: 9675915]

[29] Pierre MB, Dos Santos Miranda Costa I. Liposomal systems as drug delivery vehicles for dermal and transdermal applications. Arch Dermatol Res 2011; 303(9): 607-21.
[http://dx.doi.org/10.1007/s00403-011-1166-4] [PMID: 21805180]

[30] Torchilin VP. Recent advances with liposomes as pharmaceutical carriers. Nat Rev Drug Discov 2005; 4(2): 145-60.
[http://dx.doi.org/10.1038/nrd1632] [PMID: 15688077]

[31] Chen J, Cheng D, Li J, *et al.* Influence of lipid composition on the phase transition temperature of liposomes composed of both DPPC and HSPC. Drug Dev Ind Pharm 2013; 39(2): 197-204.
[http://dx.doi.org/10.3109/03639045.2012.668912] [PMID: 22443684]

[32] Blume G, Cevc G. Liposomes for the sustained drug release *in vivo.* Biochim Biophys Acta 1990; 1029(1): 91-7.
[http://dx.doi.org/10.1016/0005-2736(90)90440-Y] [PMID: 2223816]

[33] van Hoogevest P, Wendel A. The use of natural and synthetic phospholipids as pharmaceutical excipients. Eur J Lipid Sci Technol 2014; 116(9): 1088-107.
[http://dx.doi.org/10.1002/ejlt.201400219] [PMID: 25400504]

[34] Kaur R, Henriksen-Lacey M, Wilkhu J, Devitt A, Christensen D, Perrie Y. Effect of incorporating cholesterol into DDA:TDB liposomal adjuvants on bilayer properties, biodistribution, and immune responses. Mol Pharm 2014; 11(1): 197-207.
[http://dx.doi.org/10.1021/mp400372j] [PMID: 24171445]

[35] McMullen TPW, McElhaney RN. Physical studies of cholesterol-phospholipid interactions. Curr Opin Colloid Interface Sci 1996; 1(1): 83-90.
[http://dx.doi.org/10.1016/S1359-0294(96)80048-3]

[36] Sorkin R, Kampf N, Klein J. Effect of cholesterol on the stability and lubrication efficiency of phosphatidylcholine surface layers. Langmuir 2017; 33(30): 7459-67.
[http://dx.doi.org/10.1021/acs.langmuir.7b01521] [PMID: 28666386]

[37] Moghaddam B, Ali MH, Wilkhu J, *et al.* The application of monolayer studies in the understanding of liposomal formulations. Int J Pharm 2011; 417(1-2): 235-44.
[http://dx.doi.org/10.1016/j.ijpharm.2011.01.020] [PMID: 21251964]

[38] López-Pinto JM, González-Rodríguez ML, Rabasco AM. Effect of cholesterol and ethanol on dermal delivery from DPPC liposomes. Int J Pharm 2005; 298(1): 1-12.
[http://dx.doi.org/10.1016/j.ijpharm.2005.02.021] [PMID: 15896932]

[39] de Athayde Moncorvo Collado A, Dupuy FG, Morero RD, Minahk C. Cholesterol induces surface localization of polyphenols in model membranes thus enhancing vesicle stability against lysozyme, but reduces protection of distant double bonds from reactive-oxygen species. Biochim Biophys Acta 2016; 1858(7 Pt A): 1479-87.
[http://dx.doi.org/10.1016/j.bbamem.2016.04.002] [PMID: 27063609]

[40] Patil YP, Jadhav S. Novel methods for liposome preparation. Chem Phys Lipids 2014; 177: 8-18. [http://dx.doi.org/10.1016/j.chemphyslip.2013.10.011] [PMID: 24220497]

[41] Huang Z, Li X, Zhang T, *et al.* Progress involving new techniques for liposome preparation. Asian J Pharm Sci 2014; 9(4): 176-82. [http://dx.doi.org/10.1016/j.ajps.2014.06.001]

[42] Sorgi FL, Huang L. Large scale production of DC-Chol cationic liposomes by microfluidization. Int J Pharm 1996; 144(2): 131-9. [http://dx.doi.org/10.1016/S0378-5173(96)04733-3]

[43] Charcosset C, Juban A, Valour J, Urbaniak S, Fessi H. Preparation of liposomes at large scale using the ethanol injection method: Effect of scale-up and injection devices. Chem Eng Sci Des 2015; 94(C): 508-15. [http://dx.doi.org/10.1016/j.cherd.2014.09.008]

[44] Bangham AD, De Gier J, Greville GD. Osmotic properties and water permeability of phospholipid liquid crystals. Chem Phys Lipids 1967; 1(3): 225-46. [http://dx.doi.org/10.1016/0009-3084(67)90030-8]

[45] Zhang H. Thin-film Hydration Followed by Extrusion Method for Liposome Preparation. Liposomes: Methods and Protocols. New York, NY: Springer New York 2016; pp. 17-22.

[46] Biswas S, Dodwadkar NS, Deshpande PP, Torchilin VP. Liposomes loaded with paclitaxel and modified with novel triphenylphosphonium-PEG-PE conjugate possess low toxicity, target mitochondria and demonstrate enhanced antitumor effects *in vitro* and *in vivo.* J Control Rel 2012; 159(3): 393-402. [http://dx.doi.org/10.1016/j.jconrel.2012.01.009] [PMID: 22286008]

[47] Popa R, Vrânceanu M, Nikolaus S, Nirschl H, Leneweit G. Entrance effects at nanopores of nanocapsules functionalized with poly(ethylene glycol) and their flow through nanochannels. Langmuir 2008; 24(22): 13030-6. [http://dx.doi.org/10.1021/la8024777] [PMID: 18950201]

[48] Wang L, Pileni MP. Encapsulation of zwitterionic Au nanocrystals into liposomes by reverse phase evaporation method: Influence of the surface charge. Langmuir 2016; 32(47): 12370-7. [http://dx.doi.org/10.1021/acs.langmuir.6b01132] [PMID: 27243268]

[49] Chen G, Li D, Jin Y, *et al.* Deformable liposomes by reverse-phase evaporation method for an enhanced skin delivery of (+)-catechin. Drug Dev Ind Pharm 2014; 40(2): 260-5. [http://dx.doi.org/10.3109/03639045.2012.756512] [PMID: 23356860]

[50] Mozafari MR. Liposomes: an overview of manufacturing techniques. Cell Mol Biol Lett 2005; 10(4): 711-9. [PMID: 16341279]

[51] Jiskoot W, Teerlink T, Beuvery EC, Crommelin DJA. Preparation of liposomes *via* detergent removal from mixed micelles by dilution. The effect of bilayer composition and process parameters on liposome characteristics. Pharm Weekbl Sci 1986; 8(5): 259-65. [http://dx.doi.org/10.1007/BF01960070] [PMID: 3786108]

[52] Holzer M, Barnert S, Momm J, Schubert R. Preparative size exclusion chromatography combined with detergent removal as a versatile tool to prepare unilamellar and spherical liposomes of highly uniform size distribution. J Chromatogr A 2009; 1216(31): 5838-48. [http://dx.doi.org/10.1016/j.chroma.2009.06.023] [PMID: 19545872]

[53] Karn PR, Cho W, Park HJ, Park JS, Hwang SJ. Characterization and stability studies of a novel liposomal cyclosporin A prepared using the supercritical fluid method: comparison with the modified conventional Bangham method. Int J Nanomed 2013; 8: 365-77. [PMID: 23378759]

[54] Zhao L, Temelli F, Curtis JM, Chen L. Preparation of liposomes using supercritical carbon dioxide

technology: Effects of phospholipids and sterols. Food Res Int 2015; 77(Part 1): 63-2.
[http://dx.doi.org/10.1016/j.foodres.2015.07.006]

[55] Otake K, Shimomura T, Goto T, *et al.* Preparation of liposomes using an improved supercritical reverse phase evaporation method. Langmuir 2006; 22(6): 2543-50.
[http://dx.doi.org/10.1021/la051654u] [PMID: 16519453]

[56] Santo IE, Pedro AS, Fialho R, Cabral-Albuquerque E. Characteristics of lipid micro- and nanoparticles based on supercritical formation for potential pharmaceutical application. Nanoscale Res Lett 2013; 8(1): 386.
[http://dx.doi.org/10.1186/1556-276X-8-386] [PMID: 24034341]

[57] Lesoin L, Crampon C, Boutin O, Badens E. Preparation of liposomes using the supercritical anti-solvent (SAS) process and comparison with a conventional method. J Supercrit Fluids 2011; 57(2): 162-74.
[http://dx.doi.org/10.1016/j.supflu.2011.01.006]

[58] Zhang W, Sun Y, Li Y, Shen R, Ni H, Hu D. Preparation and influencing factors of sirolimus liposome by supercritical fluid. Artif Cells Blood Substit Immobil Biotechnol 2012; 40(1-2): 62-5.
[http://dx.doi.org/10.3109/10731199.2011.585618] [PMID: 21689052]

[59] Yang G, Zhao Y, Zhang Y, Dang B, Liu Y, Feng N. Enhanced oral bioavailability of silymarin using liposomes containing a bile salt: preparation by supercritical fluid technology and evaluation *in vitro* and *in vivo.* Int J Nanomed 2015; 10: 6633-44.
[http://dx.doi.org/10.2147/IJN.S92665] [PMID: 26543366]

[60] Hirsch M, Ziroli V, Helm M, Massing U. Preparation of small amounts of sterile siRNA-liposomes with high entrapping efficiency by dual asymmetric centrifugation (DAC). J Control Rel 2009; 135(1): 80-8.
[http://dx.doi.org/10.1016/j.jconrel.2008.11.029] [PMID: 19124051]

[61] Ingebrigtsen SG, Škalko-Basnet N, Holsæter AM. Development and optimization of a new processing approach for manufacturing topical liposomes-in-hydrogel drug formulations by dual asymmetric centrifugation. Drug Dev Ind Pharm 2016; 42(9): 1375-83.
[http://dx.doi.org/10.3109/03639045.2015.1135940] [PMID: 26710826]

[62] Massing U, Cicko S, Ziroli V. Dual asymmetric centrifugation (DAC)--a new technique for liposome preparation. J Control Rel 2008; 125(1): 16-24.
[http://dx.doi.org/10.1016/j.jconrel.2007.09.010] [PMID: 18023907]

[63] Crowe JH, Crowe LM. Factors affecting the stability of dry liposomes. Biochim Biophys Acta 1988; 939(2): 327-34.
[http://dx.doi.org/10.1016/0005-2736(88)90077-6] [PMID: 3355821]

[64] Salama AH, Aburahma MH. Ufasomes nano-vesicles-based lyophilized platforms for intranasal delivery of cinnarizine: preparation, optimization, *ex-vivo* histopathological safety assessment and mucosal confocal imaging. Pharm Dev Technol 2016; 21(6): 706-15.
[PMID: 25996631]

[65] van Winden ECA, Zhang W, Crommelin DJA. Effect of freezing rate on the stability of liposomes during freeze-drying and rehydration. Pharm Res 1997; 14(9): 1151-60.
[http://dx.doi.org/10.1023/A:1012142520912] [PMID: 9327441]

[66] Vincourt V, Nguyen L, Chaumeil JC, Dumortier G. Freeze-drying of ATP entrapped in cationic, low lipid liposomes. Cryobiology 2010; 60(3): 262-70.
[http://dx.doi.org/10.1016/j.cryobiol.2010.01.004] [PMID: 20097191]

[67] Yang S, Liu C, Liu W, *et al.* Preparation and characterization of nanoliposomes entrapping medium-chain fatty acids and vitamin C by lyophilization. Int J Mol Sci 2013; 14(10): 19763-73.
[http://dx.doi.org/10.3390/ijms141019763] [PMID: 24084723]

[68] Benjakul R, Panyarachun B, Sarisuta N. Preparation of dry reconstituted liposomal powder by freeze-

drying at room temperature. J Liposome Res 2011; 21(1): 28-37. [http://dx.doi.org/10.3109/08982101003735970] [PMID: 20387991]

[69] Peschka R, Purmann T, Schubert R. Cross-flow filtration improved detergent removal technique for the preparation of liposomes. Int J Pharm 1998; 162(1): 177-83. [http://dx.doi.org/10.1016/S0378-5173(97)00424-9]

[70] Wagner A, Vorauer-Uhl K, Kreismayr G, Katinger H. Enhanced protein loading into liposomes by the multiple crossflow injection technique. J Liposome Res 2002; 12(3): 271-83. [http://dx.doi.org/10.1081/LPR-120014762] [PMID: 12604031]

[71] Wagner A, Vorauer-Uhl K, Kreismayr G, Katinger H. The crossflow injection technique: an improvement of the ethanol injection method. J Liposome Res 2002; 12(3): 259-70. [http://dx.doi.org/10.1081/LPR-120014761] [PMID: 12604030]

[72] Wagner A, Platzgummer M, Kreismayr G, *et al.* GMP production of liposomes--a new industrial approach. J Liposome Res 2006; 16(3): 311-9. [http://dx.doi.org/10.1080/08982100600851086] [PMID: 16952884]

[73] Sebaaly C, Greige-Gerges H, Agusti G, Fessi H, Charcosset C. Large-scale preparation of clove essential oil and eugenol-loaded liposomes using a membrane contactor and a pilot plant. J Liposome Res 2016; 26(2): 126-38. [PMID: 26099849]

[74] Jaafar-Maalej C, Charcosset C, Fessi H. A new method for liposome preparation using a membrane contactor. J Liposome Res 2011; 21(3): 213-20. [http://dx.doi.org/10.3109/08982104.2010.517537] [PMID: 20860451]

[75] Szebeni J, Muggia F, Gabizon A, Barenholz Y. Activation of complement by therapeutic liposomes and other lipid excipient-based therapeutic products: prediction and prevention. Adv Drug Deliv Rev 2011; 63(12): 1020-30. [http://dx.doi.org/10.1016/j.addr.2011.06.017] [PMID: 21787819]

[76] Papahadjopoulos-Sternberg B. Freeze-fracture electron microscopy on nano- and micro-delivery vehicles for biological active compounds. Biophys J 2016; 110(3): 199a. [http://dx.doi.org/10.1016/j.bpj.2015.11.1113]

[77] Villasmil-Sánchez S, Rabasco AM, González-Rodríguez ML. Thermal and 31P-NMR studies to elucidate sumatriptan succinate entrapment behavior in phosphatidylcholine/cholesterol liposomes. Comparative 31P-NMR analysis on negatively and positively-charged liposomes. Colloids Surf B Biointerfaces 2013; 105: 14-23. [http://dx.doi.org/10.1016/j.colsurfb.2012.12.019] [PMID: 23352939]

[78] Urey C, Weiss VU, Gondikas A, *et al.* Combining gas-phase electrophoretic mobility molecular analysis (GEMMA), light scattering, field flow fractionation and cryo electron microscopy in a multidimensional approach to characterize liposomal carrier vesicles. Int J Pharm 2016; 513(1-2): 309-18. [http://dx.doi.org/10.1016/j.ijpharm.2016.09.049] [PMID: 27639623]

[79] Akashi K, Miyata H, Itoh H, Kinosita K Jr. Preparation of giant liposomes in physiological conditions and their characterization under an optical microscope. Biophys J 1996; 71(6): 3242-50. [http://dx.doi.org/10.1016/S0006-3495(96)79517-6] [PMID: 8968594]

[80] Baxa U. Imaging of Liposomes by Transmission Electron Microscopy. Characterization of Nanoparticles Intended for Drug Delivery. New York, NY: Springer 2017; pp. 73-88. [http://dx.doi.org/10.1007/978-1-4939-7352-1_8]

[81] Franken LE, Boekema EJ, Stuart MCA. Transmission electron microscopy as a tool for the characterization of soft materials: Application and interpretation. Adv Sci (Weinh) 2017; 4(5): 1600476. [http://dx.doi.org/10.1002/advs.201600476] [PMID: 28546914]

[82] Takahashi N, Higashi K, Ueda K, Yamamoto K, Moribe K. Determination of non-spherical morphology of doxorubicin-loaded liposomes by atomic force microscopy. J Pharm Sci 2017. [http://dx.doi.org/10.1016/j.xphs.2017.10.009] [PMID: 29031955]

[83] Esposito E, Mariani P, Drechsler M, Cortesi R. Structural studies of lipid-based nanosystems for drug delivery: X-ray diffraction (XRD) and Cryogenic Transmission Electron Microscopy (Cryo-TEM) Handbook of Nanopart. New York: Springer 2017. [http://dx.doi.org/10.1007/978-3-319-15338-4_39]

[84] Wei X, Patil Y, Ohana P, *et al.* Characterization of pegylated liposomal mitomycin C lipid-based prodrug (PromitilÂ®) by high sensitivity differential scanning calorimetry and cryogenic transmission electron microscopy. Mol Pharm 2017; 14(12): 4339-45. [http://dx.doi.org/10.1021/acs.molpharmaceut.6b00865]

[85] Wibroe PP, Ahmadvand D, Oghabian MA, Yaghmur A, Moghimi SM. An integrated assessment of morphology, size, and complement activation of the PEGylated liposomal doxorubicin products Doxil®, Caelyx®, DOXOrubicin, and SinaDoxosome. J Control Rel 2016; 221: 1-8. [http://dx.doi.org/10.1016/j.jconrel.2015.11.021] [PMID: 26608877]

[86] Laan AC, Denkova AG. Cryogenic transmission electron microscopy: the technique of choice for the characterization of polymeric nanocarriers. EJNMMI Res 2017; 7(1): 44. [http://dx.doi.org/10.1186/s13550-017-0290-3] [PMID: 28534134]

[87] Schilt Y, Berman T, Wei X, Barenholz Y, Raviv U. Using solution X-ray scattering to determine the high-resolution structure and morphology of PEGylated liposomal doxorubicin nanodrugs. Biochim Biophys Acta 2016; 1860(1 Pt A): 108-19. [http://dx.doi.org/10.1016/j.bbagen.2015.09.012] [PMID: 26391840]

[88] Garcia-Diez R, Gollwitzer C, Krumrey M, Varga Z. Size determination of a liposomal drug by small-angle X-ray scattering using continuous contrast variation. Langmuir 2016; 32(3): 772-8. [http://dx.doi.org/10.1021/acs.langmuir.5b02261] [PMID: 26673729]

[89] Delorme N, Fery A. Direct method to study membrane rigidity of small vesicles based on atomic force microscope force spectroscopy. Phys Rev E Stat Nonlin Soft Matter Phys 2006; 74(3 Pt 1): 030901. [http://dx.doi.org/10.1103/PhysRevE.74.030901] [PMID: 17025583]

[90] Reviakine I, Brisson A. Formation of supported phospholipid bilayers from unilamellar vesicles investigated by atomic force microscopy. Langmuir 2000; 16(4): 1806-15. [http://dx.doi.org/10.1021/la9903043]

[91] Jass J, Tjärnhage T, Puu G. From liposomes to supported, planar bilayer structures on hydrophilic and hydrophobic surfaces: an atomic force microscopy study. Biophys J 2000; 79(6): 3153-63. [http://dx.doi.org/10.1016/S0006-3495(00)76549-0] [PMID: 11106620]

[92] Takechi-Haraya Y, Sakai-Kato K, Abe Y, Kawanishi T, Okuda H, Goda Y. Observation of liposomes of differing lipid composition in aqueous medium by means of atomic force microscopy. Microscopy (Oxf) 2016; 65(4): 383-9. [http://dx.doi.org/10.1093/jmicro/dfw011] [PMID: 27020464]

[93] Takechi-Haraya Y, Sakai-Kato K, Abe Y, Kawanishi T, Okuda H, Goda Y. Atomic force microscopic analysis of the effect of lipid composition on liposome membrane rigidity. Langmuir 2016; 32(24): 6074-82. [http://dx.doi.org/10.1021/acs.langmuir.6b00741] [PMID: 27232007]

[94] Takechi-Haraya Y, Sakai-Kato K, Goda Y. Membrane rigidity determined by atomic force microscopy is a parameter of the permeability of liposomal membranes to the hydrophilic compound calcein. AAPS PharmSciTech 2017; 18(5): 1887-93. [http://dx.doi.org/10.1208/s12249-016-0624-x] [PMID: 27645470]

[95] Laouini A, Jaafar-Maalej C, Limayem-Blouza I, Sfar S, Charcosset C, Fessi H. Preparation, characterization and applications of liposomes: State of the art. J Colloid Sci Biotechnol 2012; 1(2):

147-68.
[http://dx.doi.org/10.1166/jcsb.2012.1020]

[96] Provder T. Challenges in particle size distribution measurement past, present and for the 21st century. Prog Org Coat 1997; 32(1): 143-53.
[http://dx.doi.org/10.1016/S0300-9440(97)00043-X]

[97] Grabielle-Madelmont C, Lesieur S, Ollivon M. Characterization of loaded liposomes by size exclusion chromatography. J Biochem Biophys Methods 2003; 56(1-3): 189-217.
[http://dx.doi.org/10.1016/S0165-022X(03)00059-9] [PMID: 12834977]

[98] Lundahl P, Zeng CM, Lagerquist Hägglund C, Gottschalk I, Greijer E. Chromatographic approaches to liposomes, proteoliposomes and biomembrane vesicles. J Chromatogr B Biomed Sci Appl 1999; 722(1-2): 103-20.
[http://dx.doi.org/10.1016/S0378-4347(98)00370-3] [PMID: 10068136]

[99] Xu R, Fitts A, Li X, *et al.* Quantification of small extracellular vesicles by size exclusion chromatography with fluorescence detection. Anal Chem 2016; 88(21): 10390-4.
[http://dx.doi.org/10.1021/acs.analchem.6b03348] [PMID: 27689436]

[100] Filipe V, Hawe A, Jiskoot W. Critical evaluation of Nanoparticle Tracking Analysis (NTA) by NanoSight for the measurement of nanoparticles and protein aggregates. Pharm Res 2010; 27(5): 796-810.
[http://dx.doi.org/10.1007/s11095-010-0073-2] [PMID: 20204471]

[101] Song X, Zhao Y, Hou S, *et al.* Dual agents loaded PLGA nanoparticles: systematic study of particle size and drug entrapment efficiency. Eur J Pharm Biopharm 2008; 69(2): 445-53.
[http://dx.doi.org/10.1016/j.ejpb.2008.01.013] [PMID: 18374554]

[102] Sharma G, Devi N, Thakur K, Jain A, Katare OP. Lanolin-based organogel of salicylic acid: evidences of better dermatokinetic profile in imiquimod-induced keratolytic therapy in BALB/c mice model. Drug Deliv Transl Res 2018; 8(2): 398-413.
[http://dx.doi.org/10.1007/s13346-017-0364-9] [PMID: 28224375]

[103] Smith MC, Crist RM, Clogston JD, McNeil SE. Zeta potential: a case study of cationic, anionic, and neutral liposomes. Anal Bioanal Chem 2017; 409(24): 5779-87.
[http://dx.doi.org/10.1007/s00216-017-0527-z] [PMID: 28762066]

[104] Chibowski E, Szczea A. Zeta potential and surface charge of DPPC and DOPC liposomes in the presence of PLC enzyme. Adsorp 2016; 22(4): 755-65.
[http://dx.doi.org/10.1007/s10450-016-9767-z]

[105] Grit M, Crommelin DJA. Chemical stability of liposomes: implications for their physical stability. Chem Phys Lipids 1993; 64(1-3): 3-18.
[http://dx.doi.org/10.1016/0009-3084(93)90053-6] [PMID: 8242840]

[106] Hein R, Uzundal CB, Hennig A. Simple and rapid quantification of phospholipids for supramolecular membrane transport assays. Org Biomol Chem 2016; 14(7): 2182-5.
[http://dx.doi.org/10.1039/C5OB02480C] [PMID: 26790479]

[107] Edwards KA, Baeumner AJ. Analysis of liposomes. Talanta 2006; 68(5): 1432-41.
[http://dx.doi.org/10.1016/j.talanta.2005.08.031] [PMID: 18970482]

[108] Zhang JA, Pawelchak J. Effect of pH, ionic strength and oxygen burden on the chemical stability of EPC/cholesterol liposomes under accelerated conditions. Part 1: Lipid hydrolysis. Eur J Pharm Biopharm 2000; 50(3): 357-64.
[http://dx.doi.org/10.1016/S0939-6411(00)00127-2] [PMID: 11072192]

[109] Klein RA. The detection of oxidation in liposome preparations. Biochim Biophys Acta 1970; 210(3): 486-9.
[http://dx.doi.org/10.1016/0005-2760(70)90046-9] [PMID: 5528599]

[110] Lang JK. Quantitative determination of cholesterol in liposome drug products and raw materials by

high-performance liquid chromatography. J Chromatogr A 1990; 507(Suppl. C): 157-63. [http://dx.doi.org/10.1016/S0021-9673(01)84191-1] [PMID: 2380284]

[111] Nichols JW, Deamer DW. Catecholamine uptake and concentration by liposomes maintaining p/ gradients. Biochim Biophys Acta 1976; 455(1): 269-71. [http://dx.doi.org/10.1016/0005-2736(76)90169-3] [PMID: 11002]

[112] Mayer LD, Bally MB, Hope MJ, Cullis PR. Uptake of antineoplastic agents into large unilamellar vesicles in response to a membrane potential. Biochim Biophys Acta 1985; 816(2): 294-302. [http://dx.doi.org/10.1016/0005-2736(85)90497-3] [PMID: 3839135]

[113] Cern A, Marcus D, Tropsha A, Barenholz Y, Goldblum A. New drug candidates for liposomal delivery identified by computer modeling of liposomes' remote loading and leakage. J Control Rel 2017; 252: 18-27. [http://dx.doi.org/10.1016/j.jconrel.2017.02.015] [PMID: 28215669]

[114] Bolotin EM, Cohen R, Bar LK, *et al.* Ammonium sulfate gradients for efficient and stable remote loading of amphipathic weak bases into liposomes and ligandoliposomes. J Liposome Res 1994; 4(1): 455-79. [http://dx.doi.org/10.3109/08982109409037057]

[115] Cullis PR, Mayer LD, Bally MB, Madden TD, Hope MJ. Generating and loading of liposomal systems for drug-delivery applications. Adv Drug Deliv Rev 1989; 3(3): 267-82. [http://dx.doi.org/10.1016/0169-409X(89)90024-0]

[116] Barenholz Y. Relevancy of drug loading to liposomal formulation therapeutic efficacy. J Liposome Res 2003; 13(1): 1-8. [http://dx.doi.org/10.1081/LPR-120017482] [PMID: 12725720]

[117] Gubernator J. Active methods of drug loading into liposomes: recent strategies for stable drug entrapment and increased *in vivo* activity. Expert Opin Drug Deliv 2011; 8(5): 565-80. [http://dx.doi.org/10.1517/17425247.2011.566552] [PMID: 21492058]

[118] Haran G, Cohen R, Bar LK, Barenholz Y. Transmembrane ammonium sulfate gradients in liposomes produce efficient and stable entrapment of amphipathic weak bases. Biochim Biophys Acta 1993; 1151(2): 201-15. [http://dx.doi.org/10.1016/0005-2736(93)90105-9] [PMID: 8373796]

[119] Gubernator J, Chwastek G, Korycińska M, *et al.* The encapsulation of idarubicin within liposomes using the novel EDTA ion gradient method ensures improved drug retention *in vitro* and *in vivo.* J Control Rel 2010; 146(1): 68-75. [http://dx.doi.org/10.1016/j.jconrel.2010.05.021] [PMID: 20510316]

[120] Hersh EV, Saraghi MN, Moore PA. Two recent advances in local anesthesia: Intranasal Tetracaine/Oxymetazoline and liposomal bupivacaine. Curr Oral Health Rep 2017; 4(3): 189-96. [http://dx.doi.org/10.1007/s40496-017-0144-0]

[121] Kraft JC, Freeling JP, Wang Z, Ho RJY. Emerging research and clinical development trends of liposome and lipid nanoparticle drug delivery systems. J Pharm Sci 2014; 103(1): 29-52. [http://dx.doi.org/10.1002/jps.23773] [PMID: 24338748]

[122] Allen TM, Everest JM. Effect of liposome size and drug release properties on pharmacokinetics of encapsulated drug in rats. J Pharmacol Exp Ther 1983; 226(2): 539-44. [PMID: 6875864]

[123] Maruyama K, Ishida O, Takizawa T, Moribe K. Possibility of active targeting to tumor tissues with liposomes. Adv Drug Deliv Rev 1999; 40(1-2): 89-102. [http://dx.doi.org/10.1016/S0169-409X(99)00042-3] [PMID: 10837782]

[124] Arpicco S, Lerda C, Dalla Pozza E, *et al.* Hyaluronic acid-coated liposomes for active targeting of gemcitabine. Eur J Pharm Biopharm 2013; 85(3 Pt A): 373-80. [http://dx.doi.org/10.1016/j.ejpb.2013.06.003] [PMID: 23791684]

[125] Zhang W, Wang G, See E, *et al.* Post-insertion of poloxamer 188 strengthened liposomal membrane and reduced drug irritancy and *in vivo* precipitation, superior to PEGylation. J Control Rel 2015; 203(Suppl. C): 161-9.
[http://dx.doi.org/10.1016/j.jconrel.2015.02.026] [PMID: 25701612]

[126] Cosco D, Tsapis N, Nascimento TL, *et al.* Polysaccharide-coated liposomes by post-insertion of a hyaluronan-lipid conjugate. Colloids Surf B Biointerfaces 2017; 158(Suppl. C): 119-26.
[http://dx.doi.org/10.1016/j.colsurfb.2017.06.029] [PMID: 28686903]

[127] Deshpande PP, Biswas S, Torchilin VP. Current trends in the use of liposomes for tumor targeting. Nanomedicine (Lond) 2013; 8(9): 1509-28.
[http://dx.doi.org/10.2217/nnm.13.118] [PMID: 23914966]

[128] Li W, Nicol F, Szoka FC Jr. GALA: a designed synthetic pH-responsive amphipathic peptide with applications in drug and gene delivery. Adv Drug Deliv Rev 2004; 56(7): 967-85.
[http://dx.doi.org/10.1016/j.addr.2003.10.041] [PMID: 15066755]

[129] Yao VJ, D'Angelo S, Butler KS, *et al.* Ligand-targeted theranostic nanomedicines against cancer. J Control Rel 2016; 240(Suppl. C): 267-86.
[http://dx.doi.org/10.1016/j.jconrel.2016.01.002] [PMID: 26772878]

[130] Acharya A, Das I, Chandhok D, Saha T. Redox regulation in cancer: a double-edged sword with therapeutic potential. Oxid Med Cell Longev 2010; 3(1): 23-34.
[http://dx.doi.org/10.4161/oxim.3.1.10095] [PMID: 20716925]

[131] Gu Z, Chang M, Fan Y, Shi Y, Lin G. NGR-modified pH-sensitive liposomes for controlled release and tumor target delivery of docetaxel. Colloids Surf B Biointerfaces 2017; 160: 395-405.
[http://dx.doi.org/10.1016/j.colsurfb.2017.09.052] [PMID: 28965079]

[132] Fan Y, Chen C, Huang Y, Zhang F, Lin G. Study of the pH-sensitive mechanism of tumor-targeting liposomes. Colloids Surf B Biointerfaces 2017; 151: 19-25.
[http://dx.doi.org/10.1016/j.colsurfb.2016.11.042] [PMID: 27940165]

[133] Dos Santos Ferreira D, Jesus de Oliveira Pinto BL, Kumar V, *et al.* Evaluation of antitumor activity and cardiac toxicity of a bone-targeted ph-sensitive liposomal formulation in a bone metastasis tumor model in mice. Nanomedicine (Lond) 2017; 13(5): 1693-701.
[http://dx.doi.org/10.1016/j.nano.2017.03.005] [PMID: 28343016]

[134] Movahedi F, Hu RG, Becker DL, Xu C. Stimuli-responsive liposomes for the delivery of nucleic acid therapeutics. Nanomedicine (Lond) 2015; 11(6): 1575-84.
[http://dx.doi.org/10.1016/j.nano.2015.03.006] [PMID: 25819885]

[135] Zhu L, Torchilin VP. Stimulus-responsive nanopreparations for tumor targeting. Integr Biol 2013; 5(1): 96-107.
[http://dx.doi.org/10.1039/c2ib20135f] [PMID: 22869005]

[136] Yavlovich A, Singh A, Tarasov S, Capala J, Blumenthal R, Puri A. Design of liposomes containing photopolymerizable phospholipids for triggered release of contents. J Therm Anal Calorim 2009; 98(1): 97-104.
[http://dx.doi.org/10.1007/s10973-009-0228-8] [PMID: 20160877]

[137] Fujimoto T, Ito M, Ito S, Kanazawa H. Fractional laser-assisted percutaneous drug delivery *via* temperature-responsive liposomes. J Biomater Sci Polym Ed 2017; 28(7): 679-89.
[http://dx.doi.org/10.1080/09205063.2017.1296346] [PMID: 28277004]

[138] Grüll H, Langereis S. Hyperthermia-triggered drug delivery from temperature-sensitive liposomes using MRI-guided high intensity focused ultrasound. J Control Rel 2012; 161(2): 317-27.
[http://dx.doi.org/10.1016/j.jconrel.2012.04.041] [PMID: 22565055]

[139] Ta T, Porter TM. Thermosensitive liposomes for localized delivery and triggered release of chemotherapy. J Control Rel 2013; 169(1-2): 112-25.
[http://dx.doi.org/10.1016/j.jconrel.2013.03.036] [PMID: 23583706]

[140] Fite BZ, Kheirolomoom A, Foiret JL, *et al.* Dynamic contrast enhanced MRI detects changes in vascular transport rate constants following treatment with thermally-sensitive liposomal doxorubicin. J Control Rel 2017; 256: 203-13.
[http://dx.doi.org/10.1016/j.jconrel.2017.04.007] [PMID: 28395970]

[141] Sugimoto T, Yamazaki N, Hayashi T, *et al.* Preparation of dual-stimuli-responsive liposomes using methacrylate-based copolymers with pH and temperature sensitivities for precisely controlled release. Colloids Surf B Biointerfaces 2017; 155: 449-58.
[http://dx.doi.org/10.1016/j.colsurfb.2017.04.043] [PMID: 28463812]

[142] Affram K, Udofot O, Singh M, *et al.* Smart thermosensitive liposomes for effective solid tumor therapy and *in vivo* imaging. PLoS One 2017; 12(9): e0185116.
[http://dx.doi.org/10.1371/journal.pone.0185116] [PMID: 28934281]

[143] Paul S, Nahire R, Mallik S, Sarkar K. Encapsulated microbubbles and echogenic liposomes for contrast ultrasound imaging and targeted drug delivery. Comput Mech 2014; 53(3): 413-35.
[http://dx.doi.org/10.1007/s00466-013-0962-4] [PMID: 26097272]

[144] Yudina A, de Smet M, Lepetit-Coiffé M, *et al.* Ultrasound-mediated intracellular drug delivery using microbubbles and temperature-sensitive liposomes. J Control Rel 2011; 155(3): 442-8.
[http://dx.doi.org/10.1016/j.jconrel.2011.06.006] [PMID: 21689699]

[145] Pradhan P, Giri J, Rieken F, *et al.* Targeted temperature sensitive magnetic liposomes for thermo-chemotherapy. J Control Rel 2010; 142(1): 108-21.
[http://dx.doi.org/10.1016/j.jconrel.2009.10.002] [PMID: 19819275]

[146] Zhao M, Hu J, Zhang L, *et al.* Study of amphotericin B magnetic liposomes for brain targeting. Int J Pharm 2014; 475(1-2): 9-16.
[http://dx.doi.org/10.1016/j.ijpharm.2014.08.035] [PMID: 25151436]

[147] Nahire R, Hossain R, Patel R, *et al.* pH-triggered echogenicity and contents release from liposomes. Mol Pharm 2014; 11(11): 4059-68.
[http://dx.doi.org/10.1021/mp500186a] [PMID: 25271780]

[148] Daraee H, Etemadi A, Kouhi M, Alimirzalu S, Akbarzadeh A. Application of liposomes in medicine and drug delivery. Artif Cells Nanomed Biotechnol 2016; 44(1): 381-91.
[http://dx.doi.org/10.3109/21691401.2014.953633] [PMID: 25222036]

[149] Rahimpour Y, Hamishehkar H. Liposomes in cosmeceutics. Expert Opin Drug Deliv 2012; 9(4): 443-55.
[http://dx.doi.org/10.1517/17425247.2012.666968] [PMID: 22413847]

[150] Alavi M, Karimi N, Safaei M. Application of various types of liposomes in drug delivery systems. Adv Pharm Bull 2017; 7(1): 3-9.
[http://dx.doi.org/10.15171/apb.2017.002] [PMID: 28507932]

[151] Mezei M, Gulasekharam V. Liposomes--a selective drug delivery system for the topical route of administration. Lotion dosage form. Life Sci 1980; 26(18): 1473-7.
[http://dx.doi.org/10.1016/0024-3205(80)90268-4] [PMID: 6893068]

[152] Perez AP, Altube MJ, Schilrreff P, *et al.* Topical amphotericin B in ultradeformable liposomes: Formulation, skin penetration study, antifungal and antileishmanial activity *in vitro.* Colloids Surf B Biointerfaces 2016; 139: 190-8.
[http://dx.doi.org/10.1016/j.colsurfb.2015.12.003] [PMID: 26709977]

[153] Doppalapudi S, Jain A, Chopra DK, Khan W. Psoralen loaded liposomal nanocarriers for improved skin penetration and efficacy of topical PUVA in psoriasis. Eur J Pharm Sci 2017; 96: 515-29.
[http://dx.doi.org/10.1016/j.ejps.2016.10.025] [PMID: 27777066]

[154] Jøraholmen MW, Škalko-Basnet N, Acharya G, Basnet P. Resveratrol-loaded liposomes for topical treatment of the vaginal inflammation and infections. Eur J Pharm Sci 2015; 79: 112-21.
[http://dx.doi.org/10.1016/j.ejps.2015.09.007] [PMID: 26360840]

[155] Kalat SA, Khamesipour A, Bavarsad N, *et al.* Use of topical liposomes containing meglumine antimoniate (Glucantime) for the treatment of L. major lesion in BALB/c mice. Exp Parasitol 2014; 143: 5-10.
[http://dx.doi.org/10.1016/j.exppara.2014.04.013] [PMID: 24780938]

[156] Piccioni A, Fargnoli MC, Schoinas S, *et al.* Efficacy and tolerability of 5-aminolevulinic acid 0.5% liposomal spray and intense pulsed light in wrinkle reduction of photodamaged skin. J Dermatolog Treat 2011; 22(5): 247-53.
[http://dx.doi.org/10.3109/09546634.2011.590791] [PMID: 21801099]

[157] Jain B, Singh B, Katare OP, Vyas SP. Development and characterization of minoxidil-loaded liposomal system for delivery to pilosebaceous units. J Liposome Res 2010; 20(2): 105-14. [http://dx.doi.org/10.3109/08982100903161449] [PMID: 19698000]

[158] Wadhwa S, Singh B, Sharma G, Raza K, Katare OP. Liposomal fusidic acid as a potential delivery system: a new paradigm in the treatment of chronic plaque psoriasis. Drug Deliv 2016; 23(4): 1204-13.
[http://dx.doi.org/10.3109/10717544.2015.1110845] [PMID: 26592918]

[159] Kumar R, Dogra S, Amarji B, *et al.* Efficacy of novel topical liposomal formulation of cyclosporine in mild to moderate stable plaque psoriasis: A randomized clinical trial. JAMA Dermatol 2016; 152(7): 807-15.
[http://dx.doi.org/10.1001/jamadermatol.2016.0859] [PMID: 27096709]

[160] Jøraholmen MW, Basnet P, Acharya G, Škalko-Basnet N. PEGylated liposomes for topical vaginal therapy improve delivery of interferon alpha. Eur J Pharm Biopharm 2017; 113: 132-9. [http://dx.doi.org/10.1016/j.ejpb.2016.12.029] [PMID: 28087379]

[161] Han HD, Mangala LS, Lee JW, *et al.* Targeted gene silencing using RGD-labeled chitosan nanoparticles. Clin Cancer Res 2010; 16(15): 3910-22. [http://dx.doi.org/10.1158/1078-0432.CCR-10-0005] [PMID: 20538762]

[162] Nunes PS, Rabelo AS, Souza JC, *et al.* Gelatin-based membrane containing usnic acid-loaded liposome improves dermal burn healing in a porcine model. Int J Pharm 2016; 513(1-2): 473-82. [http://dx.doi.org/10.1016/j.ijpharm.2016.09.040] [PMID: 27633280]

[163] Rajabi O, Layegh P, Hashemzadeh S, Khoddami M. Topical liposomal azithromycin in the treatment of acute cutaneous leishmaniasis. Dermatol Ther (Heidelb) 2016; 29(5): 358-63. [http://dx.doi.org/10.1111/dth.12357] [PMID: 27073044]

[164] Vázquez-González ML, Calpena AC, Domènech Ò, Montero MT, Borrell JH. Enhanced topical delivery of hyaluronic acid encapsulated in liposomes: A surface-dependent phenomenon. Colloids Surf B Biointerfaces 2015; 134: 31-9.
[http://dx.doi.org/10.1016/j.colsurfb.2015.06.029] [PMID: 26142626]

[165] Li Z, Zhang M, Liu C, *et al.* Development of Liposome containing sodium deoxycholate to enhance oral bioavailability of itraconazole. Asian J Pharm Sci 2017; 12(2): 157-64. [http://dx.doi.org/10.1016/j.ajps.2016.05.006] [PMID: 32104325]

[166] Uhl P, Helm F, Hofhaus G, *et al.* A liposomal formulation for the oral application of the investigational hepatitis B drug Myrcludex B. Eur J Pharm Biopharm 2016; 103: 159-66. [http://dx.doi.org/10.1016/j.ejpb.2016.03.031] [PMID: 27049970]

[167] Uhl P, Pantze S, Storck P, *et al.* Oral delivery of vancomycin by tetraether lipid liposomes. Eur J Pharm Sci 2017; 108: 111-8.
[http://dx.doi.org/10.1016/j.ejps.2017.07.013] [PMID: 28716758]

[168] Cui M, Wu W, Hovgaard L, Lu Y, Chen D, Qi J. Liposomes containing cholesterol analogues of botanical origin as drug delivery systems to enhance the oral absorption of insulin. Int J Pharm 2015; 489(1-2): 277-84.
[http://dx.doi.org/10.1016/j.ijpharm.2015.05.006] [PMID: 25957702]

[169] Skiba-Lahiani M, Hallouard F, Mehenni L, Fessi H, Skiba M. Development and characterization of oral liposomes of vegetal ceramide based amphotericin B having enhanced dry solubility and solubility. Mater Sci Eng C 2015; 48: 145-9.
[http://dx.doi.org/10.1016/j.msec.2014.11.069] [PMID: 25579907]

[170] Feng Y, Sun C, Yuan Y, *et al.* Enhanced oral bioavailability and *in vivo* antioxidant activity of chlorogenic acid *via* liposomal formulation. Int J Pharm 2016; 501(1-2): 342-9.
[http://dx.doi.org/10.1016/j.ijpharm.2016.01.081] [PMID: 26861689]

[171] Skibinski CG, Das A, Chen KM, *et al.* A novel biologically active acid stable liposomal formulation of docosahexaenoic acid in human breast cancer cell lines. Chem Biol Interact 2016; 252: 1-8.
[http://dx.doi.org/10.1016/j.cbi.2016.03.035] [PMID: 27041074]

[172] Abdelbary GA, Amin MM, Zakaria MY, El Awdan SA. Adefovir dipivoxil loaded proliposomal powders with improved hepatoprotective activity: formulation, optimization, pharmacokinetic, and biodistribution studies. J Liposome Res 2017; 1-16.
[PMID: 28768431]

[173] Caddeo C, Pons R, Carbone C, *et al.* Physico-chemical characterization of succinyl chitosan-stabilized liposomes for the oral co-delivery of quercetin and resveratrol. Carbohydr Polym 2017; 157: 1853-61.
[http://dx.doi.org/10.1016/j.carbpol.2016.11.072] [PMID: 27987905]

[174] Castoldi A, Herr C, Niederstraßer J, *et al.* Calcifediol-loaded liposomes for local treatment of pulmonary bacterial infections. Eur J Pharm Biopharm 2017; 118: 62-7.
[http://dx.doi.org/10.1016/j.ejpb.2016.11.026] [PMID: 27888144]

[175] Nahar K, Absar S, Patel B, Ahsan F. Starch-coated magnetic liposomes as an inhalable carrier for accumulation of fasudil in the pulmonary vasculature. Int J Pharm 2014; 464(1-2): 185-95.
[http://dx.doi.org/10.1016/j.ijpharm.2014.01.007] [PMID: 24463004]

[176] Tahara K, Tomida H, Ito Y, *et al.* Pulmonary liposomal formulations encapsulated procaterol hydrochloride by a remote loading method achieve sustained release and extended pharmacological effects. Int J Pharm 2016; 505(1-2): 139-46.
[http://dx.doi.org/10.1016/j.ijpharm.2016.03.031] [PMID: 27012982]

[177] Gandhi M, Pandya T, Gandhi R, *et al.* Inhalable liposomal dry powder of gemcitabine-HCl: Formulation, *in vitro* characterization and *in vivo* studies. Int J Pharm 2015; 496(2): 886-95.
[http://dx.doi.org/10.1016/j.ijpharm.2015.10.020] [PMID: 26453787]

[178] Chetoni P, Monti D, Tampucci S, *et al.* Liposomes as a potential ocular delivery system of distamycin A. Int J Pharm 2015; 492(1-2): 120-6.
[http://dx.doi.org/10.1016/j.ijpharm.2015.05.055] [PMID: 26183332]

[179] Tan G, Yu S, Pan H, Li J, Liu D, Yuan K, *et al.* Bioadhesive chitosan-loaded liposomes: A more efficient and higher permeable ocular delivery platform for timolol maleate. Int J Biol Macromol 2017; 94(Pt A): 355-63.
[http://dx.doi.org/10.1016/j.ijbiomac.2016.10.035]

[180] Mokhtar Ibrahim M, Tawfique SA, Mahdy MM. Liposomal diltiazem HCl as ocular drug delivery system for glaucoma. Drug Dev Ind Pharm 2014; 40(6): 765-73.
[http://dx.doi.org/10.3109/03639045.2013.783589] [PMID: 23570512]

[181] de Sá FA, Taveira SF, Gelfuso GM, Lima EM, Gratieri T. Liposomal voriconazole (VOR) formulation for improved ocular delivery. Colloids Surf B Biointerfaces 2015; 133: 331-8.
[http://dx.doi.org/10.1016/j.colsurfb.2015.06.036] [PMID: 26123854]

[182] Moustafa MA, Elnaggar YSR, El-Refaie WM, Abdallah OY. Hyalugel-integrated liposomes as a novel ocular nanosized delivery system of fluconazole with promising prolonged effect. Int J Pharm 2017; 534(1-2): 14-24.
[http://dx.doi.org/10.1016/j.ijpharm.2017.10.007] [PMID: 28987453]

[183] Zhang R, Qian J, Li X, Yuan Y. Treatment of experimental autoimmune uveoretinitis with intravitreal

injection of infliximab encapsulated in liposomes. Br J Ophthalmol 2017; 101(12): 1731-8. [http://dx.doi.org/10.1136/bjophthalmol-2016-310044] [PMID: 28986343]

[184] Kaiser JM, Imai H, Haakenson JK, *et al.* Nanoliposomal minocycline for ocular drug delivery. Nanomedicine (Lond) 2013; 9(1): 130-40. [http://dx.doi.org/10.1016/j.nano.2012.03.004] [PMID: 22465498]

[185] Razazan A, Behravan J, Arab A, *et al.* Conjugated nanoliposome with the HER2/neu-derived peptide GP2 as an effective vaccine against breast cancer in mice xenograft model. PLoS One 2017; 12(10): e0185099. [http://dx.doi.org/10.1371/journal.pone.0185099] [PMID: 29045460]

[186] Cui SH, Zhi DF, Zhao YN, *et al.* Cationic lioposomes with folic acid as targeting ligand for gene delivery. Bioorg Med Chem Lett 2016; 26(16): 4025-9. [http://dx.doi.org/10.1016/j.bmcl.2016.06.085] [PMID: 27426864]

[187] Mura P, Mennini N, Nativi C, Richichi B. *In situ* mucoadhesive-thermosensitive liposomal gel as a novel vehicle for nasal extended delivery of opiorphin. Eur J Pharm Biopharm 2018; 122: 54-61. [http://dx.doi.org/10.1016/j.ejpb.2017.10.008] [PMID: 29032194]

[188] Yusuf H, Ali AA, Orr N, Tunney MM, McCarthy HO, Kett VL. Novel freeze-dried DDA and TPGS liposomes are suitable for nasal delivery of vaccine. Int J Pharm 2017; 533(1): 179-86. [http://dx.doi.org/10.1016/j.ijpharm.2017.09.011] [PMID: 28887219]

[189] Upadhyay P, Trivedi J, Pundarikakshudu K, Sheth N. Direct and enhanced delivery of nanoliposomes of anti schizophrenic agent to the brain through nasal route. Saudi Pharm J 2017; 25(3): 346-58. [http://dx.doi.org/10.1016/j.jsps.2016.07.003] [PMID: 28344488]

[190] Xie Q, Deng W, Yuan X, *et al.* Selenium-functionalized liposomes for systemic delivery of doxorubicin with enhanced pharmacokinetics and anticancer effect. Eur J Pharm Biopharm 2018; 122: 87-95. [http://dx.doi.org/10.1016/j.ejpb.2017.10.010] [PMID: 29032193]

[191] Agarwal R, Iezhitsa I, Agarwal P, *et al.* Liposomes in topical ophthalmic drug delivery: an update. Drug Deliv 2016; 23(4): 1075-91. [http://dx.doi.org/10.3109/10717544.2014.943336] [PMID: 25116511]

[192] Nguyen TX, Huang L, Gauthier M, Yang G, Wang Q. Recent advances in liposome surface modification for oral drug delivery. Nanomedicine (Lond) 2016; 11(9): 1169-85. [http://dx.doi.org/10.2217/nnm.16.9] [PMID: 27074098]

[193] Elmowafy M, Viitala T, Ibrahim HM, *et al.* Silymarin loaded liposomes for hepatic targeting: *in vitro* evaluation and HepG2 drug uptake. Eur J Pharm Sci 2013; 50(2): 161-71. [http://dx.doi.org/10.1016/j.ejps.2013.06.012] [PMID: 23851081]

[194] Schwendener RA. Liposomes as vaccine delivery systems: a review of the recent advances. Ther Adv Vaccines 2014; 2(6): 159-82. [http://dx.doi.org/10.1177/2051013614541440] [PMID: 25364509]

[195] Ghaffar KA, Giddam AK, Zaman M, Skwarczynski M, Toth I. Liposomes as nanovaccine delivery systems. Curr Top Med Chem 2014; 14(9): 1194-208. [http://dx.doi.org/10.2174/1568026614666140329232757] [PMID: 24678703]

[196] Cipolla DC, Blanchard J. Concentrated, inhalable antibiotic formulation. inventors; Aradigm Corporation, assignee 2015.

[197] Bar-Shalom D, Skjode JS. Liposome system for ocular administration. inventors; Bioneer A/S, assignee 2011.

[198] Jensen SS, Andresen TL, Henriksen JS, Johansen PT. Cationic liposomal drug delivery system for specific targeting of human cd14+ monocytes in whole blood. inventors; Bioneer A/S, Dtu Nanotech, assignee 2013.

[199] Flood KM, Liu R, Peck KD, Zheng J. Pain reducing parenteral liposome formulation. inventors; Abbott Laboratories, assignee

[200] Proffitt RT, Adler-Moore J, Chiang S. Amphotericin B liposome preparation. inventors; Nexstar Pharmaceuticals, Inc, assignee

[201] Kim S, Kim T, Murdande S. Sustained-release liposomal anesthetic compositions. inventors; Pacira Pharmaceuticals, Inc, assignee 2012.

[202] Naeff R, Delmenico S, Spycher R, Corbo M, Flother M. Liposome-based topical vitamin D formulation. inventors; Cilag Ag, assignee 1998.

[203] Poli S, Moro L, Natali A, Coppi G. Liposome formulations of immunomodulating drugs for the topical and aerosol administrations inventors; Poli Industria Chimica Spa, assignee 1994.

[204] Venkatraman S, Natarajan JV. Stable liposomal formulations for ocular drug delivery. inventors; Nanyang Technological University, Singapore Health Services Pvt Ltd, assignee 2015.

[205] Kato M, Ohtsuki T, Egami F, Tsunoda K. Liposome eye drops. inventors; Taisho Pharmaceutical Co, Ltd, assignee 1996.

[206] Katare OP, Sinha VR, Bhatia A, Kumar R, Gupta S. Synergistic liposomal tamoxifen composition for topical application and method of preparing thereof. inventors; Lifecare Innovations Pvt Ltd, assignee 2006.

[207] Katare OP, Agarwal R, Kumar B. Novel Inter and Intra Multilamellar Vesicular Composition. inventors; Life-Care Innovations Pvt Ltd, assignee 2006.

[208] Katare OP, Agarwal R, Kumar B. Novel Inter and Intra Multilamellar Vesicular Composition. inventors; Life-Care Innovations Pvt Ltd, assignee 2008.

[209] Katare OP, Agarwal R, Kumar B. Novel Inter and Intra Multilamellar Vesicular Composition. inventors; Life-Care Innovations Pvt Ltd, assignee 2008.

[210] Katare OP, Agarwal R, Kumar B. Novel Inter and Intra Multilamellar Vesicular Composition. inventors; Life-Care Innovations Pvt Ltd, assignee 2008.

[211] Guilford FT, Venketraman V. Method of treatment using liposomally formulated reduced glutathione to cooperate with il-10 to modulate inflammatory response triggered in hiv+ andtb immune-compromised diabetic patients. inventors; Your Energy Systems, LLC, Western University Of Health Sciences, assignee 2016.

[212] Boulikas T. Therapy for human cancers using cisplatin and other drugs or genes encapsulated into liposomes. inventor Regulon, Inc, assignee 2006.

[213] Venkatraman S, Chiang Y, Boey F, Natarajan JV. Sustained timolol maleate delivery from liposomes for glaucoma therapy and ocular hypertension. inventors; Nanyang Technological University, assignees 2016.

[214] Madamsetty VS, Chaudhuri A, Mukhopadhyay D. Method for Regressing Pancreatic Tumor by a liposomal formulation along with DNA vaccines. inventors; Council of Scientific & Industrial Research (New Delhi, IN) and Mayo Clinic (Rochester, MN, US) assignee 2017.

[215] Accardo A, Ringhieri P, Morelli G. Kit for formulation of liposomal doxorubicin modified with bioactive peptides for selective targeting of receptors overexpressed by cancer cells. inventors; Invectors Srl, assignee 2017.

CHAPTER 4

Role of Dendrimers in Drug Delivery and Targeting

Sarita Rani, Neetu Sehrawat, Sudhakar Rajoria and **Umesh Gupta***

Department of Pharmacy, School of Chemical Sciences and Pharmacy, Central University of Rajasthan, Bandar Sindri, Ajmer, Rajasthan-305817, India

Abstract: In the recent past, dendrimer based nano-devices have become an attractive center for medicinal chemists as a targeted drug delivery carrier. It is promising to specifically manipulate both molecular weight and chemical composition of dendrimers by regulating dendrimer synthesis, and thereby allowing predictable tuning of their biocompatibility and pharmacokinetics. The myriad properties and functions of dendrimers provide plenty of opportunities in this regard. Presently, the fields of oncology, infectious disease, cardiovascular diseases, ocular disease, *etc.* need biodegradable polymers, macromolecules based formulations to improve the safety and efficacy of different therapeutics. Dendrimer has been extensively used for treating disorders even for anti-viral therapeutics, including AIDS-HIV . The present book chapter reports the role of dendrimers in drug delivery systems for various agents, in biology and medical science. Additionally, it covers the fundamentals of dendrimeric properties and types for cancer diagnosis and other therapy

Keywords: Dendrimers, Drug delivery, PAMAM, Therapeutics.

INTRODUCTION

Dendrimers are highly branched synthetic polymeric macromolecules with a distinct three-dimensional spherical shape, nanometric size (1-100nm), a monodispersed structure that is synthesized by the repeated sequence of reaction steps that provide a unique branching structure to the macromolecules [1, 2]. The word dendrimer is derived from the Greek word "dendron" which means tree because it branches like a tree and "meros" meaning part. Donald Tomalia, in 1985, first introduced this macromolecule [3, 4]. Dendrimers were also recognized as Cascade molecules, dendritic molecules, "arborols" or as nanometric architectures due to their monodispersed structure and nanometric configuration [1, 4, 5]. These are known as a fourth significant new class of macromolecular architecture and differ from other classes such as linear, cross-

* **Corresponding author Umesh Gupta:** Department of Pharmacy, School of Chemical Sciences and Pharmacy, Central University of Rajasthan, Bandar Sindri, Ajmer, Rajasthan-305817, India; E-mail: umeshgupta@curaj.ac.in

linked, and branched polymer due to their excellent and unique properties. The unique characteristics include a high degree of molecular uniformity, mono-dispersity as well as a highly functional surface. The fourth class of macromolecular architecture was further subdivided into four dendritic/cascade molecules, which are the random hyperbranched polymer, dendrigraft polymer, dendrons, and dendrimers [6, 7].

Dendrimers are spherical, nano-scaled macromolecules with a specific architecture composed of three different components: (i) a central core, at the center of dendrimer consisting of either a single atom or an atomic group having at least two identical chemical functions which determine the size and shape of the dendrimers, (ii) branches emanating from a core, constituted of repeating units having at least one branching junction, whose repetition is organized in a geometrical progression that results in a series of radially concentric layers called generations. The monomers attached to a core (G0) are called a first-generation (G1), and two successive generation monomers (G2) are attached to the first generations monomers (Fig. **1**). That leads to monodispersity, tree-like structure, and; (iii) many terminal functional groups which are located on the exterior of the macromolecule. These peripheral groups play a very important role in determining the dendrimer properties [8].

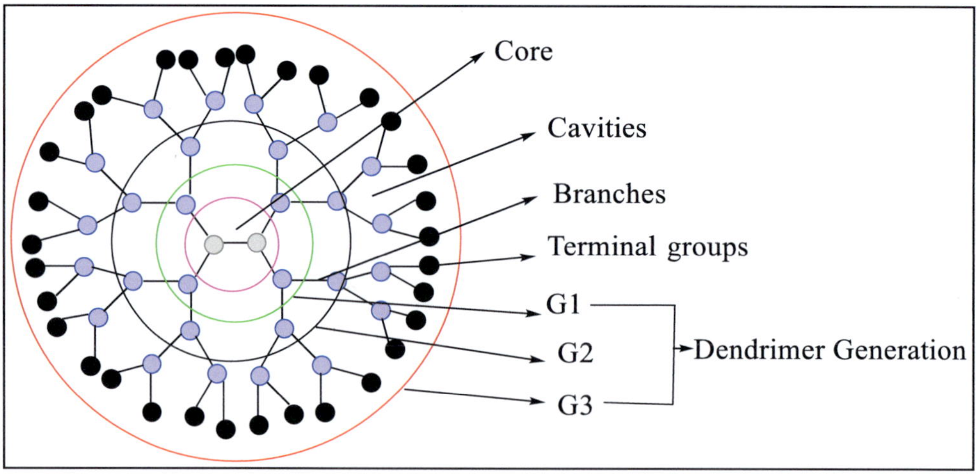

Fig. (1). Structure of dendrimers.

In recent years, dendrimers have importantly contributed to the biomedical, chemical, biological, and material sciences. Dendrimers can host both hydrophilic and hydrophobic drugs in their architecture due to their versatility. Different types of dendrimers exposed to exhibit such potential include poly (propylene imine) (PPI), polylysine dendrimers, and poly(amidoamine) (PAMAM). The later one is the first complete dendrimer family, which has been synthesized, characterized, and commercialized. The three-dimensional structure of dendrimers is promising nanocarriers for various drugs, targeting moieties [9].

Synthetic Approaches

The chemistry of dendrimers exists between molecular and polymer chemistry. These molecules are interrelated to molecular chemistry because of their stepwise controlled synthesis and are related to polymer chemistry due to their repetitive structure [10]. Two different strategies have been developed for the stepwise synthesis of dendrimers, namely, divergent and convergent (Fig. **2**). In a divergent approach, the synthesis begins from a central core to the periphery of the dendrimer, whereas in a convergent method, the synthesis of dendrimers starts from the periphery and proceeds towards the central core.

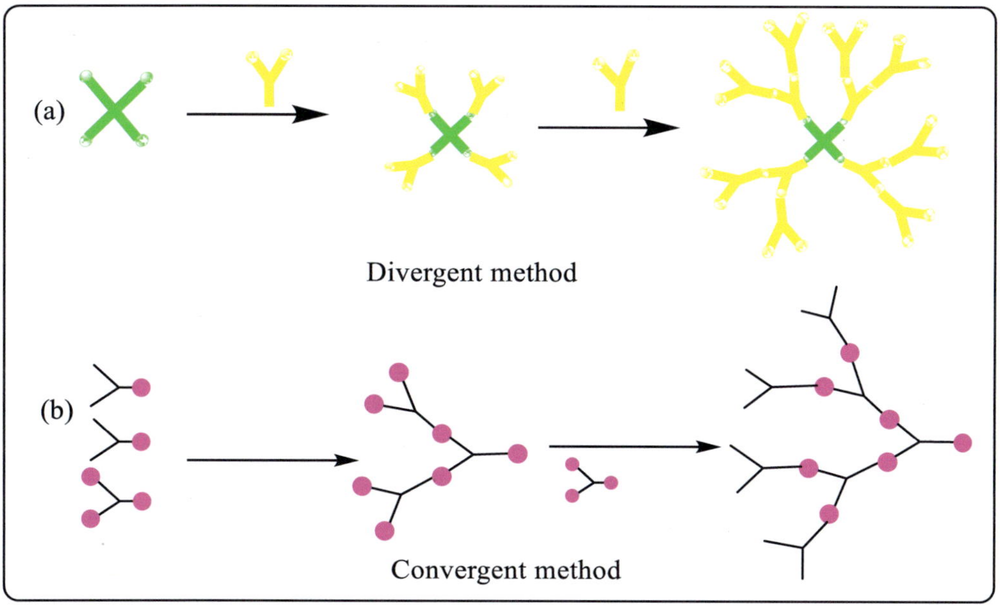

Fig. (2). Synthesis method for dendrimers **(a)** Divergent method **(b)** Convergent method.

Divergent Method

Divergent synthesis involves two steps: first, the activation of functional surface groups, and second, the addition of branching monomer units. In this method, the core reacts with two or more moles of a reagent, having at least two protecting/branching sites, abide by removal of protecting the group resulting in first-generation dendrimers. The successive liberated reactive sites lead to the formation of first-generation dendrimers. This process was repeated until the desired size dendrimer was formed. The advantage of this method is the ability to adjust the dendrimer surface molecules by changing the end groups at the outermost generation, and physical and chemical properties can be modified to particular needs [11]. The most commonly used commercially available dendrimers, which are prepared by the divergent method, include PAMAM (Tomalia, 1985) and PPI (De Brabander and Meijer, 1993) that are structurally different but have been reported to have excellent applicability in drug delivery [12].

Convergent Method

Hawker and Frechet first introduced this method in the 90s. The convergent synthesis involves two stages, (i) a repeated coupling of protected/deprotected branches to create a focal point functionalized dendron and (ii) a divergent core anchoring stage to produce several multi-dendron dendrimers. Only one type of functional group on the outermost generation is the main limitation of the divergent growth method, but the convergent method would overcome such faults [11]. The advantage of this method is that the dendrimers are synthesized with more symmetry and fewer impurities because better purification of dendrons is possible before attachment with a core. However, there is a limitation of dendrimer size because of the steric hindrance between dendrons, which is going to attach with the core [12].

Dendrimer-drug Interactions

The internal structure of dendrimers leads to host-guest interactions leading to the encapsulation of guest molecule(s) or drug. Dendrimer architecture allows many drugs or guest molecules to form complexes, both covalently and non-covalently [13]. The interactions between drug and dendrimer can be divided into two groups (i) the entrapment of drugs within the dendritic molecule, which involves the hydrophobic, electrostatic, and hydrogen bond-like interactions, (ii) the interaction of the drug with the dendritic surface, which involves covalent interactions shown in Fig. (**3**) [14, 15]. When compared to other traditional linear polymers, the exciting feature of dendrimer makes them extraordinary polymeric molecules for the novel drug delivery system [16].

Non-Covalent Interactions

The drug interacts physically with dendrimers by encapsulation into void spaces or adsorption on the surface. The forces involved in these interactions are hydrogen bonding, Vander Waal interactions, and electrostatic attractions between oppositely charged drug species and dendrimers. Small drug molecules are encapsulated into the dendrimer's interior voids space, while more abundant drug molecules adsorb on the surface of dendrimers [17]. The interior structure of the dendrimer is hydrophobic but hydrophilic externally. This micelle-like behavior of dendrimers makes them a capable carrier for the hydrophobic drug molecules. Many types of dendrimers such as PPI, PAMAM, and poly-glycerol have revealed host-guest non-covalent interactions. In the non-covalent interactions, dendrimers interact hydrophobically and electrostatically with internal tertiary nitrogens and amine groups present on the surface of dendrimers [18]. Godard and Tomalia's study shows that G1-G3 alanine dendrimers reveal an open structure although, higher generations show a densely packed surface, which is vital to produce confined internal voids that can completely encapsulate the drug molecule [19]. The hydrophilic exterior of dendrimers entirely contributes to solubility *via* non-covalent interactions, which is not possible in covalent conjugation of a drug with dendrimer since the hydrophilic functional group on the surface of the dendrimer participates in bonding, which may interfere with the surface functionality available for solubilization. Therefore, it assumed that mainly non-covalent interactions are responsible for solubilization [18].

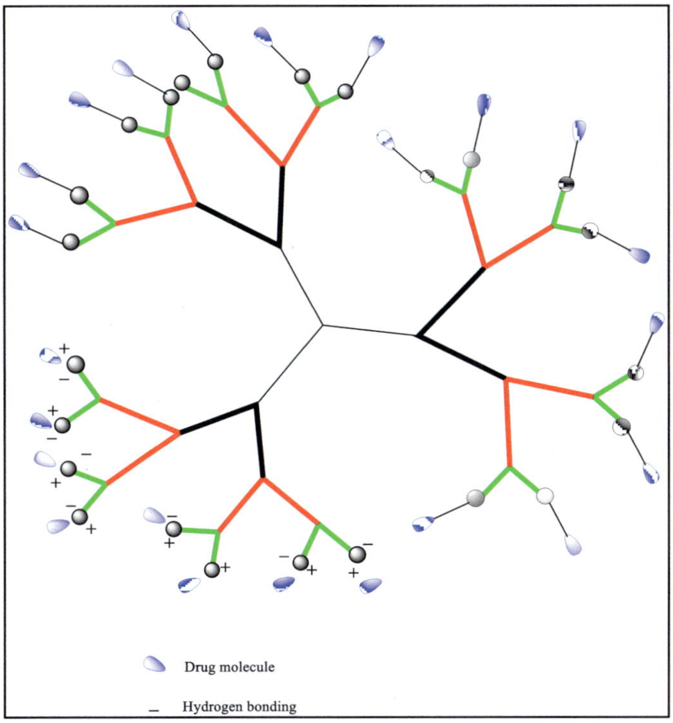

Fig. (3). Dendrimer drug loading possible approaches: Physical and chemical drug interaction such as covalent, non-covalent, hydrogen bonding, or other electrostatic interactions.

Covalent Interactions

The presence of a large number of functional groups on the dendrimeric surface and flexibility in their chemical structures provides the conjugation of drug molecules, imaging agents, and targeting ligand. These interactions of drugs and present groups of dendrimers *via* hydrolysable or biodegradable association offer the prospect for improved drug release kinetics [20]. The covalent interactions of drug moieties with dendrimers provide a stable system that is not reliant on the dynamic equilibrium or thermodynamic factors [21]. The drug can be conjugated with dendrimers by chemical bonding. There are two critical pathways: (i) direct conjugation of the drug molecules to the dendrimer surface groups (ii) drug is conjugated *via* a linker molecule [22]. Drug molecules can be conjugated to a dendrimer by either ester or amide linkage or other linkages depending upon the surface groups of a dendrimer, which can be hydrolysed by endosomal or lysosomal enzymes [23]. Covalent interaction has many advantages over drug-encapsulation (non- covalent system). The covalent attachment of the drug to the surface groups of dendrimers by chemicals bond provides the slow drug release than non-covalent interaction that is important for sustained or controlled drug delivery system.. However, problems may be arises due to the linkage large number of drug molecules to the dendrimer surface by covalent interaction, like insolubility of the product. This problem can be overcome by the attachment of soluble agents such as PEG chains [24, 25].

Properties

Dendrimers are unique hyper-branched materials, holding many attractive and exclusive properties (Fig. **4**), which make them ideal for drug delivery purposes. The physicochemical properties of dendrimers are based on their generations, surface groups, and core. The size of dendrimers is very close to many biological polymers such as proteins and DNA, which are physiologically ideal [26].

Monodispersity

Mass spectrometry, gel electrophoresis, size-exclusion chromatography, and transmission electron microscopy techniques are used for confirmation of the monodisperse nature of dendrimer. Mostly monodispersed dendrimers are obtained by the convergent method because this method allows purification at each step and abolishes cumulative effects due to failed couplings. Mass spectroscopy showed that the PAMAM dendrimers produced by the divergent approach are enormously monodisperse for the former generations (G0-G5) [27].

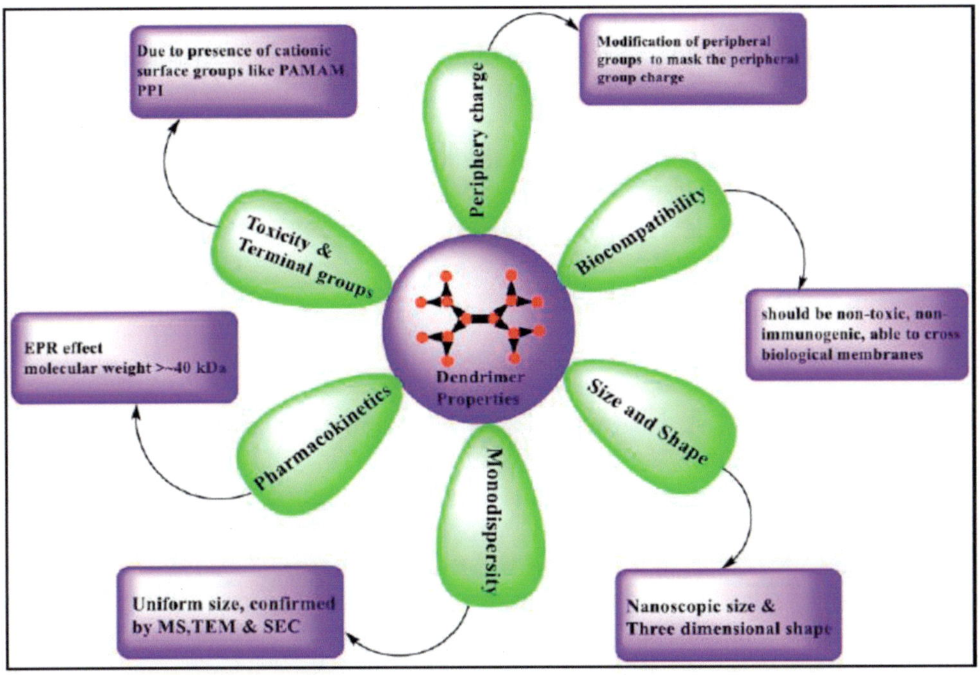

Fig. (4). Properties of dendrimers.

Size and Shape

The well-defined nanoscopic size and shape of the dendrimer is an excellent feature which helps these carriers to cross the cell membrane. The size of the dendrimer increases with an increase in the generation number. The size of the dendrimer is similar to the biological structures. The family of PAMAM dendrimer is similar in size with proteins and bio-assemblies, for example, insulin (≈30 Å), hemoglobin (≈55 Å), and cytochrome C (≈40 Å) having the same size and shape as ammonia core PAMAM dendrimer generation 3,4, and 5. Similarly, 5th and 6th generation 6G PAMAM dendrimers has a diameter that is equal to ubiquitous lipid bilayer membranes [28]. The dendrimer can be synthesized with the various core material, branching units, and surface group modifications. The size of the dendrimer is related to the three-dimensional shape; the lower generation encompasses a tendency to open structure, whereas the higher generation dendrimers have a spherical and more compact structure [29].

Biocompatibility

The dendrimer was used as a drug delivery vehicle, so these have to accomplish several biological demands of vital importance. These should be non-toxic, non-immunogenic, able to cross biological membranes, *etc.* [30]. The toxicity of dendrimers is mainly due to the presence of peripheral groups and particularly amines. The dendrimers have a cationic charge on its surface, which interacts with the negatively charged cell membrane and cause the damage of the cell membrane, which results in cell lysis [31]. The amine-terminated PAMAM and PPI dendrimers show concentration-dependent toxicity whereas neutral or anionic group terminated dendrimers show less toxicity than cationic ones. However, the toxicity of PAMAM and PPI dendrimer can be reduced by peripheral modification with negatively charged and neutral groups such as carboxylic or hydroxyl functional groups [28, 29].

Surface Charge

The surface groups of the dendrimers play an essential role in drug delivery. The toxicity of polycationic dendrimers is the main factor in their medical applications. There are many functionalization strategies developed to mask the peripheral group to overcome the problem. The functionalization of the dendrimers includes modification of their physicochemical properties, increasing the peripheral congestion, and change of peripheral groups with targeting moieties. For example, PPI an amine-terminated dendrimer, the peripheral amine groups impart toxicity but it can be overcome by the surface charge approach means *via* masking with some other molecules such as PEG, sugars, *etc.* Agashe *et al.* synthesized two types of functionalized PPI dendrimer, first they prepared t-BOC protected amino-acid coated PPI dendrimer and then prepared carbohydrate-coated PPI dendrimers [32].

Toxicity and Surface Functionality

The presence of cationic surface groups leads to the toxicity of dendrimers. As discussed above, the cationic charge on the surface of dendrimers interacts with the negatively charged cell membrane and creates nano-holes in the cell membrane, which causes the leakage of cell materials, and results in cell death [31]. It was reported that the cellular toxicity of dendrimers is due to their number of surface functional groups and the surface charge. Cationic dendrimers like PPI, PAMAM shows dose-dependent toxicity whereas negatively charged dendrimers such as sulfonated, carboxylated, and neutral dendrimers with acetyl, carboxyl, and galactose end groups show less toxicity [29, 33].

Pharmacokinetics

Pharmacokinetics is the most critical aspect of biomedical applications in medicines because of the bioavailability, toxicity, and efficacy of dendrimer-based drug relay upon their drug release profile, tissue accumulation, and elimination rate. In anticancer drug delivery, the macromolecules with prolonged circulations time shows increased accumulations in tumor tissues due to EPR (enhanced permeability and retention) effect. The dendrimer must-have neutral or negatively charged surface groups to prevent non-specific interaction with the cell membrane and must have high molecular weight to avoid rapid filtration through the glomerular filtration membrane. Poly-cationic PAMAM dendrimers show rapid clearance from the bloodstream and accumulate in the liver, kidney, or pancreas. Modification of the PAMAM surface groups leads to increased blood circulation times. Pharmacokinetic studies show that the dendrimer having a molecular weight >~40 kDa remains in the blood for a longer time than the dendrimers having low molecular weight [34].

Types of Dendrimers

In recent years, the unique hyper-branched structure of dendrimer, higher surface functional groups, globular shape, and monodispersity with different functionalities have shown great interest. Some of the dendrimers having different functionalities were described with their applications [35].

PAMAM Dendrimers

Tomalia developed the first family of dendrimers which was synthesized, characterized, and commercialized and is more examined for drug delivery in the 80s. PAMAM dendrimers can be synthesized by a divergent method from ethylenediamine (EDA) as the core. PAMAM dendrimers synthesis involves two steps (i) Michael addition of a suitable amine initiator core with methyl acrylate and (ii) amidation of the esters with the large excess of ethylenediamine (EDA). Both alkylation and amidation steps were characterized by NMR, size exclusion chromatography, and IR spectroscopy [36]. The internal structure of PAMAM dendrimers can hold guest molecules due to their distinctive functional groups [35], and used in gene transfection because of the presence of a positive charge on its surface.

PPI Dendrimers

PPI dendrimers contain poly-alkylamines having primary amines as end groups and tertiary tris-propylene amines internally. These are commercially available to fifth generations with a trading name of Astramol™ [37]. These can be synthesized by the divergent method, which is obtained by the repetition of double Michael addition of acrylonitrile to primary amines followed by heterogeneously catalyzed hydrogenation of nitrile to primary amines [38]. In the synthesis of PPI dendrimers, 1,4-diaminobutane was used as a dendrimer core. The poly (propylene imine) was introduced commercially by DSM (Dutch State

Mines) company the Netherlands and Aldrich Chemical Company Milwaukee, Wisconsin [29].

Peptide Dendrimers

Peptide dendrimers are the dendrimers containing peptide bonds [39]. These are radically branched macromolecules that include peptidyl branching core and peripheral peptides chains. These can be classified into three subclasses (i) grafted peptide dendrimer, these contain unnatural amino acids or organic groups as the branching core and peptide or proteins are attached as functional groups on the surface, (ii) peptide dendrimer, consist of mostly peptides (iii) contains amino acids in the branching core and surface functional groups with some non-peptide branching units [29, 35]. Divergent and convergent methods are used for the synthesis of peptide dendrimers. Peptide dendrimers are used as contrast agents for magnetic resonance imaging (MRI), magnetic resonance angiography (MRA), fluorogenic imaging [34], proteins, liposomal mimetics, vehicles for drug and gene delivery, and also as a biomaterial in life science and biomedical applications [35].

Glycodendrimers

The word 'glycodendrimer' is used to describe dendrimers that are composed of carbohydrates such as glucose, mannose, galactose. The primary group of glycodendrimers has saccharide residue on their surface, but containing a sugar moiety as a central core. Glycodendrimers are classified into three types: (i) carbohydrate coated (ii) carbohydrate centered and (iii) carbohydrate-based. Glycodendrimers with surface carbohydrate units have been used in the study of protein-carbohydrate interactions [29, 35].

Tecto/Core-shell Dendrimers

The core generation number is more significant than the surrounding shell dendrimer, called the core-shell tecto dendrimer [40]. Core-shell/tecto dendrimers represent a polymeric structure obtained as a product of controlled covalent attachment of the dendrimer building block [40]. Tecto dendrimers are composed of a cored dendrimer, which may or may not contains the therapeutic agents surrounded by the dendrimers. The surrounded dendrimers are of several types to perform a function that is necessary for a smart therapeutic nanodevice [29, 35]. The core-shell/tecto dendrimers inhibit the growth of melanoma cells at a concentration that is safe for the Keratinocytes epithelial cells [41].

Chiral Dendrimers

Chiral dendrimer offers the possibility to investigate the impact of chirality in the macromolecular system. Their properties are based on their well-defined, highly ordered structure with nanoscopic dimensions. Chiral dendrimers differ from achiral dendrimers by the following features: (i) the functional group's arrangement on the surface could be chiral, (ii) chiral structures in chiral dendrimers should be detected by optical measurement, (iii) their shape could be chiral, not spherical [42]. The chirality in the dendrimers is based on the construction of constitutionally different but chemically similar branches to a chiral core [29, 43]. These are suitable for the addition of useful functionalities aimed at an application as a drug delivery system and light-harvesting materials [35].

Triazine Dendrimers

The triazine dendrimers used in a drug delivery system have the payback from their synthetic versatility and well-defined structure. They can be synthesized and designed to display an orthogonally working surface that will help in the attachment of drugs, ligands. Triazine dendrimers are used for medical applications such as drug delivery, non-viral DNA, and RNA delivery. The synthesis of triazine is based on the following substitution of trichlorotriazine with amine nucleophiles [44].

Poly-L-lysine Dendrimers

Poly-L-lysine is a cationic polypeptide with amino acids as repetition units. The polypeptide chain of PLL is ideal for *in vivo* gene delivery. These are the first polymer developed for substitute viral vector for gene delivery to avoid the immunogenicity and oncogenicity [45]. Poly-L-lysine dendrimer shows a systemic antiangiogenic activity that may be used for solid tumors therapy. PLL Dendrimer Inhibits Tubule Formation by Murine Endothelial Cells (SVEC4-10) *in vitro* at Nontoxic Concentrations and Inhibits Neovascularization in the CAM Assay [46].

Applications of Dendrimers

Currently, the demands of dendrimers have gained more attention from scientists. Dendrimers are interesting for pharmaceutical applications because of their unique characteristics. Dendrimers are hyperbranched macromolecules and have a lot of space in the internal cavities with multiple functional groups on the surface [47]. The following are the major biomedical applications about dendrimers.

The claims are solely related to biomedicals however the non-biomedical applications of dendrimers are beyond the scope of this chapter.

Dendrimers in Cancer Therapy

For most of the cancer treatment, chemotherapy becomes an essential aspect, but chemotherapeutic agents still show poor specificity to reach the tumorous site. The nanoscale delivery vehicles are capable of controlled release of chemotherapeutic agents directly into the tumor cell. Ligands such as folic acid, antibodies, nucleic acid along with polymer functionalization are used to target the tumor tissue [48]. Unique properties of nano-drug delivery system discriminate against them from other chemotherapeutic agents: (i) nanoscale system can be attached to targeting moieties, which gives high affinity and specificity for target cells, (ii) nanoscale devices can carry a drug moiety. By applying both active and passive targeting strategies, the nanocarriers can attain an increased intracellular concentration of drugs and minimizing toxicity in healthy cells, simultaneously enhancing anticancer effects and reducing systemic toxicity. Dendrimers are a nanoscopic compound that comprises a series of branches around the central core. The unique structure of dendrimers allows for the multivalent attachment of imaging probes, targeting moieties so it can also be used as a diagnostic tool for cancer therapy [49, 50]. The hydrophobic interior of the dendrimer allows for the physical encapsulation of the hydrophobic drug molecules [51]. Majoros *et al.* conjugated fluorescein isothiocyanate (FITC), folic acid (FA), and methotrexate to the G5 PAMAM dendrimers. The dendritic device synthesized was capable of targeting the overexpressed folate receptors found in many types of cancer cells [52]. Lai research group conjugated doxorubicin to the PAMAM dendrimers through the amide or the hydrazone bond [53]. Wang *et al.* also conjugated FITC and folic acid to G5 PAMAM dendrimers and encapsulated another anticancer drug, 2-methoxyestradiol [51]. The applicability of dendrimers has majorly been explored for cancer treatment. The day by day applicability of dendrimers can establish this nanocarrier for better delivery and targeting of anticancer drugs.

Dendrimer in Infectious Diseases

The microorganisms such as bacteria, viruses, fungi, protozoa cause infectious diseases, which leads to mortality and morbidity. Treatment of viral infections especially HIV and influenza virus was delayed by drug resistance and adverse side effects [54]. One of the most crucial tasks in infection therapy is the design of nanocarriers which can transport antimicrobial agents to the target site and controlled release. Both gram-positive and gram-negative bacteria have different constituents in their outer membranes, which are readily deprotonated to give the negative charges, increasing their hydrophilicity [55]. Dendrimer biocides contain quaternary ammonium salt as functional end groups. Quaternary ammonium salts are recognized antimicrobial agents that interrupt bacterial cell walls. It is reported that poly-cationic dendrimers give the high concentration of positively charged amine groups which are bind to the negatively charged bacterial cell walls, increases bacterial membrane permeability, and allows more dendrimers for the entry of bacteria that cause the leakage of potassium ions and a complete breakdown of the bacterial cell wall [56, 57]. Poly-anionic dendrimers displayed excellent antiviral activities against both HIV and herpes virus by targeting the viral life cycle to prevent it's binding to host cells [54]. The surface amine group of the dendrimer is modified with sulfate groups shows that sulfated linear-dendritic polymer blocked the HIV-1 infection *in vitro* and reduced the viral load in late-stage AIDS patients [58].

Dendrimers in Ocular Diseases

Dendrimers act as a safe and effective drug delivery cargo. Dendrimers show physiochemical properties like pH, osmolality, and viscosity, which are well suited for the formulations of ocular diseases [59]. Dendrimers' surface functional groups are modified or conjugated with different types of guest molecules to assist the multivalent interactions with the biological membranes. Cationic dendrimers can interact with a negatively charged biological membrane to increase their permeability. At physiological pH, the cornea is a negative charge, and this characteristic feature of dendrimer could be an advantage of an ocular drug delivery system. Vandamme and Brobeck demonstrated that PAMAM dendrimers could enhance the corneal residence time of pilocarpine and tropicamide [60, 61]. Yao *et al.* prepared puerarin dendrimer complexes to increase the residence time of Puerarin because Puerarin drains from the ocular surface rapidly which results in less residence time, short absorption time low bioavailability [62].

Dendrimer in Dermatological Diseases

The large surface area and secure use of skin make it an attractive route of drug delivery but the structure of skin restrict the transport of molecules through it. So, various chemical and physical methods are developed to increase drug penetration. The unique feature of dendrimer includes the presence of the surface functional group in large quantities which results in the multivalent interactions with drug and biological membranes. The membrane permeability, bio-distribution, and toxicity of dendrimers are dependent on their surface functional groups, size, and charge on the surface groups [63, 64]. PAMAM dendrimers enhanced the transdermal drug delivery system of various drugs including indomethacin, ketoprofen, 5-fluorosis, 8-methoxy psoralen. Cationic dendrimers show higher skin penetration than the neutral or anionic dendrimers because of the binding of negatively charged skin, interacting with the intercellular lipids, and inducing holes in the cell membranes which causes the leakage of a cell [65, 66]. The lower generation of dendrimers shows higher permeation than the more elevated generation dendrimers. Dendrimers are also increasing the skin penetration of lipophilic and hydrophilic drugs [65].

Dendrimer in CNS Disorders

One of the main challenges in the delivery of drugs to the CNS in the presence of the Blood-brain barrier (BBB). Most of the drugs, nucleic acid, proteins, and imaging agents fail to cross the BBB. Nanocarriers are used for the delivery of drugs across the BBB. A dendrimer is a highly branched, three-dimensional structure having low polydispersity and high functionality [67]. The presence of multiple surface groups and the hydrophobic core allows high drug loading and multifunctional group attachment. All these properties of dendrimers make versatile nano-carrier for transporting the drug across the BBB [68]. Dendrimers can transport drugs across various cell membranes or biological barriers by an endocytosis-mediated cellular internalization.

Dendrimers in Cardiovascular Diseases/Hypertension

Dendrimers are hyper-branched, monodispersed, three-dimensional macromolecules utilized in cardiovascular chemotherapy. Amine terminal functionalities of dendrimer can enter into the cell due to the presence of a positive charge on its surface, whereas the lipid membrane of a cell is negatively charged. Dendrimer can cross several barriers by active and passive targeting. Dendrimers can deliver hydrophobically, as well as, hydrophilic drug molecules, due to the presence of hydrophobic core and terminal charged functional groups. Many researchers delivered bioactive like drugs, propranolol, Quinidine, genes, and nitrous oxide in cardiovascular diseases (CVD) treatment [68].

Dendrimers in Ulcerative Colitis Diseases

Ulcerative colitis is a disease of the gastrointestinal tract characterized by inflammation of the mucosal barrier. The various inflammatory cytokines such as interleukin 1β and interleukin six are involved in the pathogenesis of ulcerative colitis. Pure and surface functionalized dendrimers present *in vivo* anti-inflammatory activities by inhibition of pro-inflammatory cytokine secretion, which result in the prevention of immigration of leukocytes across the endothelial cells to the inflammatory sites. These properties of dendrimers make them a much more fascinating drug delivery carrier for the biomedical sciences [69].

CONSENT FOR PUBLICATION

Not applicable.

CONFLICT OF INTEREST

The author declares no conflict of interest, financial or otherwise.

ACKNOWLEDGEMENTS

Declared none.

REFERENCES

[1] Kesharwani P, Tekade RK, Gajbhiye V, Jain K, Jain NK. Cancer targeting potential of some ligand-anchored poly(propylene imine) dendrimers: a comparison. Nanomedicine (Lond) 2011; 7(3): 295-304.
[http://dx.doi.org/10.1016/j.nano.2010.10.010] [PMID: 21070888]

[2] Gupta U, Dwivedi SKD, Bid HK, Konwar R, Jain NK. Ligand anchored dendrimers based nanoconstructs for effective targeting to cancer cells. Int J Pharm 2010; 393(1-2): 185-96.
[http://dx.doi.org/10.1016/j.ijpharm.2010.04.002] [PMID: 20382210]

[3] Otto DP, Villiers MM. Poly (amidoamine) PAMAM dendrimers as a pharmaceutical excipient. J Pharmaceut Sci 2017; 107: 75-83.
[http://dx.doi.org/10.1016/j.xphs.2017.10.011] [PMID: 29045886]

[4] Buhleier E, Wehner W, Vogtle F. "Cascade"- and "nonskid-chain-like" syntheses of molecular cavity topologies. Synth 1978; 155-8. http://pascal-francis.inist.fr/vibad/index.php?action=getRecordDetail&idt=PASCAL7860240769

[5] Kesharwani P, Jain K, Jain NK, *et al.* Dendrimer as the nanocarrier for drug delivery. Prog Polym Sci 2014; 39: 268-307.
[http://dx.doi.org/10.1016/j.progpolymsci.2013.07.005]

[6] Tomalia DA, Naylor AM, William A, *et al.* Starburst dendrimers: molecular-level control of size, shape, surface chemistry, topology, and flexibility from atoms to macroscopic matter. Angew Chem Int Ed Engl 1990; 29: 138-75.
[http://dx.doi.org/10.1002/anie.199001381]

[7] Tomalia DA, Frechet JMJ. Discovery of dendrimers and dendritic polymers: a brief historical perspective. J Polym Sci A Polym Chem 2002; 40: 2719-28.
[http://dx.doi.org/10.1002/pola.10301]

[8] Caminade AM, Laurent R, Majoral JP. Characterization of dendrimers. Adv Drug Deliv Rev 2005; 57(15): 2130-46.
[http://dx.doi.org/10.1016/j.addr.2005.09.011] [PMID: 16289434]

[9] Noriega-Luna B, Godínez LA, Rodríguez FJ, *et al.* Applications of dendrimers in drug delivery agents diagnosis therapy and detection. J Nanomater 2014; 39: 19.
[http://dx.doi.org/10.1155/2014/507273]

[10] Abbasi E, Aval SF, Akbarzadeh A, *et al.* Dendrimers: synthesis, applications, and properties. Nanoscale Res Lett 2014; 9(1): 247.
[http://dx.doi.org/10.1186/1556-276X-9-247] [PMID: 24994950]

[11] Hawker CJ, Frechet J. Control of surface functionality in the synthesis of dendritic macromolecules using the convergent-growth approach. Macromol 1990; 23: 4726-9.
[http://dx.doi.org/10.1021/ma00223a036]

[12] Gupta V, Nayak SK. Dendrimers: A review on synthetic approaches. J Appl Pharm Sci 2015; 5: 117-22.
[http://dx.doi.org/10.7324/JAPS.2015.50321]

[13] Jain NK, Gupta U. Application of dendrimer-drug complexation in the enhancement of drug solubility and bioavailability. Expert Opin Drug Metab Toxicol 2008; 4(8): 1035-52.
[http://dx.doi.org/10.1517/17425255.4.8.1035] [PMID: 18680439]

[14] D'Emanuele A, Attwood D. Dendrimer-drug interactions. Adv Drug Deliv Rev 2005; 57(15): 2147-62.
[http://dx.doi.org/10.1016/j.addr.2005.09.012] [PMID: 16310283]

[15] Maingi V, Kumar MVS, Maiti PK. PAMAM dendrimer-drug interactions: effect of pH on the binding and release pattern. J Phys Chem B 2012; 116(14): 4370-6.
[http://dx.doi.org/10.1021/jp211515g] [PMID: 22420638]

[16] Cheng Y, Wang J, Rao T, He X, Xu T. Pharmaceutical applications of dendrimers: promising nanocarriers for drug delivery. Front Biosci 2008; 13: 1447-71.
[http://dx.doi.org/10.2741/2774] [PMID: 17981642]

[17] Svenson S. Dendrimers as versatile platform in drug delivery applications. Eur J Pharm Biopharm 2009; 71(3): 445-62.
[http://dx.doi.org/10.1016/j.ejpb.2008.09.023] [PMID: 18976707]

[18] El-Sayed M, Ginski M, Rhodes C, Ghandehari H. Transepithelial transport of poly(amidoamine) dendrimers across Caco-2 cell monolayers. J Control Rel 2002; 81(3): 355-65.
[http://dx.doi.org/10.1016/S0168-3659(02)00087-1] [PMID: 12044574]

[19] Medina SH, El-Sayed ME. Dendrimers as carriers for delivery of chemotherapeutic agents. Chem Rev 2009; 109(7): 3141-57.
[http://dx.doi.org/10.1021/cr900174j] [PMID: 19534493]

[20] Sharma A, Gautam SP, Gupta AK. Surface modified dendrimers: synthesis and characterization for cancer targeted drug delivery. Bioorg Med Chem 2011; 19(11): 3341-6.
[http://dx.doi.org/10.1016/j.bmc.2011.04.046] [PMID: 21570304]

[21] Najlah M, Freeman S, Attwood D, D'Emanuele A. Synthesis, characterization and stability of dendrimer prodrugs. Int J Pharm 2006; 308(1-2): 175-82.
[http://dx.doi.org/10.1016/j.ijpharm.2005.10.033] [PMID: 16384673]

[22] Svenson S, Tomalia DA. Dendrimers in biomedical applications--reflections on the field. Adv Drug Deliv Rev 2005; 57(15): 2106-29.
[http://dx.doi.org/10.1016/j.addr.2005.09.018] [PMID: 16305813]

[23] Singh J, Jain K, Mehra NK, Jain NK. Dendrimers in anticancer drug delivery: mechanism of interaction of drug and dendrimers. Artif Cells Nanomed Biotechnol 2016; 44(7): 1626-34.
[http://dx.doi.org/10.3109/21691401.2015.1129625] [PMID: 26747336]

[24] Wolinsky JB, Grinstaff MW. Therapeutic and diagnostic applications of dendrimers for cancer treatment. Adv Drug Deliv Rev 2008; 60(9): 1037-55.
[http://dx.doi.org/10.1016/j.addr.2008.02.012] [PMID: 18448187]

[25] Betley TA, Hessler JA, Mecke A, *et al.* Tapping mode atomic force microscopy investigation of poly(amidoamine) core-shell tecto(dendrimers) using carbon nanoprobes. Langmuir 2002; 18: 3127-33.
[http://dx.doi.org/10.1021/la025538s]

[26] Bharali DJ, Khalil M, Gurbuz M, Simone TM, Mousa SA. Nanoparticles and cancer therapy: a concise review with emphasis on dendrimers. Int J Nanomed 2009; 4: 1-7.
[PMID: 19421366]

[27] Tomalia DA. Birth of a new macromolecular architecture: dendrimers as quantized building blocks for

nanoscale synthetic polymer chemistry. Prog Polym Sci 2005; 30: 294-324.
[http://dx.doi.org/10.1016/j.progpolymsci.2005.01.007]

[28] Esfand R, Tomalia DA. Poly(amidoamine) (PAMAM) dendrimers: from biomimicry to drug delivery and biomedical applications. Drug Discov Today 2001; 6(8): 427-36.
[http://dx.doi.org/10.1016/S1359-6446(01)01757-3] [PMID: 11301287]

[29] Turnbull WB, Stoddart JF. Design and synthesis of glycodendrimers. J Biotechnol 2002; 90(3-4): 231-55.
[PMID: 12071227]

[30] Boas U, Heegaard PM. Dendrimers in drug research. Chem Soc Rev 2004; 33(1): 43-63.
[http://dx.doi.org/10.1039/b309043b] [PMID: 14737508]

[31] Jain K, Kesharwani P, Gupta U, Jain NK. Dendrimer toxicity: Let's meet the challenge. Int J Pharm 2010; 394(1-2): 122-42.
[http://dx.doi.org/10.1016/j.ijpharm.2010.04.027] [PMID: 20433913]

[32] Agashe HB, Dutta T, Garg M, Jain NK. Investigations on the toxicological profile of functionalized fifth-generation poly (propylene imine) dendrimer. J Pharm Pharmacol 2006; 58(11): 1491-8.
[http://dx.doi.org/10.1211/jpp.58.11.0010] [PMID: 17132212]

[33] Malik N, Wiwattanapatapee R, Klopsch R, *et al.* Dendrimers: relationship between structure and biocompatibility *in vitro*, and preliminary studies on the biodistribution of 125I-labelled polyamidoamine dendrimers *in vivo*. J Control Rel 2000; 65(1-2): 133-48.
[http://dx.doi.org/10.1016/S0168-3659(99)00246-1] [PMID: 10699277]

[34] Lee CC, MacKay JA, Fréchet JM, Szoka FC. Designing dendrimers for biological applications. Nat Biotechnol 2005; 23(12): 1517-26.
[http://dx.doi.org/10.1038/nbt1171] [PMID: 16333296]

[35] Nanjwade BK, Bechra HM, Derkar GK, Manvi FV, Nanjwade VK. Dendrimers: emerging polymers for drug-delivery systems. Eur J Pharm Sci 2009; 38(3): 185-96.
[http://dx.doi.org/10.1016/j.ejps.2009.07.008] [PMID: 19646528]

[36] Tomalia DA, Baker H, Dewald J, *et al.* A new class of polymers: starburst-dendritic macromolecules. Polym J 1985; 17: 117.
[http://dx.doi.org/10.1295/polymj.17.117]

[37] Boas U, Christensen JB, Heegaard PM, *et al.* Dendrimers in medicine and biotechnology: new molecular tools. Royal Soc Chem 2006.
[http://dx.doi.org/10.1039/9781847552679]

[38] Brabander-van E, Meijer EW. Poly (propylene imine) dendrimers: large-scale synthesis by heterogeneously catalyzed hydrogenations. Angew Chem 1993; 32: 1308-11.
[http://dx.doi.org/10.1002/anie.199313081]

[39] Sadler K, Tam JP. Peptide dendrimers: applications and synthesis. J Biotechnol 2002; 90(3-4): 195-229.
[PMID: 12071226]

[40] Li J, Swanson DR, Qin DT, *et al.* Characterizations of core-shell tecto-(dendrimer) molecules by tapping mode atomic force microscopy. Langmuir 1999; 15: 7347-50.
[http://dx.doi.org/10.1021/la990517s]

[41] Schilrreff P, Mundiña-Weilenmann C, Romero EL, Morilla MJ. Selective cytotoxicity of PAMAM G5 core--PAMAM G2.5 shell tecto-dendrimers on melanoma cells. Int J Nanomed 2012; 7: 4121-33.
[PMID: 22904625]

[42] Seebach D, Rheiner PB, Greiveldinger G, *et al.* Chiral dendrimers dendrimers. Chem Eur J 1998; 125-64.
[http://dx.doi.org/10.1007/3-540-69779-9_4]

[43] Ritzén A, Frejd T. Synthesis of a chiral dendrimer based on polyfunctional amino acids. Chem Commun (Camb) 1999; 207-8. [http://dx.doi.org/10.1039/a809195a]

[44] Lim J, Simanek EE. Triazine dendrimers as drug delivery systems: from synthesis to therapy. Adv Drug Deliv Rev 2012; 64(9): 826-35. [http://dx.doi.org/10.1016/j.addr.2012.03.008] [PMID: 22465784]

[45] Boyd BJ, Kaminskas LM, Karellas P, Krippner G, Lessene R, Porter CJ. Cationic poly-L-lysine dendrimers: pharmacokinetics, biodistribution, and evidence for metabolism and bioresorption after intravenous administration to rats. Mol Pharm 2006; 3(5): 614-27. [http://dx.doi.org/10.1021/mp060032e] [PMID: 17009860]

[46] Al-Jamal KT, Al-Jamal WT, Akerman S, *et al.* Systemic antiangiogenic activity of cationic poly--lysine dendrimer delays tumor growth. Proc Natl Acad Sci USA 2010; 107(9): 3966-71. [http://dx.doi.org/10.1073/pnas.0908401107] [PMID: 20150514]

[47] Biswas S, Torchilin VP. Dendrimers for siRNA Delivery. Pharmaceuticals (Basel) 2013; 6(2): 161-83. [http://dx.doi.org/10.3390/ph6020161] [PMID: 24275946]

[48] Gu FX, Karnik R. Targeted nanoparticles for cancer therapy. Nano Today 2007; 2: 14-21. [http://dx.doi.org/10.1016/S1748-0132(07)70083-X]

[49] Misra R, Acharya S, Sahoo SK. Cancer nanotechnology: application of nanotechnology in cancer therapy. Drug Discov Today 2010; 15(19-20): 842-50. [http://dx.doi.org/10.1016/j.drudis.2010.08.006] [PMID: 20727417]

[50] Peer D, Karp JM, Hong S, Farokhzad OC, Margalit R, Langer R. Nanocarriers as an emerging platform for cancer therapy. Nat Nanotechnol 2007; 2(12): 751-60. [http://dx.doi.org/10.1038/nnano.2007.387] [PMID: 18654426]

[51] Wang Y, Guo R, Cao X, Shen M, Shi X. Encapsulation of 2-methoxyestradiol within multifunctional poly(amidoamine) dendrimers for targeted cancer therapy. Biomaterials 2011; 32(12): 3322-9. [http://dx.doi.org/10.1016/j.biomaterials.2010.12.060] [PMID: 21315444]

[52] Majoros IJ, Thomas TP, Mehta CB, Baker JR Jr. Poly(amidoamine) dendrimer-based multifunctional engineered nanodevice for cancer therapy. J Med Chem 2005; 48(19): 5892-9. [http://dx.doi.org/10.1021/jm0401863] [PMID: 16161993]

[53] Lai PS, Lou PJ, Peng CL, *et al.* Doxorubicin delivery by polyamidoamine dendrimer conjugation and photochemical internalization for cancer therapy. J Control Rel 2007; 122(1): 39-46. [http://dx.doi.org/10.1016/j.jconrel.2007.06.012] [PMID: 17628166]

[54] Kim Y, Park EJ, Na DH, *et al.* Recent progress in dendrimer-based nanomedicine development. Arch Pharma Res 2018; 1-12.

[55] González B, Colilla M, Díez J, *et al.* Mesoporous silica nanoparticles decorated with polycationic dendrimers for infection treatment. Acta Biomater 2018; 68: 261-71. [http://dx.doi.org/10.1016/j.actbio.2017.12.041] [PMID: 29307796]

[56] Zhang L, Pornpattananangku D, Hu CM, Huang CM. Development of nanoparticles for antimicrobial drug delivery. Curr Med Chem 2010; 17(6): 585-94. [http://dx.doi.org/10.2174/092986710790416290] [PMID: 20015030]

[57] Huh AJ, Kwon YJ. Nanoantibiotics: a new paradigm for treating infectious diseases using nanomaterials in the antibiotics resistant era. J Control Rel 2011; 156(2): 128-45. [http://dx.doi.org/10.1016/j.jconrel.2011.07.002] [PMID: 21763369]

[58] Shaunak S. Perspective: Dendrimer drugs for infection and inflammation. Biochem Biophys Res Commun 2015; 468(3): 435-41. [http://dx.doi.org/10.1016/j.bbrc.2015.07.033] [PMID: 26168733]

[59] Baig T. Dendrimer as a carrier for ocular drug delivery. J Drug Dis Ther 2010; 2: 20.

[60] Yao W, Sun K, Mu H, *et al.* Preparation and characterization of puerarin-dendrimer complexes as an ocular drug delivery system. Drug Dev Ind Pharm 2010; 36(9): 1027-35.
[http://dx.doi.org/10.3109/03639041003610799] [PMID: 20545508]

[61] Reimondez-Troitiño S, Csaba N, Alonso MJ, de la Fuente M. Nanotherapies for the treatment of ocular diseases. Eur J Pharm Biopharm 2015; 95(Pt B): 279-93.
[http://dx.doi.org/10.1016/j.ejpb.2015.02.019] [PMID: 25725262]

[62] Vandervoort J, Ludwig . A ocular drug delivery: nanomedicine applications. NanoMed 2007; 2(1): 11-21.
[http://dx.doi.org/10.2217/17435889.2.1.11]

[63] Venuganti VV, Sahdev P, Hildreth M, Guan X, Perumal O. Structure-skin permeability relationship of dendrimers. Pharm Res 2011; 28(9): 2246-60.
[http://dx.doi.org/10.1007/s11095-011-0455-0] [PMID: 21633876]

[64] Yang Y, Sunoqrot S, Stowell C, *et al.* Effect of size, surface charge, and hydrophobicity of poly(amidoamine) dendrimers on their skin penetration. Biomacromolecules 2012; 13(7): 2154-62.
[http://dx.doi.org/10.1021/bm300545b] [PMID: 22621160]

[65] Winnicka K, Wroblewska M, Sosnowska K, Car H, Kasacka I. Evaluation of cationic polyamidoamine dendrimers' dermal toxicity in the rat skin model. Drug Des Devel Ther 2015; 9: 1367-77.
[http://dx.doi.org/10.2147/DDDT.S78336] [PMID: 25834395]

[66] Xu L, Zhang H, Wu Y. Dendrimer advances for the central nervous system delivery of therapeutics. ACS Chem Neurosci 2014; 5(1): 2-13.
[http://dx.doi.org/10.1021/cn400182z] [PMID: 24274162]

[67] Leiro V, Duque SS, Lopes CD, *et al.* Dendrimers as powerful building blocks in central nervous system disease: Headed for Successful. NanoMed Adv Func Mat 2017.
[http://dx.doi.org/10.1002/adfm.201700313]

[68] Gothwal A, Kesharwani P, Gupta U, *et al.* Dendrimers as an effective nanocarrier in cardiovascular disease. Curr Pharm Des 2015; 21(30): 4519-26.
[http://dx.doi.org/10.2174/1381612820666150827094341] [PMID: 26311317]

[69] Wang Y, Shen W, Shi X, *et al.* Alpha-Tocopheryl Succinate-Conjugated G5 PAMAM dendrimer enables effective inhibition of ulcerative colitis. Adv Healthc Mater 2017; 6(14): 14.
[http://dx.doi.org/10.1002/adhm.201700276] [PMID: 28474434]

CHAPTER 5

Chronicle of Nanomicelles in Drug Delivery: From Bench to Bedside

Neha Bajwa[1], Preet Amol Singh[1], Vaskuri G.S. Sainaga Jyothi[2], Rupinder Kaur Sodhi[3], Ashish Baldi[1], Satish Sardana[4], Ramesh Chandra[5,6] and Jitender Madan[2,*]

[1] *Department of Pharmaceutical Sciences and Technology, Maharaja Ranjit Singh Punjab Technical University, Bathinda, Punjab, India*

[2] *Department of Pharmaceutics, National Institute of Pharmaceutical Education and Research, Hyderabad, Telangana, India*

[3] *Department of Pharmacology, Chandigarh College of Pharmacy, Mohali, Punjab, India*

[4] *Amity Institute of Pharmacy, Amity University, Gurgaon (Haryana) India*

[5] *Dr. B.R. Ambedkar Centre for Biomedical Research, University of Delhi, Delhi, India*

[6] *Department of Chemistry, University of Delhi, Delhi, India*

Abstract: Currently, nanomicelles are used pharmaceutically for solubilizing hydrophobic drugs, improving bioavailability, targeting anticancer drugs, and many more applications are associated with the use of nanomicelles. By definition, nanomicelles are colloidal dispersions containing a hydrophobic core and a hydrophilic shell that self-assemble itself into the nanosize range of 10-100 nm. The major limiting factor for formulating a clear aqueous solution of a drug is its hydrophobic nature. Nanomicelles improve the solubilization of hydrophobic drugs by entrapping the drugs within a mixed micellar hydrophobic core with hydrophilic chains protruding outwards, forming a clear aqueous formulation. Pharmaceutically nanomicelles serve as outstanding carriers as they can avert or moderate drug degradation by dropping adverse side-effects, thereby augmenting drug permeation through biological barriers with very minimum or no irritation at all, which ultimately enhance bioavailability. The major factors that affect the micelle formation are the hydrophobic part size in the amphiphilic molecule, amphiphiles concentration, temperature and solvent. The assembling process starts only when a certain minimum concentration is crossed by the amphiphilic molecules, called Critical Micelle Concentration (CMC). Therefore, in this chapter, a critical history of nanomicelles is provided from its fundamental theory to preclinical and clinical achievements.

Keywords: Drug release mechanism, Nanomicelles, Ocular drug delivery, Peptide drug delivery, Pulmonary drug delivery, Theory, Vaccine adjuvant.

* **Corresponding author Jitender Madan:** Department of Pharmaceutics, National Institute of Pharmaceutical Education and Research, Hyderabad, Telangana, India; E-mail: jitenderpharmacy@gmail.com

Surendra Nimesh, Nidhi Gupta and Ramesh Chandra (Eds.)

INTRODUCTION

The arena of nanomicelles is growing its roots from academic to an industrial level by implementing different techniques and preclinical and clinical tools. With the increase in interest in this technology, pioneering research is being carried out in nanomicelles, which aids in drug solubilization, drug targeting, and targeting of nucleic acids [1]. Nanomicelles are colloidal dispersions containing a hydrophilic shell and hydrophobic core that self-assemble itself into the nanosize range of 10-100 nm. Currently, nanomicelles are adopted pharmaceutically for solubilizing hydrophobic drugs [2]. The major factor limiting the formulation of a clear aqueous solution of a drug is its hydrophobicity. Nanomicelles improve hydrophobic drug solubility by loading the drug in the micelle core, which is hydrophobic in nature and is surrounded by hydrophilic chains protruding outwards, forming a clear solution [3]. Pharmaceutically nanomicelles serve as exceptional carriers as they can prevent or reduce degradation of drug, lower adverse effects and enhance permeation of drug through biological barriers with very minimum or no irritation at all, which ultimately enhance bioavailability [4]. Micelles are the aggregates of molecules containing both hydrophobic and hydrophilic segments. These molecules, with both hydrophilic and hydrophobic parts, are called amphiphilic molecules. These molecules self assemble in the suitable solvent and form micelles which are called nanomicelles with a size in the nano range. The micelles aggregates in the hydrophilic solvent in such a way that the hydrophobic part is in the core and the hydrophilic part is outward. This limits the exposure of the hydrophobic part to the hydrophilic environment. If the amphiphilic molecules are exposed to a hydrophobic solvent, the micelles are formed with the hydrophobic region on the exterior and the hydrophilic chains in the core which are called reverse nanomicelles. Normal nanomicelles are employed in encapsulation, solubilization and delivery of hydrophobic drugs, whereas reverse nanomicelles are employed in encapsulation and delivery of water-soluble drugs. Nanomicelles have been established as excellent drug delivery systems. Furthermore, nanomicelles minimize the degradation of drugs, reduce the adverse effects and show improvement in drug bioavailability. Moreover, nanomicelles provide unique advantages of submicron size, aqueous clear/transparent drug compartment, encapsulation and solubilization of hydrophobic drugs, and enable high permeation across different barriers in skin. In addition, either a surfactant or polymeric system can be used to form nanomicelles [5].

Principle of Micelles Formation

Amphiphilic molecules or surfactants in the presence of solvent undergo an abrupt change in the physicochemical properties. The hydrophilic head and the lipophilic chains of these amphiphilic molecules are oriented in the solvent to form spherical clusters called micelles. The molecules are assembled and formed the hydrophobic core and hydrophilic surface within the micelle. Around 50-200 molecules called monomers are assembled to form a micelle, which is called aggregation number. The micelles are spherical with a radius of 1-3 nm, unlike the length of the surfactant monomer and fall in the colloidal range [6, 7]. The main motive for the self-assembly of monomers is the reduction of free energy, and the diminution of it is the outcome of the exclusion of hydrophobic head from the polar environment by forming a hydrophobic core with the hydrophilic tails towards outward. The chief factors that affect the formation of a micelle are a hydrophobic part, monomer size, the concentration of monomers, solvent and temperature. The monomers are self-assembled only at a particular concentration level called CMC (Critical Micelle Concentration). At lower concentrations, monomers exist distinctly with a smaller size giving an impression of a sub-colloidal range. Below CMC, the monomers adsorb at the interface of two phases, and the interface gets saturated with monomers when the concentration of monomers reaches the CMC value. Further increase of monomers beyond the CMC causes the assemble of monomers in the air-water interface into a spherical cluster in the bulk phase called a micelle. In this way, a micelle is formed by reducing the free energy. The temperature at which the monomers starts forming micelle is called critical micellization temperature (CMT) [6, 8, 9]. Mainly two forces are involved in micelle formation; one is an attractive force that aids in the self-assembly of monomers, and the other force is a repulsive force, which averts limitless micelle grow to reach macroscopic state [10, 11]. Amphiphilic copolymers show association in the polar solvent, which has both hydrophobic and hydrophilic polymers characteristics. At concentrations below CMC, the polymers are individual, but when the concentration reaches the CMC value, then the polymeric chains assemble and form micelles. At CMC, the amphiphilic copolymer forms micelles with a considerable amount of solvent inside the core, and hence the micelles are aggregates with a size greater than the polymeric micelles produced at concentrations above CMC [12]. The CMC concentration of amphiphilic copolymers is lower than that of the low molecular weight surfactant. For example, polystyrene-poly (N-isopropylacrylamide) (PS-PNIPA) and poly(ethylene oxide)-poly (beta-benzyl-L-aspartate) (PEO-PBLA) have a CMC value of 0.0005-0.002% [13, 14]. Although, few amphiphilic copolymers displayed a high CMC value of 0.01±10%, for example, poloxamers [15 - 17]. But amphiphilic copolymers of high CMC value are not preferred for drug delivery because of the lack of stability in the polar solvent and shows dissociation when diluted. Amphiphilic copolymers form micelles of two different categories based on the link of the hydrophobic chain to the hydrophilic backbone. The hydrophobic chain-linked randomly to the hydrophilic polymer formed micelles of smaller size compared to micelles formed from the hydrophobic chain linked to the terminal of the hydrophilic chain [18]. The size of the micelle is estimated by the hydrophobic forces in the core that hold the hydrophobic chains and repulsion between the chains, limiting their size. These two forces result in micelles of different sizes with random and end-altered copolymers. In the end-altered polymers, the hydrophobic groups are linked together to form micelles, which form the core, and the water is omitted from the core, and there is no direct contact between the hydrophobic core and the hydrophilic part [18, 19]. In the randomly linked polymer, both the hydrophilic and hydrophobic chains are intertwined together, permitting exposure of micelle core to the surrounding media where the hydrophilic chains creating the shell are less movable [19]. In this, the hydrophobic core may cause polymeric micelles secondary aggregation [12, 13, 20]. This secondary aggregation is a hypothesis in interpreting the formation of large micelles (>100 nm) of polyethylene oxide and poly(aspartic acid) bearing conjugated doxorubicin [20].

Amphiphilic Copolymers

Amphiphilic copolymers that form micelles can either block copolymers of graft copolymers. A graft copolymer has a backbone made up of one polymer and side grafts of different polymeric chains. These copolymers consist of properties of both backbone and graft polymers. A well-defined graft copolymer is synthesized by 'click' reaction, which integrates the polymeric chains onto the backbone [21]. The block copolymers form spherical micelles in aqueous solutions by self-assembling amphiphilic di-block or tri-block copolymers. However, the copolymers exist as individual molecules if the hydrophilic chain length is too long, and the copolymers may develop different shapes if the hydrophobic block is too lengthy.

Amphiphilic graft or block copolymers behave in a similar way to that of the conventional amphiphiles and form polymeric micelles in hydrophilic solvent above CMC. As compared to conventional micelles, there exists a covalent linkage between the monomers in the interior of the hydrophobic core. This covalent linkage turns away the dynamic moment of polymeric chains between the micellar pseudo-phase and the bulk of the solution and thus imparts rigidity to the micelle [22]. Polymeric micelles have an aggregation number of several hundred. An important factor that governs the polymeric micelles' size includes amphiphilic block copolymer molecular weight (M_w), amphiphiles aggregation number, preparation method and the composition of hydrophilic and hydrophobic chains [23].

The amphiphilic block copolymers aggregate into spherical micelles, cylindrical micelles or vesicles, or polymersomes in the aqueous medium. Hydrophilic-hydrophobic balance, which is defined by the hydrophilic volume fraction (*f*), is the main factor, which governs the morphology of the micelles. Polymeric vesicles are formed with *f* value of approximately 35% and spherical micelles with *f* greater than 45% [24, 25]. Micelles with multifaceted morphologies such as toroids, multicompartment, screw-cut micelles, *etc.*, are generated by the amphiphiles of more complicated designs such as miktoarm star copolymers, or by changing the processing parameters for the association of polymers. Controlled synthesis of amphiphilic block copolymers of desired structure can be accomplished by fluctuating total M_w, chemical structure and the block ratio, and thus, the size and shape of the micelles can be fabricated. These polymeric micelles contain a hydrophobic core, which solubilizes the hydrophobic drug resolving the solubility issue of hydrophobic drugs, enhancing the oral bioavailability [26, 27].

Polymeric micelles are safer than conventional surfactant micelles as the surfactant lyses the cell membrane. The micelles obviate the filtration and clearance by the spleen in the body due to its nanoscopic size and the hydrophilic shell [28]. This extends the circulation duration of the drug and also the shell stabilizes the micelle and controls the biodistribution. The nano-size range of micelles aids in diminishing the embolism risk in blood capillaries as compared to other drug carriers. The nanosize also supports intravenous administration, good absorption in the intestinal tract, faster clearance and lowers toxicity. As the drug is encapsulated in the micelle, chemical structure alteration of the drug is not recommended [29]. Targeted delivery can be achieved by these polymeric micelles owing to their submicron size and high drug loading. Receptor-mediated endocytosis of drugs and genes can be realized by functionalizing the block copolymers with sugars and peptides. Tagging the monoclonal antibody to the polymeric or surfactant micelles results in immunomicelles, which display high binding specificity and targetability [30]. Stimuli selective copolymers incorporation into polymeric micelles are being discovered for controlled release, which can lead to the production of '*Intelligent Vehicles*' [31].

Micellization of Amphiphilic Block Copolymers

Amphiphilic molecules have a unique structure and owing to this, they assemble at the interface of two and are thus termed surfactants. The amphiphilic molecules in the polar solvent assemble in such a way that the exposure of the hydrophobic blocks is minimized to a hydrophilic environment. With an increase in the concentration of amphiphiles, the interface gets saturated with the monomers leading to unfavorable interactions of hydrophobic parts with the polar solvent and results in increased free energy. Therefore, these amphiphilic molecules assemble together and form a cluster reducing the entropy at the CMC level, which is termed as a micelle. Above the level of CMC, polymeric micelles are stable, but on dilution, the concentration falls underneath CMC and leads to dissociation of monomers from the micelle. This disassembly of amphiphiles depends on their structure and the interactions between the chains [32]. Therefore, amphiphilic copolymers form micelles with greater stability when compared to the surfactants like Cremophor EL or polysorbates, due to their distinctly lower CMC value and physical stability, which is the added advantage. Micelles fall under the category of colloids but nevertheless can be considered as solid particles [33]. The amphiphiles that form micelles are in dynamic equilibrium with the molecules in the external solvent and are established by sedimentation velocity, small-angle X-ray scattering and light scattering. Micelles with a size range of 10-100 nm and the core of the micelle are hydrophobic nature, which aids in solubilizing the poorly soluble drug [32, 34 - 37]. The hydrophobic core is encircled by the hydrophilic chains of the amphiphilic copolymers. The hydrophilic outer core is surrounded by water molecules and adopts a "splayed" appearance, forming different conformations such as polymeric "brush" [38]. This hydrophilic chain conformation on the micelle reduces opsonization sterically in the body and thus prevents the engulfing of micelles by macrophages and reduces clearance.

Characterization of Polymeric Micelles

Characterization of micelles is very important and needs to be carried out appropriately. Micelle formation takes place as a consequence of attractive force, which is responsible for the assembling of molecules and the growth of micelle to macroscopic size is controlled b the repulsive force [39, 40]. The amphiphilic copolymer in a solvent shows self-assembly of molecules based on the affinity hydrophilic or hydrophobic polymer towards the solvent.

Shape of Polymeric Micelles

The shape of the micelle determines the thermodynamic properties of self-assembly and the intra-micelle forces produced within the molecules inside the micelles [41]. These two major factors (intra-micelle and thermodynamic forces), in addition to the interactions between different micelles, govern the structure and type of micelle formed at equilibrium. Hence understanding the fundamentals that determine the inter-dependence between size and morphology of the micelles attained by self-assembling is crucial. Also, other decisive factors like temperature, block length, concentration, composition and amphiphilic block copolymer architecture play a vital role. Polymeric micelles often have a spherical shape with narrow size distribution [42]. Thus, the inner hydrophobic core is always safeguarded from the aqueous medium by the hydrophilic polymeric chains. Spherical micelles are obtained by the association of block copolymers, which is driven by entropic mechanism, but under certain conditions, the micelle shape may vary [43 - 47]. Based on the length of the blocks in the copolymer, the structure is affected. If the hydrophilic block length is larger, then the micelles are with a large shell and very small inner core and therefore called 'Star-micelles'. In contrast, micelle with small corona generated by a large hydrophobic block is referred to as 'Crew-cut micelles' [48]. Shape of the micelles varies from spherical to vesicular to other lesser common structures under certain circumstances [49 - 54]. Apart from spherical and vesicular structures, other complex morphologies include cylinders or bilayers, may also form depending on the block copolymers. The cylindrical micelles can be created by specifically designed amphiphilic block copolymers [55 - 59].

Size of Polymeric Micelles

A fascinating feature of polymeric micelles is their small size, approx. 10-100 nm. Not only it allows extravasation of carriers, but it also allows the sterilization of the formulation usually carried out by filtration and reduces embolism risks inside the capillaries with respect to larger conventional drug carriers [60]. Size of micelles rarely surpasses 100 nm, however, the factors that influence the size include M_w, aggregation number and relative proportion of hydrophobic and hydrophilic chains [61 - 63]. Micelles of PEO±PBLA prepared in dimethyl-formamide (DMF) solvent and dialyzed in water, produced micelles larger than that of micelles obtained by the conventional method [64]. Calculating the size of micelles can be used to estimate the polymeric micelles interaction with the biological medium. For example, micelles of the amphiphilic triblock copolymer (PEO±PPO±PEO) was found to preserve their intact size in the solution of bovine serum albumin and antibodies as well as displayed negligible effect of interaction with plasma proteins on particle size [65]. Finding out the size of the micelle can be used in the thermo-responsive micelles characterization.

Polydispersity index and size of the micelles can be estimated directly by DLS (dynamic light scattering) method. The sphericity of the polymeric micelles can be determined by DLS [66, 67]. It is also possible with DLS to terminate the micelle structure and to bring a change in the size from 50 nm to 3 nm by adding a low M_w surfactant such as 1% W/V solution of sodium dodecyl sulfate [68]. Apart from DLS, the size of the micelles can be calculated by AFM (atomic force microscopy) [69, 70], TEM (transmission electron microscopy) [71] and SEM (scanning electron microscopy) [72]. SEM is conventionally implied for characterization of colloidal carriers because of its easy sample preparation and high resolution, however, the sample should withstand the high vacuum during analysis. Moreover, the sample should be conducive for the visualization of particles and this is accomplished by coating with gold. But the coating of the samples leads to an increase in thickness and should be taken into consideration while estimating the particle size. New imaging tools like AFM gives the image of the sample without any sample preparation at atmospheric pressure [73]. By AFM, it was found that micelles of PNIPA-b-PSt (poly-*N*-isopropylacrylamid--block-polystyrene) had a disc-shaped structure with 5 nm heights and a diameter of 20 nm that was approximately nearer to the 24 nm size measured by DLS [69].

Polydispersity of polymeric micelles is evaluated occasionally in ultracentrifugation velocity studies [74, 75]. The size of the polymeric micelles can be controlled easily. Micelles in the size of 10-100 nm are acceptable for the administration through the bloodstream, and this size eludes the clearance from the kidney and also from reticuloendothelial systems (liver and spleen) [76, 77] and also it shows selective extravagation to the target tissue from the bloodstream. In this way, solid tumours are targeted by using the EPR (Enhanced permeation and retention) effect [78 - 81], and many tests turned positive by this EPR. Other than polymeric micelles, liposomes and nanospheres are also effective for implementing the EPR targeting. However, polymeric micelles have two strong advantages over the liposomes and nanospheres. Smaller size range of less than 50 nm can be achieved by polymeric micelles, whereas with liposomes and nanospheres, the particles will in the range of 10-200 nm. The second benefit is having control of size by controlling the chemical structure of the polymer without altering the preparation process. So the size of the micelle particle can be foreseen to be appropriate with that of theoretical calculation [82 - 84] from the chain length of the block copolymer and its chemical structure. The suitable size for targeting through the bloodstream is established to be 10-200 nm [76, 77, 85]. However, particles having a size more than 400 nm are extensively captured [86]; although they can cross the walls of blood capillaries. As polymeric micelles are most frequently falls in size range of 20-100 nm, so they are more suitable for drug targeting owing to their smaller size. Furthermore, polymeric micelles are less prone to chronic toxicity as the micelles disassemble into a single polymer *in vivo,* which can be excreted from the kidney. Owing to the EPR effect, solid tumors can be targeted by polymeric micelles. Polymeric micelles are diffused from the bloodstream into the tumor tissue through convection mechanisms. The appropriate size of polymeric micelles to target the tumor tissue is vital, but still, it is not explored. Technically, two difficulties have been associated to perform this study, which includes variations in the circulation time and calculation of the amount deposited in the bone marrow [87, 88]. However, it is difficult to make an exact comparison between size effects on selective targeting [88]. Calculation of delivered amount to bone marrow is vital since it is a significant location for inducing hematopoietic toxicity by antitumour agents. Nevertheless, this calculation is not regularly performed because of the complexity associated with bone marrow collection [89].

Critical Micelles Concentration (CMC) Determination

The CMC can be estimated by numerous methods [90, 91]. Actually, the change in the physical property (*e.g.*, conductivity, osmotic pressure and interfacial tension), which occurs near the CMC level, can be utilized for estimating the CMC value [92]. Generally, the plot of such properties can be used to detect the micellization where the change in the graph gives the onset of micellization process [93]. However, the CMC of polymeric micelles is very small to estimate by these methods. Aggregation number and the M_w of micelles is extensively determined by light scattering. Nevertheless, the scattering method can identify the onset of micellization only if the CMC matches with the sensitivity of the instrument [10]. GPC (Gel permeation chromatography) can be used to determine the CMC since micelles as the single chains exhibit different elution volumes [94]. However, polymers may adsorb on to the column surface and may create a problem [95], particularly at a level near to CMC concentration where are micelles are in loose aggregates. A fluorescent probe was used for the determination of micellar CMC [10, 96 - 98], of which pyrene is used widely. It is highly hydrophobic, condensed aromatic hydrocarbon and is sensitive to the surrounding environment polarity [62]. Pyrene is soluble in the water below CMC, but when the micelles are formed, it partitions favorably towards the core of the micelle, which is hydrophobic in nature [62]. As a result, various changes occur, which include the red shift of the band in the excitation spectra, increase in fluorescence intensity and change in vibrational frequency of the emission spectra. However, a small variation in the fluorescence spectrum can occasionally be ascribed to the individual polymer chains or premicellar aggregates or presence of hydrophobic impurities [99]. Variations in anisotropy of fluorescent probes have also been linked to micellization initiation [100, 101]. It has been established that the micellization initiation for numerous copolymers is mainly influenced by hydrophobic polymer chain length, whereas the effect of the chain length of hydrophilic polymer on the CMC is less noticeable [10, 98, 102]. The CMC of the micelle methoxy(Me) PEO-PDLLA (poly(DL-lactic acid)) remained constant in water, 5% w/v dextrose solutions and 0.9% w/v saline solution. Though, it was observed that PNIPA(poly (*N*-isopropylacrylamide) (PNIPAM) copolymerized with methacrylic acid and octadecylacrylate has a CMC which was constant in all different media like PBS (phosphate buffered saline) and water, since micelles have a negative zeta potential [97]. Electrolytes, in this case has payed off the amplified ionization of the copolymer in PBS vis-a-vis' water so the CMC remained constant [93]. It was observed that loading of 10% paclitaxel into Me-PEO-PDLLA micelles showed no effect on the value of CMC [100]. The fluorescent probes have been maximally utilized for the determination of CMC [103 - 106]. CMC estimation can be attested using UV spectroscopy method [107]. Other processes, such as SAXS (Small Angle X-ray Scattering) or SLS/DLS (Static and Dynamic Light Scattering), which are widely utilised for the surfactants, can principally be used for polymeric micelles. Nevertheless, they have their usage only in limited number of applications owing to low signal intensity [108].

Stability Studies of Polymeric Micelles

The degree of disassembly of micelles into individual polymers and their affinity towards the plasma proteins is as crucial as that of the size of the micelle and CMC. The integrity of the polymeric micelle should be maintained to allow drug reach the action site after the administration of micelles. For example- stability of polymeric micelles of doxorubicin in *in vitro* was interrelated to the *in vivo* antitumor activity against murine colon adenocarcinoma [109]. Lipid moieties can enhance the stability of micelles as these two fatty acid acyls in the micelle enhance the interaction with the polymeric chains inside the micelle. Certainly, upon dilution of diacyllipid-PEO conjugates, the polymeric chains showed no dissociation in the chromatogram [110]. Polymeric micelles are not excreted by kidney as they are in micelle form, once they reached the target site, they dissociate into individual polymers and get excreted from the kidney (Fig. **1**). Furthermore, the rate of disassembly is also crucial for rate of release of drug because the drugs should demonstrate instant release upon disassembly of micellar structure. Many factors are unidentified regarding the polymericmicelle's *in vivo* disassembly. For instance, impact of disassembly on pharmacokinetic behavior of polymeric micelles, rigidity, hydrophobicity and ratio of hydrophobic/hydrophilic chain length are needed to be addressed during development of polymeric micelles [111].

Synthesis of Drug Conjugated Polymeric Micelles

For the fabrication of prodrug, natural and synthetic polymers are being used. For prodrug conjugation natural biopolymers like dextran and chitosan have been used. Dextran is a polysaccharide with simple sugars as its monomeric units. The monomeric units are linked through α-1,6 linkages, with hydroxylatedcyclohexyl unit as well as normally formed *via* enzymes from definite strains of Streptococcus or Leuconostoc. However, dextran can be customized to conjugate with drug or protein with its primary or secondary hydroxyl groups using a spacer arm. Few reports have elucidated the potential application of dextran *via* oral route [112] in colon targeted delivery. Studies on human and animal illustrated that distribution and elimination of dextran are based on the net charge and the M_w. Dextran is being explored for the delivery of antitumor agents as a polymeric vesicle [113]. Dextrans were conjugated with corticosteroids were evaluated [114]. Pharmacokinetics of T-0128, a camptothecin analogue-carboxymethyl dextran conjugate, was tested in rats [115]. Drugs in conjugation with dextran was reported to reduce the toxicity and prolong the drug effect, moreover few studies have also been carried out in humans [116]. The polymeric dextran backbone provides the multiple -OH groups as possible functional sites for the conjugation of drug. For intratumoral delivery [117] *via* an acid-labile linkage, Dextran with M_w of 70,000 Da was conjugated to doxorubicin. In this way, dextran has been proved as a suitable carrier for its biodegradation and biocompatible properties [118].

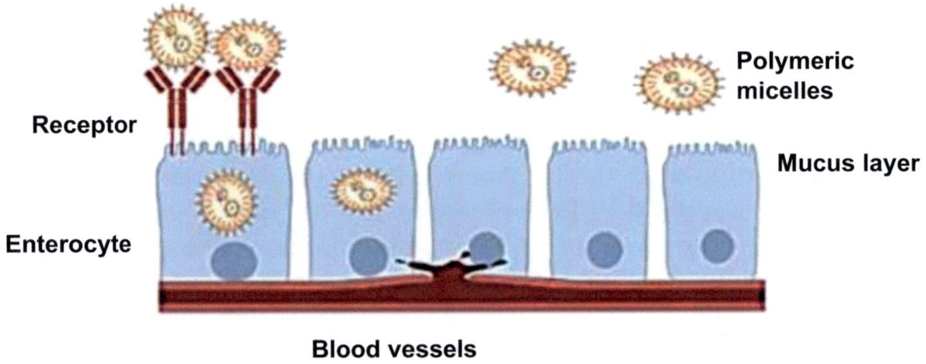

Fig. (1). Schematic representation of receptor mediated uptake of polymeric nanomicelles.

Drug Incorporation in Polymeric Micelles

One significant feature of micelles is the control of drug loading in the micelle core. This amount controls the stability of micelles and determines the targeting ability at a particular site [119, 120]. The extended circulation time in the body requires a high stability of micelles with their solid-state inner cores. In case of polymeric micelles, the drug cannot be incorporated by simple mixing to the inner core or the drug in an aqueous medium. Dialysis and emulsion process are required for the loading of drug in to the solid inner core of the micelle. For efficient targeting, it is very important to analyze this incorporation process to obtain the stable drug incorporation. Polymeric micelles having hydrophobic solid-state inner core have no capability to incorporate drugs, the alternatively hydrophobic block poly(D,L-lactide), in a polymeric micelle system can incorporate drugs due to the solid state inner core having glass transition temperature of about 54°C [121]. Hence, dialysis method is used for loading of drugs in solid state inner core. In the dialysis process, an organic solvent dissolves the drug and the block copolymer forms micelles during dialysis as the water content increases. It was proves that the quantity of incorporated adriamycin was effectively controlled through the feed adriamycin and triethanolamine.

Drug Solubilization Capacity of the Polymeric Micelles

Generally, in a micellar system, the amount of solubilised drug increases with rise in the temperature [122]. Sulfonyl urea which is a second generation drug is used in non-insulin dependent diabetes mellitus. Determination of solubility of drug with respect to surfactant concentration found to be different at different temperature from 25 to 37°C and the drug solubility with the ionic surfactant was found to be high at 37°C as compared to 25°C. This may be due to the thermal agitation which increased the micellar solubilization. For instance, the solubility of gliclazide was improved with ionic surfactant and decreased with non-ionic surfactants. In another case, micelles/water molar partition coefficient was decreased for the non-polar steroidal drugs in PEO surfactants solution when the temperature rose from 10-50 °C [123]. The amount of drug found in the PEO micelle inner core is anticipated to rise as the temperature is increased owing to growth of micelles [124]. One another strategy is to combine [125] micelles solubilization with other strategies. Recently, solubilization property of a surfactant was improved with ascorbic acid that not only improved solubility but also protects it from degradation. Example of the surfactant used is ascorbyl decanoate. Addition of surfactant in the hydrophobic micelles will enhance the hydrophobic drug solubility and protects it from degradation.

Drug partitioning in Polymeric Micelles

Drug in the core of the micelle is always remain in dynamic equilibrium between the hydrophobic core and the bulk of the solution called 'partitioning' If the rate of moment of drug from the bulk of solution to the micelle is equal to the moment from micelle to the bulk of the solution, then it is called equilibrium and the equilibrium constant called '*Partition Coefficient*' [126]. However, circumstance of equilibrium is rarely observed in the body. Partitioning coefficient is used to evaluate the amount of drug in micelles administered to patients and the drug release from the micelles into the body fluids. Various methods are used to determine the partition coefficient of micelles. One of the most handy methods is fluorescent technique that allows determination of partition coefficient [127, 128]. In the drugs which are non-fluorescent containing chromophor, partition coefficient is determined by the electronic spectra. Measuring the diffusion rate of drug by semi-permeable membrane through micelles is used to determine the partition coefficient [129, 130]. Partition coefficient determined by the diffusion-based technique and fluorescence are more reliable.

Drug Release from Polymeric Micelles

In most of the drug carrier systems, drug exhibits the pharmacological action after the drug release from the carrier. If drug release from the micelles is too slow then pharmacological actions are not well expressed. Incorporation of drug into polymeric micelles occurs in two ways; physical entrapment and chemical conjugation. In case of chemical conjugation [34, 84], the drug release rate is usually too slow. Hence, conjugation of hydrophobic drug to the polymeric system results in slow release of drug. This information indicates extra hydrophobic inner core environment which inhibits competently hydrolytic drug release reactions. The hydrophobic inner core mechanism reduced access of enzymes (hydrolytic enzyme) or reactants (*e.g.* hydroxyl ion), representing selection of a relatively labile bond (*e.g.* ester bond) which is suitable to conjugate with the drug. For instance, camptothecin derivative was incorporated in polymeric micelles through chemical conjugation [131], which is currently in clinical trials. This conjugation was fabricated with an ester bond between carboxyl group and OH group of a carrier polymer. Hence, a large number of polymeric micelles entrapping different categories of drugs were customized owing to lacking of conjugation chemistry between drug and polymer. Currently, sufficient novel tools and techniques are not available for obtaining a proper or ideal release rate from the polymeric system. Moreover, analysis of drug content, polymer composition and incorporating process for the adequate release rate were carried out only by hit and trial methods but not in a systematic manner. It is well known that the core inside the micelles should be in the solid form for drug integration. If liquid-state core is used for drug integration (*e.g.* by poloxamer, PEO-PPO-PEO block copolymer), moment of drug molecules from the hydrophobic inside core to the bulk of the solution will be very fast upon dilution of polymeric micelles and consequently instant drug release would be obtained. Therefore, solid state micelle core is essential for the proper release of drug and the hydrophobic drug can go into the blood in significant amount owing to binding with albumin. Yamamoto *et al.* reported [132] that a polymeric micelle in absence of hydrophobic and rigid inside core showed lack of stability. They recommended the polymeric chain configuration is also responsible for incorporating the stability to the micelle. More studies are warranted to evaluate the role of physicochemical parameters that influence the drug stability and optimization of drug release rate in a systematic manner. The drug release from polymeric micelles usually follows sustained and controlled drug release as well as pH sensitive drug release.

Sustained and Controlled Drug Release

The micelles can support a sustain release of drug when an amphiphilic polymer is produced by a combination of polymer and polyamino-acid. This type of micelles release the entrap drug in a molecularly dispersed form inside the upper GIT (Gastrointestinal tract) area where absorption is maximum. This drug release pattern in the GIT can be designed which increases drug absorption or oral bioavailability. For example-In PEG-polyglutamic acid co-polymer system, drug release was analyzed up to 150 h and only 60% drug release was observed during 150 h. Mice were injected with micelles and compared with that of plain drug. After 25 h, about 10% of the drug reached plasma from the micelles whereas with the plain drug it is about 0.1%. This enhance in plasma concentration is credited to decreased clearance of micelles as compared to free drug. This corresponds to buildup of the drug in kidney, spleen and liver is less as compared to that in the tumor tissue [133]. Amphiphilic block copolymers are more efficient carriers as they remain stable after dilution and they potentially extend the delivery of drug [134]. Amphiphilic copolymers like PLGA-PEG H40 (Dendritic boltron H40 having 40 hydroxyl groups]-PLA-mPEG (dendritic boltron H40 core-poly-lactic acid-methoxy polyethylene glycol) contains CMC of less than 100 μg/ml. However, at larger concentration, dilution of the micelle is possible without any segregation. With such polymers, the development of sustained release micelles is possible having capability of drug release for more than 2-3 months. Sustained delivery of drug is necessary for chronic therapy. Sustained release of drug is also predicted to be effective in treating communicable and non-communicable diseases. Present strategies to attain sustain release of drug with respect to micelles include use of new polymers, synthesis of pro-drug and assemble of micelles as layer by layer. The disadvantages associated with this system are sustained release only for shorter period, instability over longer duration, inadequate fitness for hydrophilic drugs and optimization of the micelles with each drug [135]. In addition to the amphiphilic copolymers and pharmaceutical surfactants, lipids, oligopeptides and polysaccharides can also be used to fabricate the polymeric micelles along with the polymer especially to minimize the dissociation of micelle upon dilution and to achieve prolong release of drug [134]. Micelles can be stored as solid particles by forming the reverse micelle where the particles will regain as micelles in *in vivo*. Other applications include preparing the pH responsive micelles and enhancing the solubility of hydrophobic drugs and they need not sustain the drug release. The smaller size of the micelles enables targeting of drug efficiently as compared with the conventional particulate systems. However, sufficient development is to be made to realize the sustain delivery as compared to the large particulate delivery systems and implants. Multilayered micelles can attain superior sustained delivery when compared to other techniques. Micelles are layered one above the other with the aid of H-bond formation between the H-bond acceptor and donor in the micelle shell. The H-bonding can be customized to break under preferred environment for the drug to release from the micelles [135].

pH Sensitive Drug Release

A variety of pH gradient exists throughout the human GIT. The pH varies to a great extent in the stomach from acidic (1.0-2.5) to pH of 5.1-7.8 in the small intestine [136]. Stomach has a mean pH value of 1.1 ± 0.15 in fasted healthy subjects and 3.6 ± 0.4 in fed state. The acidic nature of stomach is a result of gastric acid secretion [137]. Hydrochloric acid is the main component in the gastric acid is responsible for the pH of 1-2 and the acid is maintained by the H^{+}/K^{+} ATPase proton pump [138]. The cell lining of duodenum produces large amount of bicarbonate which neutralizes the gastric acid and hence the pH is higher in the small intestine. The pH of 5.8 ± 0.3 is observed in proximal duodenum in healthy individuals and rises to 6.6 ± 0.5 in jejunum and 7.5 ± 0.5 in the terminal ileum. However, a drop in pH in the proximal colon to 6.4 ± 0.6 and a rises in pH to 7.0 ± 0.7 in the distal colon was observed [139 - 141]. The pH of GIT is affected by many factors such as pathological condition, age, food and gender. So the use pH-sensitive polymeric micelles in the oral drug delivery is also being investigated apart from intravenous (*i.v*) administration [142]. Oral delivery of drug is the major route of administration to patients owing to its compliance, convenience, simplicity, convenience, patient acceptance and cost-effectiveness particularly in the treatment of chronic disease. pH-responsive polymeric micelles have advantages in the oral delivery of drug where the drug is encapsulated in the core of micelle reducing its exposure to harsh acidic conditions and pre-systemic enzymatic degradation. Moreover, pH-sensitive polymeric micelles minimizes the precipitation upon dilution in stomach and also reduces initial burst release. While in small intestinal pH (pH>5.0), micelles releases the drug from the micelle in molecular form in the upper GIT region where the maximum absorption occurs. Such a design of drug delivery in the GIT increases drug absorption and oral bioavailability [143].

Pharmaceutical Applications of Polymeric Micelles

Polymeric micelles are primarily fabricated to augment the performance of drug by enhancing the solubility of hydrophobic drug, targeting the drug to certain cells or organs and sustaining the release of drug.

Polymeric Micelles as Solubilizing Agents for Poorly Soluble Drugs

Currently, new chemical entities synthesized in the drug discovery process was found that around 90% are poorly soluble in water and belongs to BCS

(Biopharmaceutical Classification System) class II or IV category [144, 145]. Synthesis of these drugs has great implication in the pharmaceutical research but with the poor solubility, these compounds cannot show their effectiveness. Because of the poor solubility and low bioavailability, these compounds remain in pipelines in spite of having great therapeutic efficacy. So to improve the solubility of these compounds, many strategies are being used in the pharmaceutical development but are often not effective and may increase the complications linked with the administration of drug such as embolisms due to drug segregation [146]. Current studies are looking towards the polymeric micelles as they enhance the solubility, enhance dissolution of drug and thereby enhance bioavailability. The outcomes displayed that micelles favorably enhanced the solubility of drug by many folds of approximately 10-8400 [147, 148] by providing a hydrophobic reservoir to the drug. Drugs in the micelle are trapped either by chemical conjugation or by physical means or by ionic interaction based on the drug nature and the properties of polymer [149]. One of the example of polymeric micelle formulations is paclitaxel micelle with poly (d,l-lactide)-mPEG-diblock copolymer which enhanced the solubility about 5000-folds [150].

Sustained Release Profile of Polymeric Micelles

Polymeric micelles in comparison to conventional surfactants, forms more stable micelles, showed slow dissociation, exists low CMC values, holds drugs for longer duration and ultimately, obtain a higher gathering of the drug at the site of target [151]. The capability of micelles to sustain the drug release needs definite properties such as high thermodynamic stability (low CMC) to the dilution of micelles, low mobility of chains in the core and kinetic stability based on high viscous. Nevertheless, drugs enclosed in micelles have low diffusion coefficients which is suitable to sustain the release of drug [152]. Loading of drug can be increased by varying the chemical structure of the hydrophobic block which increases the hydrophobicity and firmness of the micellar core and prevents the wastage of drug. In clinical trial of NK105, micelles of paclitaxel, the core producing block has been partly altered by esterification with 4-phenyl-1-butanol and the enclosed drug displayed larger AUC_{last} than paclitaxel alone when given to cancerous mice [153]. This 25-fold increase in AUC_{last} in comparison to plain paclitaxel also showed that polymeric micelle carriers permit efficient sustained release of drug to the tumor [153]. Along with increasing the solubility of poorly soluble drugs and augmenting its release profile, polymeric micelles improve the drug bioavailability by various mechanisms mainly by shielding the drug from the acidic conditions throughout the GIT and easing the transport as well as pH-responsive micelles regulates the release of drug at the target site. Furthermore, mucoadhesive properties of micelles extend the residence period of drug in GIT. Micelles augment absorption of drug from the mucosa by enhancing the permeability of the membrane or by preventing efflux of drug in the mucosa (P-gp inhibitors). Pluronic-PAA (poly(acrylic acid) micellar formulations have been utilized for fabricating several drug delivery systems by oral route as the exclusive structure of provides all the elements for efficient delivery of drug [154, 155].

Polymeric Micelles Drug Steering Application

Drug targeting through polymeric micelles can be done by the enhanced permeability and retention (EPR) effect and by tagging the specific ligand molecule to the surface of the micelle, or coupling the monoclonal antibody results in active targeting using immunomicelles, stimuli-sensitivity are few strategic inventions [156].

Enhanced Permeation and Retention Effect

Due to nanoscopic size of polymeric micelle, they gather between the interstitial spaces by leasing out blood from capillaries (especially of solid tumors) and also get transported to various systems that don't have an effective barrier functions like cytoplasmic organelles, intact tissues, inflammatory sites and infected areas [157]. Reduced side-effects of the drugs can detected as the polymeric micelles are not able to surpass the walls of the normal blood vessels. There is no well-developed lymphatic drainage mechanism of micelles which causes improvised holding of micelles inside the tumor cells as micelles are not effectively cleared and cause tumor neovasculature. This characteristic of polymeric micelles permits the extended circulation time upon administration [158]. With these characteristic features, it is likely to obtain passive targeting of drug by utilizing polymeric micelles. The enhanced permeability of tumors which is called EPR effect and the greater secretion of vascular permeability factors induce the micelle extravasations within the tumor tissue. Generally secreted vascular factors are vascular endothelial growth factor nitric oxide, bradykinins, prostaglandins, collagenase enzyme and peroxynitrite [159]. Vetvicka *et al.,* fabricated micelles to increase circulation period and to increase the effectiveness of the EPR effect [160]. Poly (ethylene oxide)-block-poly(allyl glycidyl ether)-doxorubicin micelles were prepared which had longer circulation time and delivered the drug effectively to the cancerous tissue (acidic pH). Maitaniet and his group formulated polymeric micelles of camptothecin with poly(ethylene glycol)-poly(aspartate ester) block copolymers [161]. The micellar system enhanced the water solubility and stability of camptothecin. The stability of miceller system depends on amount of benzyl esters and polymeric chain of PEG. By EPR effect the micelles were effectively transported to tumor site.

Stimuli Sensitive Micelles

For effective drug targeting the drug should not be released in circulation by micelle. The micelles should be accumulated in the target tissue with internal (pH, particular enzyme, *etc.*) and external causes (temperature, light, ultrasound or magnetic field) and then the drug should be released. Numerous stimuli can cause response in the micelle which includes disturbance in the structure, variation in the volume, shape hydration state, conformation of chains, permeation rate, swelling/collapsing and hydrophilic/hydrophobic surface. Response of micelles to the stimuli can be either by physiological or by external cause is stated as 'Environmental sensitivity' or 'Stimuli-sensitivity' [162]. Release of drug from these micelle systems are known as 'Smart Delivery' or 'Intelligent Delivery'.

Acid-sensitive Polymeric Micelles

pH of the biological fluids varies with normal and pathological conditions. So this variation of pH with disease condition can be utilized for the targeted drug delivery with the pH-sensitive or acid-sensitive polymeric micelles. In inflammatory and tumor tissues, a slightly acidic pH is noticed (pH approximately 5.5-6.5) due to deposition of lactic acid. It is a slightly more acidic in comparison to the pH of normal tissues and blood which is approximately 7.4. Micelles also utilize the endocytosis to enter the cells and may associate with cell organelles such as lysosomes, endosomes, *etc.* The pH of these organelles is around 5.5 [163]. This laid the foundation in the advancement of pH-responsive polymeric micelles. *e.g.*, oligo/poly(nucleic acids) which are negatively charged entered into the cells by forming complex with cationic polymers. Once they transferred into endosomes, deprotonation occurs disturbing the membrane of endosomes releasing the nucleic acid in the cytosol [164]. Addition of titrable group and a linkage susceptible to acidic pH are the two advancements utilized for preparing the pH sensitive systems. The titrable groups widely used are carboxylic acid, amine *etc.* which cause variation in the solubility of polymer upon protonation. This protonation disturbs the structure micelle. The integrity of the micelle may be disturbed or the micelle structure may demolished by these acid-liable linkages like benzoic imine when exposed to acidic environment [165].

Thermosensitive Polymeric Micelles

The thermosensitive micelles experience a change in structure due to release of drug with rise in temperature. Thermosensitive polymers show an abrupt variation in the solution state and volume phase transition at a firm temperature known as critical solution temperature (CST). Temperature above which the polymer is completely soluble is called Upper critical solution temperature (UCST) and the temperature below which the polymer is insoluble is called lower critical solution temperature (LCST). So, LCST is the key factor in this strategy [166]. Temperature variations can be internal (raise in temperature at inflammatory site) or can be external. By applying ultrasound or by local application of high frequency resulting in the oscillation of magneto-sensitive micelles accumulated at the target site produces heat inside target tissues. Liu *et al.* established the utilisation of poly(Nisopropylacrylamide-co-acrylamide)-b-poly(D,L-lactide) copolymer with docetaxel for targeting cancer [167]. They found that raise in temperature largely improved effective targeting micelles and also aided in reducing the drug toxicity.

Targeting the Micelles with Ligand

A remarkable approach to deliver the polymeric micelles to the target tissue is tagging the cell-specific ligand to the micelle (Fig. **1**). Thus, association of cell specific ligands, *e.g*., peptides, monoclonal antibodies and sugars to the micelles has been demonstrated to improve the delivery of drug to the intended tissue. In case of targeting tumor, tumor-specific peptides are selected as ligands to construct the micelleswhich improves the tumor specificity and *in vivo* stability. In a study reported by Lavasanifar *et al.* arginine-glycine-aspartic acid (RGD) peptide was complexed with poly(ethylene oxide)-block-poly(caprolactone) micelles which identified the bound molecules over-expressed on metastatic cancer cells [168]. Hence, micelles were found to be suitable for delivery of drug to the targeted metastatic tumor cells with the aid of ligand tagging. Torchilin*et et al.* [166] utilised the over-expression of peripheral benzodiazepine receptor (PBR) for targeting cancerous cells [169]. Selective ligands of PBR enable the effective seizure of cell cycle. Therefore, paclitaxel loaded polyethylene glycol-phosphatidylethanolamine PBR-targeted micelles was developed and established its superiority in targeting the cancer cells.

Active Targeting with the Help of Immunomicelles

Association of antibodies to micelle surface offers wide range of opportunities for active targeting. Numerous studies have been performed tagging the antibody covalently to the micelle 'immunomicelles' [166]. To validate the efficiency of utilising immunomicelles in targeting tumor, Torchillin *et al.* [166] prepared micelles of camptothecin and paclitaxel with polymers polyethylene glycol-phosphatidyl ethanolamine and vitamin E. Furthermore, antinucleosome monoclonal antibody 2C5 (mAb 2C5) was tagged to the prepared polymeric micelle where the antibody directs toward the cancer cells. These immunomicelles showed high cytotoxicity in *in vitro* studies of various tumor cell lines [170].

Application of Polymeric Micelles in Drug Delivery Through Different Routes of Administration

Polymeric micelles have gained much interest in multifunctional nanotechnology-based delivery system. Due to this, the preclinical and clinical merits of polymeric micelles as delivery vehicles have been demonstrated through different routes of administration like oral drug delivery (Table **1**), ocular drug delivery (Table **2**), colon drug delivery (Table **3**), pulmonary drug delivery (Table **4** and Fig. **1**), topical and transdermal drug delivery (Table **5**), vaginal drug delivery and gene delivery (Table **6**) as well as vaccine and peptide drug delivery (Table **7**) systems.

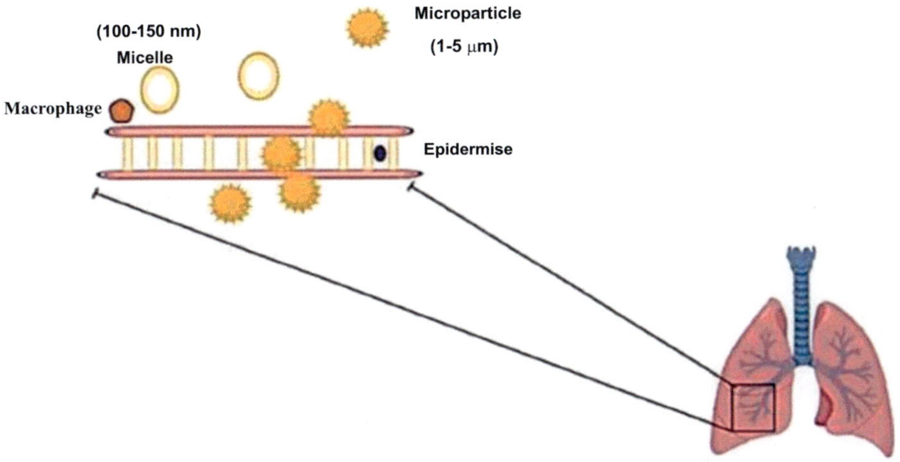

Fig. (2). Schematic representation of pulmonary deposition of polymeric nanomicelles.

Table 1. Nanomicelles administered through oral route.

Drug	Polymer	Size	Remarks	Reference
Scopoletin	Soluplus	59.4 nm	Sco micelles improved 4.38-and 8.43-fold, area under the curvein comparison to free scopoletin	[171]
Amphotericin B	Lecithin	187.20 nm	Ambicelles improved the bioavailability when administered iv and orally to 2.18- and 1.50-folds respectively and improved solubility to 5 mg/ml from 0.001 mg/ml	[172]

(Table 1) cont.....

Drug	Polymer	Size	Remarks	Reference
Silybin	Hyaluronic acid-deoxyoxycholic acid	130 nm	Silybin loaded micelles obtainedsignificantly higher AUC andCmax values	[173]
Insulin	Dodecylamine, Polyglutamic Acid, trimethylchitosan	106.8 nm	*in vivo* studies performed on ratselicited a prolonged hypoglycemicresponse and a relative bioavailabilityof 7.05%.	[174]
Probucol	MethoxyPEG-distearylphosphatidylethanolamine	200 nm	Micellar formulation significantlyincreased the uptake and transportof probucol	[175]
Morin	Pluronic f127	91.97 nm	Absolute bioavailability of morinincreased from 0.4% to 11.2%	[176]
Berberine	Soy phosphatidylcholine	36 nm	Oral bioavailabilty of BER loadedreverse micelles was increased up.4 folds & maximum bloodconcentration increased 2.1-folds	[177]
Valspodar	EpsilonCaprolactone,Ethylene oxide	62 nm	Micelles provided higher (77%)plasma AUC	[178]
Paclitaxel	Poly(ethyleneglycol)(PEG)	105 nm	Micelles resulted in efficientpaclitaxel solubilization &stabilization	[179]

Table 2. Nanomicelles in ocular drug delivery.

Drug	Polymer	Size	Remarks	Reference
Dexamethsone	e-caprolactone mPEG	25-30 nm	Nanomicelles increased permeability of dexamethasone by 2 folds	[180]
Itraconazole	Pluronics	79.99 nm	Micelles showed improvedtranscorneal permeationwhen compared with eyedrop and pure drug suspension	[181]

(Table 2) cont.....

Drug	Polymer	Size	Remarks	Reference
Gatifloxacin	Pluronics F127	176 nm	Micellar formulation increased ocular availability with no corneal damage. Solubility increased to18.67-folds	[182]
Curcumin	PluronicP123,TPGS	10.83 nm	*Ex-vivo* studies showed that the cumulative drug permeation increased 1.32 folds than the curcumin solution	[183]
Triamcinoacetonide	PNIPAAm	217.5 nm	Drug loaded micelles significantly increased bioavailabilty of corticosteroid triamcinoloneacetonide	[184]
Cyclosporin A	PEG, Sympatens AS, Sympatens ACS	9-12 nm	In porcine models exceptionally high concentrations of cyclosporine A were obtained with micellar CsA solution	[185]
Metipranolol	Pluronic-Chitosan	123-232 nm	*In vivo* studies indicated pluronicmicelles have sustained releasebehaviour and good pharmacological response	[186]
Diclofenac	MPEG-PCL	100 nm	diclofenac loaded MPEG-PCL micelles showed 2 fold increase in AUC than diclofenac PBS solutioneye drops	[187]
Ethoxzolamide	poloxamines	17-120 nm	Micelles induced 50-fold increasein apparent solubility	[188]
Netilmicin Sulphate	PHEA-PEG-C_{16}	19-23 nm	Micelles loaded with drug showed greater bioavailability than water suspension of same drug	[189]

Table 3. Nanomicelles in colon cancer drug delivery.

Drug	Polymer	Size	Remarks	Reference
Paclitaxel	SMA	195 nm	30 mg/kg and 60 mg/kg dose of SMA-paclitaxel micelles reduced tumor weight by 54% and 69%, respectively	[190]
Dimethyl Fumarate	sPCEG-azo	153.6 nm	Drug loaded micelles showed accumulative release of 60% in 2h & of 100% in 10 h	[191]
Fluorouracil	Pluronic P85	10-100 nm	Drug loaded micelles showed strong pesticide effect at lower doses and reduced the migration and invasion of cancer cells	[192]
Curcumin	PEGMEMA	34 & 80 nm	Loaded micelles showed enhanced stability and were rapidly internalized and exocytosed	[193]
Quercetin	MPEG-PCL	34.8 nm	Enhanced anti-colon cancer effectwas obtained in quercetin loaded micelles	[194]

(Table 3) cont.....

Drug	Polymer	Size	Remarks	Reference
Fisetin	MPEG-PCL	22±3 nm	MPEG was more competent in reducing the rate of tumor growth compared to plain drug	[195]
Doxorubicin	MPEG-PCL	25 nm	Micelles showed enhanced anticolon cancer activity	[196]

Table 4. Nanomicelles in pulmonary drug delivery.

Drug	Polymer	Size	Remarks	Reference
Celastrol	PEG	121.53 nm	The celastrol micelles had long elimination life & area under curve increased to 2.4 folds	[197]
Curcumin	MPEG-PLA	34.9 nm	Micelles showed constant release of the drug and therefore its cellular uptake was increased	[198]
Baohuoside I	DDAB,DTBM	22-30 nm	Micelles improved cellular uptake and had a greater hypersensitizing effect on cells	[199]
Ginsenoside Compound K	AP/TPGS	30.28 nm	The solubility of compound wasfrom 35.2±4.2 to 1463.2±153.3μg/ml	[200]
OleanolicAcid	PluronicP105	95.7 nm	*In vivo* studies showed that micellescould accumulate in the tumors	[201]
Acetylthevetin B	Chitosan-Pluronic P123	40-50 nm	15.31 fold& 9.49 fold accumulation of drug in tumors and lungs increased	[202]
Cyclosporin A	poly(MPC-co-BMA)	<0.5 nm	The micellizing drug solid dispersion of the drug showed better dissolution behaviour than amorphousform	[203]
Niclosamide	PluronicP123/F127	<100 nm	These micelles showed high cytotoxicity for cancer cells at lower concentrations than the free drug	[204]
Cabazitaxel	MPEG-E-CL	28.77 nm	The anti-tumor activity of cabazitaxel loaded polymericmicelles was 50% more	[205]
7-ethyl-14Aminocamptothecin	MPEG-PLA	20.7 nm	The micelles increased the solubility 800 folds than the free drug	[206]

(Table 4) cont.....

Drug	Polymer	Size	Remarks	Reference
Fasudil	DSPE-PEG	14 nm	The cellular uptake was increased to 1.7 folds and half life increased to 5 folds	[207]

Table 5. Nanomicelles in topical and transdermal delivery.

Drug	Polymer	Size	Remarks	Reference
Cyclosporin A	MPEG-dihexPLA	25-52 nm	The micelles enhanced the solubility of Cyclosporin by 518 times	[208]
Clotrimazole, Fluconazole, EconazoleNitrate	MPEG-hexPLA	70-165 nm	*In vivo* studies showed13 folds higher drug deposition	[209]
Ketoconazole	MPEG-E-PCL	58.66 nm	*In vivo* studies showed higherrates of skin deposition of ketoconazole than the marketed formulation	[210]
Lidocaine, Prilocaine	PluronicF127	32.5 nm	*In vivo* studies showed enhancedskin permeability of the anaesthetics	[211]
Resolvin	HydrogenatedCastor oil, Octoxynol-40	≤12 nm	Micellar formulation was more effective and showed more patient compliance	[212]
Glabridin	PMCP	188±31.7 nm	The absorption of glabrid in was approximately 4 times higher than the normal oil in water micelles	[213]
Arbutin	Urah Micellar System	90 nm	Results showed that micellar arbutincream increased both delivery andcellular melanin suppression	[214]
Podophyllotoxin	Glycyrrhizicacid	10 nm	Micelles achieved a higher drug distribution level and as well aslighter skin inflammation	[215]
Tacrolimus	MPEG-dihexPLA	10-50 nm	Studies showed that that there was improved cutaneous drug levelsdue to improved drug levels instratum corneum for psoriasistreatment	[216]
Sumatriptan	PluronicF127	10-100 nm	Micelles increased the stability ofthe preparation by increasing the viscosity and also increased the solubility (67.39 to 74.21)	[217]

Table 6A. Nanomicelles in vaginal drug delivery.

Drug	Polymer	Size	Remarks	Reference
Progesterone	Pluronic F127, F-68, Brij 35, Myrj 52	10-100 nm	The micelles increased theper-meability by 2 folds	[218]

Table 6B. Nanomicelles in gene delivery.

Gene	Polymer	Size	Remarks	Reference
miR-34a Modulator Rubone	PEG-PCD	-	Gene delivery by micelles along with a co drug wascarried out effectively, as the drug loading capacities andcontrolled drug release were enhanced	[219]
Genep53	Pullalan	160.8 nm	The micelles showed good drug loading capacity & the micelles also exhibited higher cytotoxicity and induced increased apoptosisin tumor cells	[220]
HIV-1gag	PSA	117.77 nm	Immune responses in BALB/c mice were noticeably induced on administration of micelles	[221]
P-gpsiRNA	PEG,PCL,PEI	141.1 nm	The obtained nanomicelles could prevent renal clearance, RNasedegradation & aggregation incirculation	[222]
iRGD	DSPE-PEG2000	10 nm	*In vivo* studies showed superior tumor penetration efficacy	[223]
Anti-VEGFsiRNA	MPEG-PCL	100-200 nm	Micelles containing anti-VEGF si RNA showed a high anti-tumor activity in tumor bearing mice	[224]
Anti-VEGF siRNA	PEI-SA	149.6 nm	The post transcriptional gene silencing was greatly improved by the polymeric micelles	[225]
Human adrenomedullin gene	PEG	5-100 nm	The micelles showed 100 fold increased gene expression inanimal studies	[226]

Table 7. Nanomicelles in vaccine and peptide delivery.

Vaccines	Polymer	Size	Remarks	Reference
Cancer Vaccines	PEG-PE	10-100 nm	The micelles mediated delivery of vaccines greatly enhances the vaccination process	[227]
Hepatitis BDNA vaccine	Man-CS-Phe	130 nm	The polymeric micelles protect the DNA from nuclear degradation & also induced multi-fold higher serum antibody titer	[228]
Curcumin conjugated vaccine	PEG	45.3 nm	This vaccine combination therapy boosted *in vivo* cytotoxic T-lymphocyte response and showed 7 fold increase in interferon production	[229]
OVA vaccine	PEG2000,PCL2000, PEI2000	50-80 nm	The micelles increased OVA antibody production by 190 folds	[230]
Group AS treptococcal vaccination	Peptideamphiphilemicelles	5-15 nm(diameter)25-125 nm(length)	The micelles induced a strong IgG1 antibody immune response in mice	[231]

(Table 7) cont.....

Vaccines	Polymer	Size	Remarks	Reference
Dendritic Cell vaccine	PEG-pGlu	70-80 nm	The micelles significantly enhanced the dendritic cell vaccine efficacy against tumors	[232]
Mucosal Vaccine	y-PGA	20 nm	The intranasal administration of micelle vaccine induced a high immune response both incellular and humoral immunity	[233]
Salmonella enterica serovar, Abortusovisvaccine	PCL or Gantrez	<300 nm(PCL)267 nm(Gantrez)	The micelle vaccine induced significant protection against	[234]
Thymopeptides	Sodiumdeoxycholate/phospholipids	149 nm	Nanomicelles demonstrated significant advantage in absorption characteristics as compared to other substances	[235]

Application of Polymeric Micelles in Diagnostics

Numerous micelles of contrasting agents are being generated for medical purpose (X-ray computed tomography (CT), magnetic resonance imaging (MRI), nuclear imaging and ultrasonography) which is used for both solely imaging/ diagnostic methods and for the virtual regulation delivery of drug by micelles [236]. Nanoparticles are far ahead of the small contrast molecular agents in terms of extended circulation period and target imaging and possibility of cell tracking. Molecular imaging agents and ligands of cancer specific can be linked to the nanoparticle which increases the bio-stability, bio-distribution and thus improves the contrast between the cancerous and healthy tissues [237]. Chelated paramagnetic metals such as manganese (Mn), iron oxide, dysprosium (Dy) and gadolinium (Gd) particles are used to fabricate the magnetic resonance contrast agents [236]. pH responsive micelles were fabricated as MRI contrast agents to recognize cancer. Amphiphilic block copolymers, comprising of methoxy poly(ethylene glycol)-b-poly(l-lactic acid)-diethylenetriaminopentaacetic acid and methoxy poly(ethylene glycol)-b-poly(l-histidine) (PEG-p(l-His)) and dianhydride-gadolinium chelate (PEG-p (l-LA)-DTPA-Gd) produced a micelle which displayed its effectiveness as a T1 MR contrast agent in the low pH cancer tissues. It was concluded that small tumors of ~3mm *in vivo* at 1.5T can be detected in a short time interval [238]. A recent study on the theranostic micelle was carried out which showed concurrent imaging of cancer cells with high contrast and superior photothermal therapy. The polymeric micelles showed good biodistribution and prolonged circulation holding the carbocyanine dyes in the tumor and gave near infrared fluorescence in both hyper and hypovascular cancerous cell for a period of up to 8 days. Furthermore, the micelles showed photothermal damage to tumor cells by destabilizing the organelles which resulted in positive tumor necrosis [239]. Torchilin *et al.* used iodine as CT contrasting agent in the PLL-PEG micelles with 34% w/w of iodine load and a size of 80 nm. Subsequent i.v. injection to rats resulted in sustained improvement of blood pool (aorta and heart), spleen and liver for a period of 3 h [240]. SPECT/ CT imaging procedure are also being used to examine the biodistribution and pharmacokinetics of DNA loaded in micelles [241].

Loading Polymeric Micelles with Diagnostic Agents

Therapeutics delivery by polymeric micelles has been well developed, however; delivery of contrast agents through micellar structure is a quite new field [242]. Apart from diagnosis, it also permits to capture the drug delivery site and the deposition inside the body with astonishing anatomical resolution. MR positive (T1) contrast agents are represented by the chelated paramagnetic metal moieties (Dy, GD, or Mn ions). Various pathways are used to produce contrast micelles and now mixed micelles were achieved from taurocholate and monoolein with Mn-mesoproporphyrin, which was found to be an effective contrasting agent for T1-weighted MRI to detect oral hepatobiliary disease [243]. Chelated metals generally are hydrophilic in nature. Diethylenetriamine pentaaceticacid (DTPA) complexed with Gd3, for instance, has a water molecule that can be linked to the sphere of metal coordination [244]. For incorporation into micelle, the contrasting agents should be amphiphilic in nature. Numerous chelating probes have been reported which includes (DTPA-PE) [245], DTPA-stearylamine, DTPA-SA [246], acylated to the chelating ions of Gd and Mn [247]. These agents are hydrophilic and are to be coupled with a hydrophobic lipid chain. The lipid chain is affixed to the micelle core which is hydrophobic or liposome bilayer while the chelate is attached to the shell of the micelle or to the liposome surface. Loading of chelates enables the examination of pharmacokinetic parameters and bio-distribution of drug.

The amphiphilic micelles PEG(5 kDa)-PE were loaded with chelating probes (paramagnetic Gd-DTPA-PE and radioactive ^{111}In-DTPASA) and utilised *in vivo* for scintigraphy and MR imaging. Particle size is the major character that enables the micelles of PEG lipid best suited for imaging applications for diagnostic purpose. Due to limitation in the lipid bilayer curvature, it is impossible to produce liposomes of smaller size than the usually size of 70-100 nm [248]. Diagnostic particulates of small size are needed for diagnostic applications. For instance, during percutaneous lymphography, uptake of polystyrene nanospheres was noticed in the primary lymph node prior to subcutaneous injection and the uptake was dramatically increased with reducing the size of the particles specifically when the size of the nanospheres was less than 100 nm [249]. In studies on lymphatic imaging [250], micelles of PEG(5 kDa)-PE with a size of 20 nm were loaded with amphiphilic chelating probes 111In-DTPA-SA and Gd(111In)-DTPA-PE and an attempt was made to utilise these contrasting agents in percutaneous lymphography by means of MR imaging and g-scintigraphy in rabbits. The popliteal lymph node was pictured after a few seconds of administration of these agents. A little massage at the site of injection can improves accumulation at lymph nodes and thrust the labeled material to the thoracic duct *via* lymphatic passage. It is obvious that micelles being in smaller size display greater gathering in primary lymph node. Massage at the site of injection eradicates accumulation at lymph node perhaps, because of dilation of lymph vessel. From the obtained data, it can be demonstrated that the micelles owing to their surface properties and size, they can be transported easily from the site of injection through the lymphatic system to the blood circulation along with the lymph. So it can be concluded that micelles are good at capturing the primary lymph nodes and are considered as lymphangiographic in nature. Their working principle is the conception of lymph flowing through various elements of the lymphatics. The working of other lymphotropic contrasting agents is principally uptake by the nodal macrophages [251].

Clinical Applications of Nanomicelles and Regulatory Approval

Due to erratic pharmacokinetic biodistribution elements, administration of therapeutic moieties is a challenge. The nanomicelles drug delivery system has gained success in preclinical to clinical studies of various drugs for regulatory approval. Recently next-generation of polymeric micelles has been developed for improved therapeutic index. A summary of drugs that gained clinical success is presented in Table **8**.

Table 8. Clinical applications of nanomicelles and regulatory approval.

Drug	Product Name/ Company	Composition	Size(nm)	Characterstics	Status	Therapeutic Indication	Reference
Paclitaxel	Genexol- PM®/ Samyang Biopharmac euticals	mPEG-b- PDLLA (~3750 Da)	20-50	This formulation permits administration of higher PTX doses than Taxol® without associated increase in toxicities (MTD of 390 mg/m2). In phase I studies,	Marketed in South Korea since 2007	Breast cancer NSCLC Ovarian cancer	[252, 253]
				it showed lower AUC and Cmax than conventional PTX and a shorter plasma half-life (11.0-12.7 h).			
Paclitaxel	Nanoxel®/ Dabur Pharma Ltd	PVP-b- PNIPAAM	80-100	It is activated by the low pH level of the tumor microenvironment: in acidic conditions, PNIPAAM is degraded and the release of the free drug occurs. It has linear pharmacokinetics and can be administered without premedication.	DCGI approval in 2008	Breast cancer, NSCLC And AIDS related KS	[254]
Paclitaxel	NK105 /NanoCarrie r™	PEGP(aspartate)	85 (20- 430)	It is administered as i.v. infusion for 1-h. At the recommended phase II dose (150mg/m^2), NK105 showed more than 15-fold higher plasma AUC and a 26- fold lower clearance than Taxol®.	Phase III	Stomach Cancer Breast Cancer	[255]
Paclitaxel	Paxceed®/ Angiotech Pharmaceuti cals Inc	mPEGPDLLA	N.S.	PTX was observed to suppress the cell growth and prevents functions of certain types associated with RA.	Phase II/comple Ted	Rheumatoid Arthritis	[256]

(Table 8) cont.....

Drug	Product Name/ Company	Composition	Size(nm)	Characterstics	Status	Therapeutic Indication	Reference
Paclitaxel	Paclical/ Oasmia Pharmaceutical AB	a CremophorEL-free formulation of paclitaxel	N.S.	Pcal recommended as iv infusion for 1-h without anti-allergic premedication. It comparable with Paclitaxel in efficacy and safety without standard use of premedication and a shorter infusion time	Phase III	Ovarian cancer	[257, 258]
SN-38 (active metabolite of irinotecan hydrochloride)	NK012	7-Ethyl-10-hydroxy-camptothecin (SN-38) releasing nanodevice	N.S.	Treatment against metastatic colorectal cancer was more efficient and was comparable with that of monotherapy reported previously. Side effects like myelo suppression and GI toxicity is mild and this formulation is much preferred.	Phase II	Colorectal cancer, lung and ovarian cancer	
Doxorubicin	SP1049C	Pluronic L61 and F127 polymeric micelle	N.S.	Improved circulation time of about 3.1 folds compared to plain drug and $t_{1/2}$ was found to be 2.4 ± 2.1 h for SP1049C whereas the plain drug showed 1.4 ± 0.6 h.	Phase II/III	Lung Cancer	[259]
Cisplatin	NC-6004	PEG-PGA polymeric micelle	N.S.	NC-6004 is a platinum complex with improved antitumor activity, reduced nephrotoxicity and lymphatic drug delivery	Phase II/III	Solid tumor, gastrointestinal and genitourinary cancer	[260]
Camptothecin analog	Lipotecan	Polymeric micelle	N.S.	Lipotecan administered doses up to 50 mg/m^2 is safe, tolerable and has demonstrated broad anticancer activity to support Phase 2 disease-specific investigations.	Phase I/II	Renal and liver cancer	[261]

(Table 8) cont.....

Drug	Product Name/ Company	Composition	Size(nm)	Characterstics	Status	Therapeutic Indication	Reference
Oxaliplatin	NC-4016	1,2-diaminocyclohexane platinum (II)-incorporating micelles (NC-4016)	N.S.	NC-4016 has anticancer activity superior to oxaliplatin	Phase I	Solid tumors	[262]
Epirubicin	NC-6300	PEG-*b*PAH polymeric micelle	40-80 nm	NC-6300 showed anticancer effect and decreased the cardiotoxicity of epirubicin	Phase I	Solid tumors	[263]
Doxorubicin	NK911	PEG-PAA polemric micelle	40 mm	Micelles formulation showed 3.5-folds increase in circulation time compared to plain drug, the elimination half life, $t_{1/2}$ (2.8 ± 0.3 h) was better than that of plain drug (1.4 ± 0.6 h).	Phase I	Solid tumors	[264]

N.S.: not specified.

CONCLUSION

In conclusion, polymeric micelles have emerged a vital pharmaceutical drug delivery system owing to its lucrative merits as compared to other nanocarriers. Moreover, synthesis of polymeric micelles is relatively simple and can be scaled up at industrial level. The interesting feature of polymeric micelles is that its ease of loading of lipophilic drugs and consequently improve their bioavailability. Importantly, polymeric nanomicelles can be effectively delivered through various routes of administration such as oral route, pulmonary route and vaginal route. Therefore, owing to these reasons polymeric micelles gained clinical success in a short period of time.

CONSENT FOR PUBLICATION

Not applicable.

CONFLICT OF INTEREST

The authors declare no conflict of interest, financial or otherwise.

ACKNOWLEDGEMENTS

Declared none.

REFERENCES

[1] Mourya VK, Nazma I, Nawale RB, Kulthe SS. Polymeric micelles: General considerations and their applications. Ind J Pharm Edu Res 2011; 45: 128-38.

[2] Trivedi R, Kompella UB. Nanomicellar formulations for sustained drug delivery: strategies and underlying principles. Nanomed (Lond) 2010; 5(3): 485-505. [http://dx.doi.org/10.2217/nnm.10.10] [PMID: 20394539]

[3] Velagaleti PR, Anglade E, Khan IJ, Gilger BC, Mitra AK. Topical delivery of hydrophobic drugs using a novel mixed nanomicellar technology to treat diseases of the anterior and posterior segments of the eye. Drug Del Technol 2010; 10(4): 42-7.

[4] Cholkar K, Patel A, Vadlapudi AD, Mitra AK. Novel nanomicellar formulation approaches for anterior and posterior segment ocular drug delivery. Recent Pat Nanomed 2012; 2(2): 82-95. [http://dx.doi.org/10.2174/1877912311202020082] [PMID: 25400717]

[5] Kataoka K, Harada A, Nagasaki Y. Block copolymer micelles for drug delivery: design, characterization and biological significance. Adv Drug Deliv Rev 2001; 47(1): 113-31. [http://dx.doi.org/10.1016/S0169-409X(00)00124-1] [PMID: 11251249]

[6] Zana R, Ed. Dynamics of surfactant self-assemblies micelles: microemulsions, vesicles, and lyotropic phases. Boca Raton: CRC Press 2005; pp. 2-16. [http://dx.doi.org/10.1201/9781420028225]

[7] Sinko P, Ed. Martin's physical pharmacy and pharmaceutical sciences. Maryland, USA: Lippincott Williams and Wilkins 2007; pp. 469-97.

[8] Adams ML, Lavasanifar A, Kwon GS. Amphiphilic block copolymers for drug delivery. J Pharm Sci 2003; 92(7): 1343-55. [http://dx.doi.org/10.1002/jps.10397] [PMID: 12820139]

[9] Moroi Y. Micelles: Theoretical and applied aspects. Springer international ed. New York: Springer 2005; pp. 41-50.

[10] Astafieva I, Zhong XF, Eisenberg A. Critical micellization phenomena in block polyelectrolyte solutions. Macromol 1993; 26: 7339-52. [http://dx.doi.org/10.1021/ma00078a034]

[11] Mukerjee P. Formation and some properties of micelles. Ber Bunsenges Phys Chem 1978; 82: 931-7. [http://dx.doi.org/10.1002/bbpc.19780820944]

[12] Hanafy NAN, El-Kemary M, Leporatti S. Micelles structure development as a strategy to improve smart cancer therapy. Cancers (Basel) 2018; 10(7): 238. [http://dx.doi.org/10.3390/cancers10070238] [PMID: 30037052]

[13] La SB, Okano T, Kataoka K. Preparation and characterization of the micelle-forming polymeric drug indomethacin-incorporated poly(ethylene oxide)-poly(beta-benzyl L-aspartate) block copolymer micelles. J Pharm Sci 1996; 85(1): 85-90. [http://dx.doi.org/10.1021/js950204r] [PMID: 8926590]

[14] Cammas S, Suzuki K, Sone C, Sakurai Y, Kataoka T, Okano T. Thermo-responsive polymer nanoparticles with a core-shell micelle structure as site-specific drug carriers. J Control Rel 1997; 48: 157-64. [http://dx.doi.org/10.1016/S0168-3659(97)00040-0]

[15] Prasad KN, Luong TT, Florence AT, *et al.* Surface activity and association of ABA polyoxyethylene-polyoxypropylene block copolymers in aqueous solution. J Colloid Interface Sci 1979; 69: 225-32. [http://dx.doi.org/10.1016/0021-9797(79)90151-6]

[16] Kabanov AV, Batrakova EV, Melik-Nubarov NS, *et al.* New class of drug carriers: micelles of poly(oxyethylene)-poly(oxypropylene) block copolymers as microcontainers for drug targeting form blood in brain. J Control Rel 1992; 22: 141-58.

[http://dx.doi.org/10.1016/0168-3659(92)90199-2]

[17] Turro NJ, Chung CJ. Photoluminescent probes for water-water soluble polymers, Pressure and temperature effect on a polyol surfactant. Macromol 1984; 17: 2123-6. [http://dx.doi.org/10.1021/ma00140a043]

[18] Winnik FM, Davidson AR, Hamer GK, Kitano H. Amphiphilic poly(N-isopropylacrylamides) prepared by using a lipophilic radical initiator: synthesis and solution properties in water. Macromol 1992; 25: 1876-80. [http://dx.doi.org/10.1021/ma00033a006]

[19] Chung JE, Yokoyama M, Aoyagi T, Sakurai Y, Okano T. Effect of molecular architecture of hydrophobically modified poly(N-isopropylacrylamide) on the formation of thermoresponsive core-shell micellar drug carriers. J Control Rel 1998; 53(1-3): 119-30. [http://dx.doi.org/10.1016/S0168-3659(97)00244-7] [PMID: 9741919]

[20] Osada K, Christie RJ, Kataoka K. Polymeric micelles from poly(ethylene glycol)- poly(amino acid) block copolymer for drug and gene delivery. J R Soc Interface 2009; 6(Suppl 3): S325-39.

[21] Liu YL, Lin GC, Wu CS. Preparation of polysulfone-gpoly(N-isopropylacrylamide) graft copolymers through atom transfer radical polymerization and formation of temperature-responsive nanoparticles. J Polym Sci A Polym Chem 2008; 46: 4756-65. [http://dx.doi.org/10.1002/pola.22809]

[22] Hickok RS, Wedge SA, Hansen LA, Morris KF, Billiot FH, Warner IM. Pulsed field gradient NMR investigation of solubilization equilibria in amino acid and dipeptide terminated micellar and polymeric surfactant solutions. Magn Reson Chem 2002; 40: 755-61. [http://dx.doi.org/10.1002/mrc.1099]

[23] Kulthe SS, Choudhari YM, Inamdar NN, Mourya V. Polymeric micelles: authoritative aspects for drug delivery. Des Mono Poly 2012; 15: 465-521. [http://dx.doi.org/10.1080/1385772X.2012.688328]

[24] Erhardt R, Boker A, Zettl H, *et al.* Janus micelles. Macromol 2001; 34: 1069-75. [http://dx.doi.org/10.1021/ma000670p]

[25] Riess G, Hurtrez G, Bahadur P. Block Copolymers. In: Mark HS, Kroschwitz JI, Eds. Encyclopedia of Polymer Science and Engineering. New York: Wiley press 1985; pp. 324-34.

[26] Varma MV, Panchagnula R. Enhanced oral paclitaxel absorption with vitamin E-TPGS: effect on solubility and permeability *in vitro, in situ* and *in vivo.* Eur J Pharm Sci 2005; 25(4-5): 445-53. [http://dx.doi.org/10.1016/j.ejps.2005.04.003] [PMID: 15890503]

[27] Bromberg L. Polymeric micelles in oral chemotherapy. J Control Rel 2008; 128(2): 99-112. [http://dx.doi.org/10.1016/j.jconrel.2008.01.018] [PMID: 18325619]

[28] Peng X, Zhang L. Formation and morphologies of novel self-assembled micelles from chitosan derivatives. Langmuir 2007; 23(21): 10493-8. [http://dx.doi.org/10.1021/la701325v] [PMID: 17877377]

[29] Mahmud A, Xiong XB, Aliabadi HM, Lavasanifar A. Polymeric micelles for drug targeting. J Drug Target 2007; 15(9): 553-84. [http://dx.doi.org/10.1080/10611860701538586] [PMID: 17968711]

[30] Papagiannaros A, Upponi J, Hartner W, Mongayt D, Levchenko T, Torchilin V. Quantum dot loaded immunomicelles for tumor imaging. BMC Med Imaging 2010; 10: 22. [http://dx.doi.org/10.1186/1471-2342-10-22] [PMID: 20955559]

[31] Klaikherd A, Nagamani C, Thayumanavan S. Multi-stimuli sensitive amphiphilic block copolymer assemblies. J Am Chem Soc 2009; 131(13): 4830-8. [http://dx.doi.org/10.1021/ja809475a] [PMID: 19290632]

[32] Allen C, Maysinger D, Eisenberg A. Nano-engineering block copolymer aggregates for drug delivery.

Colloids Surf B Biointerfaces 1999; 16: 3-27.
[http://dx.doi.org/10.1016/S0927-7765(99)00058-2]

[33] Martin AN. Colloids. In: Martin AN, Alfred N, Eds. Physical Pharmacy: Physical Chemical Principles in the Pharmaceutical Sciences. 4th ed. Baltimore: Williams and Wilkins press 1993; pp. 393-22.

[34] Jones M, Leroux J. Polymeric micelles - a new generation of colloidal drug carriers. Eur J Pharm Biopharm 1999; 48(2): 101-11.
[http://dx.doi.org/10.1016/S0939-6411(99)00039-9] [PMID: 10469928]

[35] Kwon GS, Kataoka K. Block copolymer micelles as long- circulating drug vehicles. Adv Drug Deliv Rev 1995; 16: 295-09.
[http://dx.doi.org/10.1016/0169-409X(95)00031-2]

[36] Kwon GS. Diblock copolymer nanoparticles for drug delivery. Crit Rev Ther Drug Carrier Syst 1998; 15(5): 481-512.
[http://dx.doi.org/10.1615/CritRevTherDrugCarrierSyst.v15.i5.20] [PMID: 9822868]

[37] Lavasanifar A, Samuel J, Kwon GS. Poly(ethylene oxide)-block-poly(L-amino acid) micelles for drug delivery. Adv Drug Deliv Rev 2002; 54(2): 169-90.
[http://dx.doi.org/10.1016/S0169-409X(02)00015-7] [PMID: 11897144]

[38] Vonarbourg A, Passirani C, Saulnier P, Benoit JP. Parameters influencing the stealthiness of colloidal drug delivery systems. Biomaterials 2006; 27(24): 4356-73.
[http://dx.doi.org/10.1016/j.biomaterials.2006.03.039] [PMID: 16650890]

[39] Maibaum L, Dinner AR, Chandler D. Micelle formation and the hydrophobic effect. J Phys Chem B 2004; 108: 6778-81.
[http://dx.doi.org/10.1021/jp037487t]

[40] Price C. Micelle formation by block copolymer in organic solvents. Pure Appl Chem 1983; 55: 1563-72.
[http://dx.doi.org/10.1351/pac198355101563]

[41] Hernandez JR, Checot F, Gnanou Y, Lecommandoux S. Toward smart nano-objects by self-assembly of block co-polymers in solution. Prog Polym Sci 2005; 30: 691-24.
[http://dx.doi.org/10.1016/j.progpolymsci.2005.04.002]

[42] Chu B, Zhou Z. Physical chemistry of poly(oxyalkylene) block copolymer surfactants. In: Nace VN, Ed. Non-ionic Surfactants: Polyoxyalkylene Block Copolymers. New York: Marcel Dekker press 1996; Vol. 60: pp. 68-144.

[43] Daoud M, Cotton JP. Star shaped polymers: a model for the conformation and its concentration dependence. J Phys (Paris) 1982; 43: 531-8.
[http://dx.doi.org/10.1051/jphys:01982004303053100]

[44] Halperin A. Polymeric micelles: a star model. Macromol 1987; 20(11): 2943-6.
[http://dx.doi.org/10.1021/ma00177a051]

[45] Shusharina NP, Nyrkova IA, Khoklov AR. Diblock copolymers with a charged block in a selective solvent. Micellar Struct Macromol 1996; 29: 3167-74.
[http://dx.doi.org/10.1021/ma9515745]

[46] Wu C, Gao J. A Simple scaling for the core–shell nanostructure formed by self-assembly of block copolymers in a selective solvent. Macromol 2000; 33: 645-6.
[http://dx.doi.org/10.1021/ma990288g]

[47] Shusharina NP, Alexandridis P, Linse P, Balijepalli S, Gruenbauer HJM. Phase behavior and structure of an ABC triblock copolymer dissolved in selective solvent. Eur Phys J E Soft Matter 2003; 10(1): 45-54.
[http://dx.doi.org/10.1140/epje/e2003-00001-6] [PMID: 15011078]

[48] Moffitt M, Khougaz K, Eisenberg A. Micellization of ionic block copolymers. Acc Chem Res 1996;

29: 95-102.
[http://dx.doi.org/10.1021/ar940080]

[49] Chen XL, Jenekhe SA. Supramolecular self-assembly of three-dimensional nanostructures and microstructures: Microcapsules from electroactive and photoactive rod-coil-rod triblock copolymers. Macromol 2000; 33: 4610-2.
[http://dx.doi.org/10.1021/ma000100p]

[50] Discher BM, Hammer DA, Bates FS, Discher DE. Polymer vesicles in various media. Curr Opin Colloid Interface Sci 2000; 5: 125-31.
[http://dx.doi.org/10.1016/S1359-0294(00)00045-5]

[51] Burke S, Shen H, Eisenberg A. Multiple vesicular morphologies from block copolymers in solution. Macromol Symp 2001; 175: 273.
[http://dx.doi.org/10.1002/1521-3900(200110)175:1<273::AID-MASY273>3.0.CO;2-P]

[52] Lee JCM, Bermudez H, Discher BM, *et al.* Preparation, stability, and *in vitro* performance of vesicles made with diblock copolymers. Biotechnol Bioeng 2001; 73(2): 135-45.
[http://dx.doi.org/10.1002/bit.1045] [PMID: 11255161]

[53] Stoenescu R, Meier W. Vesicles with asymmetric membranes from amphiphilic ABC triblock copolymers. Chem Commun (Camb) 2002; 24(24): 3016-7.
[http://dx.doi.org/10.1039/b209352a] [PMID: 12536794]

[54] Soo PL, Eisenberg A. Preparation of block copolymer vesicles in solution. J Polym Sci 2004; 42(Pt B): 923.

[55] Liu G, Qiao L, Guo A. Diblock copolymer nanofibers. Macromol 1996; 29: 5508.
[http://dx.doi.org/10.1021/ma9604653]

[56] Liu G. Nanofibers. Adv Nanofibers Mater 1997; 9: 437.
[http://dx.doi.org/10.1002/adma.19970090517]

[57] Yu KE, Eisenberg A. Bilayer morphologies of self-assembled crew-cut aggregates of amphiphilic PS-*b*-PEO diblock copolymers in solution. Macromol 1998; 31: 3509.
[http://dx.doi.org/10.1021/ma971419l]

[58] Stewart S, Liu G. Hollow nanospheres from polyisoprene-*block*-poly(2-cinnamoylethyl methacrylate)-*block*-poly(*tert*-butyl acrylate). Chem Mater 1999; 11: 1048.
[http://dx.doi.org/10.1021/cm981009r]

[59] Stewart S, Liu G. Block copolymer nanotubes. Angew Chem Int Ed Engl 2000; 39(2): 340-4.
[http://dx.doi.org/10.1002/(SICI)1521-3773(20000117)39:2<340::AID-ANIE340>3.0.CO;2-H]
[PMID: 10649402]

[60] Kwon GS, Okano T. Polymeric micelles as new drug carriers. Adv Drug Deliv Rev 1996; 21: 107-16.
[http://dx.doi.org/10.1016/S0169-409X(96)00401-2]

[61] Trubetskoy VS, Torchilin VP. Polyethylene glycol based micelles as carriers of therapeutic and diagnostic agents. S T P Pharma Sci 1996; 6: 79-86.
[http://dx.doi.org/10.1208/aapsj0902015]

[62] Shin IG, Kim SY, Lee YM, Cho CS, Sung YK. Methoxy poly(ethylene glycol)/epsilon-caprolactone amphiphilic block copolymeric micelle containing indomethacin. I. Preparation and characterization. J Control Rel 1998; 51(1): 1-11.
[http://dx.doi.org/10.1016/S0168-3659(97)00164-8] [PMID: 9685899]

[63] Yokoyama M, Kwon GS, Okano T, Sakurai Y, Seto T, Kataoka K. Preparation of micelle-forming polymer-drug conjugates. Bioconjug Chem 1992; 3(4): 295-301.
[http://dx.doi.org/10.1021/bc00016a007] [PMID: 1390984]

[64] La SB, Okano T, Kataoka K. Preparation and characterization of the micelle-forming polymeric drug indomethacin-incorporated poly(ethylene oxide)-poly(beta-benzyl L-aspartate) block copolymer

micelles. J Pharm Sci 1996; 85(1): 85-90.
[http://dx.doi.org/10.1021/js950204r] [PMID: 8926590]

[65] Kabanov AV, Batrakova EV, Melik-Nubarov NS, *et al.* New class of drug carriers: micelles of poly(oxyethylene)-poly(oxypropylene) block copolymers as microcontainersfor drug targeting from blood in brain. J Control Rel 1992; 22: 141-58.
[http://dx.doi.org/10.1016/0168-3659(92)90199-2]

[66] Kataoka K, Togawa H, Harada A, Yasugi K, Matsumoto T, Katayose S. Spontaneous formation of polyion complex micelles with narrow distribution from antisense oligonucleotide and cationic block copolymer in physiological saline. Macromol 1996; 29: 8556-7.
[http://dx.doi.org/10.1021/ma961217+]

[67] Nagasaki Y, Okada T, Scholz C, Iijima M, Kato M, Kataoka K. The reactive polymeric micelle based on an aldehyde-ended poly(- ethylene glycol)/poly(lactide) block copolymer. Macromol 1998; 31: 1473-9.
[http://dx.doi.org/10.1021/ma971294k]

[68] Yokoyama M, Miyauchi M, Yamada N, *et al.* Characterization and anticancer activity of the micelle-forming polymeric anticancer drug adriamycin-conjugated poly(ethylene glycol)-poly(aspartic acid) block copolymer. Cancer Res 1990; 50(6): 1693-700.
[PMID: 2306723]

[69] Baalousha M, Lead JR. Rationalizing nanomaterial sizes measured by atomic force microscopy, flow field-flow fractionation, and dynamic light scattering: sample preparation, polydispersity, and particle structure. Environ Sci Technol 2012; 46(11): 6134-42.
[http://dx.doi.org/10.1021/es301167x] [PMID: 22594655]

[70] Kohori F, Sakai K, Aoyagi T, Yokoyama M, Sakurai Y, Okano T. Preparation and characterization of thermally responsive block copolymer micelles comprising poly(N-isopropylacrylamide-b-DL-lactide). J Control Rel 1998; 55(1): 87-98.
[http://dx.doi.org/10.1016/S0168-3659(98)00023-6] [PMID: 9795019]

[71] Yu BG, Okano T, Kataoka K, Kwon G. Polymeric micelles for drug delivery: solubilization and haemolytic activity of amphotericin B. J Control Rel 1998; 53(1-3): 131-6.
[http://dx.doi.org/10.1016/S0168-3659(97)00245-9] [PMID: 9741920]

[72] Kim SY, Shin IG, Lee YM, Cho CS, Sung YK. Methoxy poly(ethylene glycol) and epsilon-caprolactone amphiphilic block copolymeric micelle containing indomethacin. II. Micelle formation and drug release behaviours. J Control Rel 1998; 51(1): 13-22.
[http://dx.doi.org/10.1016/S0168-3659(97)00124-7] [PMID: 9685900]

[73] AlleÂmann E, Leroux JC, Gurny R. Biodegradable nanoparticles of poly(lactic acid) and poly(lactic-co-glycolic acid) for parenteral administration. In: Lieberman H, Rieger M, Banker G, Eds. Pharmaceutical Dosage Forms: Disperse Systems. New York: Marcel Dekker press 1998; Vol. 3: pp. 163-93.

[74] Yokoyama M, Okano T, Sakurai Y, Kataoka K. Improved synthesis of adriamycin-conjugated poly(ethylene oxide)-poly(aspartic acid) block copolymer and formation of unimodal micellar structure with controlled amount of physically entrapped adriamycin. J Control Rel 1994; 32: 269-77.
[http://dx.doi.org/10.1016/0168-3659(94)90237-2]

[75] Hagan SA, Coombes GA, Garnett MC, *et al.* Polylactide- poly(ethylene glycol) copolymers as drug delivery systems. 1. Characterization of water dispersible micelle-forming systems. Langmuir 1996; 12: 2153-61.
[http://dx.doi.org/10.1021/la950649v]

[76] Ishida O, Maruyama K, Sasaki K, Iwatsuru M. Size-dependent extravasation and interstitial localization of polyethyleneglycol liposomes in solid tumor-bearing mice. Int J Pharm 1999; 190(1): 49-56.
[http://dx.doi.org/10.1016/S0378-5173(99)00256-2] [PMID: 10528096]

[77] Litzinger DC, Buiting AM, van Rooijen N, Huang L. Effect of liposome size on the circulation time and intraorgan distribution of amphipathic poly(ethylene glycol)-containing liposomes. Biochim Biophys Acta 1994; 1190(1): 99-107.
[http://dx.doi.org/10.1016/0005-2736(94)90038-8] [PMID: 8110825]

[78] Matsumura Y, Maeda H. A new concept for macromolecular therapeutics in cancer chemotherapy: mechanism of tumoritropic accumulation of proteins and the antitumor agent smancs. Cancer Res 1986; 46(12 Pt 1): 6387-92.
[PMID: 2946403]

[79] Maeda H. Macromolecular therapeutics in cancer treatment: the EPR effect and beyond. J Control Rel 2012; 164(2): 138-44.
[http://dx.doi.org/10.1016/j.jconrel.2012.04.038] [PMID: 22595146]

[80] Maeda H. The link between infection and cancer: tumor vasculature, free radicals, and drug delivery to tumors *via* the EPR effect. Cancer Sci 2013; 104(7): 779-89.
[http://dx.doi.org/10.1111/cas.12152] [PMID: 23495730]

[81] Maeda H, Seymour LW, Miyamoto Y. Conjugates of anticancer agents and polymers: advantages of macromolecular therapeutics *in vivo.* Bioconjug Chem 1992; 3(5): 351-62.
[http://dx.doi.org/10.1021/bc00017a001] [PMID: 1420435]

[82] Hamad E, Qutubuddin S. Theory of micelle formation by amphiphilic side-chain polymers. Macromolecules 1990; 23: 4185-91.
[http://dx.doi.org/10.1021/ma00221a001]

[83] Munch MR, Gast AP. Block copolymers at interfaces. 1. Micelle formation. Macromolecules 1988; 21: 1360-6.
[http://dx.doi.org/10.1021/ma00183a030]

[84] Xu R, Winnik MA, Riess G, *et al.* Micellization of polystyrenepoly(ethylene oxide) block copolymers in water. 5. A test of the star and mean-field models. Macromolecules 1992; 25: 644-52.
[http://dx.doi.org/10.1021/ma00028a025]

[85] Li Y, Kwon GS. Methotrexate esters of poly(ethylene oxide)-block-poly(2-hydroxyeth-l-L-aspartamide). Part I: Effects of the level of methotrexate conjugation on the stability of micelles and on drug release. Pharm Res 2000; 17(5): 607-11.
[http://dx.doi.org/10.1023/A:1007529218802] [PMID: 10888314]

[86] Yuan F, Dellian M, Fukumura D, *et al.* Vascular permeability in a human tumor xenograft: molecular size dependence and cutoff size. Cancer Res 1995; 55(17): 3752-6.
[PMID: 7641188]

[87] Tuzar Z, Kratochvil P. Block and graft copolymer micelles insolution. Adv Colloid Interface Sci 1976; 6: 201-32.
[http://dx.doi.org/10.1016/0001-8686(76)80009-7]

[88] Weber SE. Polymer micelles: an example of self-assembling polymers. J Phys Chem B 1998; 102: 2618-26.
[http://dx.doi.org/10.1021/jp980386o]

[89] Cabral H, Matsumoto Y, Mizuno K, *et al.* Accumulation of sub-100 nm polymeric micelles in poorly permeable tumours depends on size. Nat Nanotechnol 2011; 6(12): 815-23.
[http://dx.doi.org/10.1038/nnano.2011.166] [PMID: 22020122]

[90] Mittal KL, Ed. Micellization, Solubilization, and Microemulsions. New York: Plenum Press 1976.

[91] Hiemenz PC. Principles of Colloid and Surface Chemistry. New York: Marcel Dekker 1986.

[92] Nakamura K, Ryuichi E, Takeda M. Surface properties of styrene-ethylene oxide block copolymers. J Polym Sci, Polym Phys Ed 1976; 14: 1287-95.
[http://dx.doi.org/10.1002/pol.1976.180140712]

[93] Lieberman HA, Rieger MM, Banker GS, Eds. Pharmaceutical Dosage Forms: Disperse Systems. New York: Marcel Dekker press 1996; Vol. 2.

[94] Weissig V, Lizano C, Torchilin VP. Micellar delivery system for dequalinium. Proc Int Symp Control Release Bioact Mater 1998; 25: 415-6.

[95] Yokoyama M, Sugiyama T, Okano T, Sakurai Y, Naito M, Kataoka K. Analysis of micelle formation of an adriamycin-conjugated poly(ethylene glycol)-poly(aspartic acid) block copolymer by gel permeation chromatography. Pharm Res 1993; 10(6): 895-9.
[http://dx.doi.org/10.1023/A:1018921513605] [PMID: 8321859]

[96] Turro NJ, Chung CJ. Photoluminescent probes for water-water soluble polymers. Pressure and temperature effect on a polyol surfactant. Macromol 1984; 17: 2123-6.
[http://dx.doi.org/10.1021/ma00140a043]

[97] Fluksman A, Benny O. A robust method for critical micelle concentration determination using coumarin-6 as a fluorescent probe. Anal Methods 2019; 11: 3810-8.
[http://dx.doi.org/10.1039/C9AY00577C]

[98] Kalyanasundaram K, Thomas JK. Environmental effects on vibronic band intensities in pyrene monomer fluorescence and their application in studies of micellar systems. J Am Chem Soc 1977; 99: 2039-44.
[http://dx.doi.org/10.1021/ja00449a004]

[99] Chen WY, Alexandridis P, Su CK, Patrickios CS, Hertler WR, Hatton TA. Effect of block size and sequence on the micellization of ABC triblock methacrylic acid polyampholytes. Macromol 1995; 28: 8604-11.
[http://dx.doi.org/10.1021/ma00129a020]

[100] Zhang X, Jackson JK, Burt HM. Development of amphiphilicdiblock copolymers as micellar carriers of taxol. Int J Pharm 1996; 132: 195-206.
[http://dx.doi.org/10.1016/0378-5173(95)04386-1]

[101] Zhang X, Jackson JK, Burt HM. Determination of surfactant critical micelle concentration by a novel fluorescence depolarization technique. J Biochem Biophys Methods 1996; 31(3-4): 145-50.
[http://dx.doi.org/10.1016/0165-022X(95)00032-M] [PMID: 8675957]

[102] Gao Z, Eisenberg A. Model of micellization for block copolymers in solutions. Macromol 1993; 26: 7353-60.
[http://dx.doi.org/10.1021/ma00078a035]

[103] Taillefer J, Jones MC, Brasseur N, Van Lier JE, Leroux JC. Preparation and characterization of pH-responsive polymeric micelles for the delivery of anticancer photosensitizing drugs. J Pharm Sci 2000; 89: 52-62.
[http://dx.doi.org/10.1002/(SICI)1520-6017(200001)89:1<52::AID-JPS6>3.0.CO;2-D] [PMID: 10664538]

[104] Dong DC, Winnik MA. The P*y* scale of solvent polarities. Solvent effects on the vibronic fine structure of pyrene flourescence and empirical correlations with E_T and Y values. Photochem Photobiol 1982; 35: 17.
[http://dx.doi.org/10.1111/j.1751-1097.1982.tb03805.x]

[105] Calderara F, Hruska Z, Hurtrez G, Lerch JP, Nugay T, Riess G. Investigation of polystyrene-poly(ethylene oxide) block copolymer micelle formation in organic and aqueous solutions by nonradiative energy transfer experiments. Macromol 1994; 27(5): 1210.
[http://dx.doi.org/10.1021/ma00083a020]

[106] Kim MS, Hyun H, Cho YH, *et al.* Synthesis and characterization of PEG-PCL di-block copolymers. Polym Bull 2005; 55: 149.
[http://dx.doi.org/10.1007/s00289-005-0410-5]

[107] Dominguez A, Fernandez A, Gonzalez N, Iglesias E, Montenegro L. Determination of critical micelle

concentration of some surfactants by three techniques. J Chem Educ 1997; 74: 1227.
[http://dx.doi.org/10.1021/ed074p1227]

[108] Hamley IW. The Physics of Block Copolymers. Oxford Science Publication 1998; p. 131.

[109] Yokoyama M, Kwon GS, Okano T, Sakurai Y, Naito M, Kataoka K. Influencing factors on *in vitro* micelle stability of adriamycin-block copolymer conjugates. J Control Rel 1994; 28: 59-65.
[http://dx.doi.org/10.1016/0168-3659(94)90153-8]

[110] Trubetskoy VS, Torchilin VP. Use of polyethylene-lipid conjugates as long-circulating carriers for delivery of therapeutic and diagnostic agents. Adv Drug Deliv Rev 1995; 16: 311-20.
[http://dx.doi.org/10.1016/0169-409X(95)00032-3]

[111] Yokoyama M. Polymeric micelles as drug carriers: their lights and shadows. J Drug Target 2014; 22(7): 576-83.
[http://dx.doi.org/10.3109/1061186X.2014.934688] [PMID: 25012065]

[112] Sund M, Xu LL, Rahman A, Qian BF, Hammarström ML, Danielsson A. Reduced susceptibility to dextran sulphate sodium-induced colitis in the interleukin-2 heterozygous (IL-2) mouse. Immunology 2005; 114(4): 554-64.
[http://dx.doi.org/10.1111/j.1365-2567.2005.02123.x] [PMID: 15804292]

[113] Takakura Y, Hashida M. Macromolecular carrier systems for targeted drug delivery: pharmacokinetic considerations on biodistribution. Pharm Res 1996; 13(6): 820-31.
[http://dx.doi.org/10.1023/A:1016084508097] [PMID: 8792417]

[114] McLeod AD, Friend DR, Tozer TN. Synthesis and chemical stability of glucocorticoid-dextran esters: potential prodrugs for colon- specific drug delivery. Int J Pharm 1993; 92: 105-14.
[http://dx.doi.org/10.1016/0378-5173(93)90269-L]

[115] Harada M, Murata J, Sakamura Y, Sakakibara H, Okuno S, Suzuki T. Carrier and dose effects on the pharmacokinetics of T-0128, a camptothecin analogue-carboxymethyl dextran conjugate, in non-tumor- and tumor-bearing rats. J Control Rel 2001; 71(1): 71-86.
[http://dx.doi.org/10.1016/S0168-3659(00)00372-2] [PMID: 11245909]

[116] Mehvar R. Dextrans for targeted and sustained delivery of therapeutic and imaging agents. J Control Rel 2000; 69(1): 1-25.
[http://dx.doi.org/10.1016/S0168-3659(00)00302-3] [PMID: 11018543]

[117] Munechika K, Sogame Y, Kishi N, *et al.* Tissue distribution of macromolecular conjugate, adriamycin linked to oxidized dextran, in rat and mouse bearing tumor cells. Biol Pharm Bull 1994; 17(9): 1193-8.
[http://dx.doi.org/10.1248/bpb.17.1193] [PMID: 7531053]

[118] Vercauteren R, Schacht E, Duncan R. Effect of the chemical modification of dextran on the degradation by rat-liver lysosomal-enzymes. J Bioact Compat Polym 1992; 7: 346-57.
[http://dx.doi.org/10.1177/088391159200700404]

[119] Yokoyama M, Fukushima S, Uehara R, *et al.* Characterization of physical entrapment and chemical conjugation of adriamycin in polymeric micelles and their design for *in vivo* delivery to a solid tumor. J Control Rel 1998; 50(1-3): 79-92.
[http://dx.doi.org/10.1016/S0168-3659(97)00115-6] [PMID: 9685875]

[120] Yokoyama M, Okano T, Sakurai Y, Fukushima S, Okamoto K, Kataoka K. Selective delivery of adriamycin to a solid tumor using a polymeric micelle carrier system. J Drug Target 1999; 7(3): 171-86.
[http://dx.doi.org/10.3109/10611869909085500] [PMID: 10680973]

[121] Brandrup J, Immergut EH, Grulke EA, Eds. Polymer Handbook. 4th ed. New York: John Wiley and Sons press 1999; p. VI/202.

[122] Alkhamis KA, Allaboun H, Al-Momani WY. Study of the solubilization of gliclazide by aqueous micellar solutions. J Pharm Sci 2003; 92(4): 839-46.
[http://dx.doi.org/10.1002/jps.10350] [PMID: 12661069]

[123] Barry BW, El Eini DI. Solubilization of hydrocortisone, dexamethasone, testosterone and progesterone by long-chain polyoxyethylene surfactants. J Pharm Pharmacol 1976; 28(3): 210-8. [http://dx.doi.org/10.1111/j.2042-7158.1976.tb04133.x] [PMID: 6695]

[124] Florence AT, Attwood D. Physicochemical principles of Pharmacy. 3rd ed., London: The Mac Millan Press 2003.

[125] Palma S, Manzo RH, Allemandi D, Fratoni L, Lo Nostro P. Drug solubilization in ascorbyl-decanoate micellar solution. Coll Surf A 2003; 212: 163-73. [http://dx.doi.org/10.1016/S0927-7757(02)00299-6]

[126] Leo A, Hansch C, Elkins D. Partition coefficients and their uses. Chem Rev 1971; 71: 525. [http://dx.doi.org/10.1021/cr60274a001]

[127] Wilhelm M, Zhao CL, Wang Y, *et al.* Poly (styrene-ethylene oxide) block copolymer micelle formation in water: a fluorescence probe study. Macromol 1991; 24: 1033. [http://dx.doi.org/10.1021/ma00005a010]

[128] Kabanov AV, Nazarova IR, Astafieva IV, *et al.* Micelle Formation and Solubilization of Fluorescent Probes in Poly(oxyethylene-b-oxypropylene-b-oxyethylene) Solutions. Macromol 1995; 28: 2303. [http://dx.doi.org/10.1021/ma00111a026]

[129] Saito Y, Miura K, Tokuoka Y, Kondo Y, Abe M, Sato T. Volatility and solubilization of synthetic fragrances by pluronic P-85. J Dispers Sci Technol 1996; 17: 567. [http://dx.doi.org/10.1080/01932699608943525]

[130] Melik-Nubarov NS, Kozlov MY. Evaluation of partition coefficients of low molecular weight solutes between water and micelles of block copolymer of ethylene oxide based on dialysis kinetics and fluorescence spectroscopy. Coll Polym Sci 1998; 276: 381. [http://dx.doi.org/10.1007/s003960050256]

[131] Koizumi F, Kitagawa M, Negishi T, *et al.* Novel SN-38-incorporating polymeric micelles, NK012, eradicate vascular endothelial growth factor-secreting bulky tumors. Cancer Res 2006; 66(20): 10048-56. [http://dx.doi.org/10.1158/0008-5472.CAN-06-1605] [PMID: 17047068]

[132] Yamamoto T, Yokoyama M, Opanasopit P, Hayama A, Kawano K, Maitani Y. What are determining factors for stable drug incorporation into polymeric micelle carriers? Consideration on physical and chemical characters of the micelle inner core. J Control Rel 2007; 123(1): 11-8. [http://dx.doi.org/10.1016/j.jconrel.2007.07.008] [PMID: 17716772]

[133] Nishiyama N, Okazaki S, Cabral H, *et al.* Novel cisplatin-incorporated polymeric micelles can eradicate solid tumors in mice. Cancer Res 2003; 63(24): 8977-83. [PMID: 14695216]

[134] Wiradharma N, Tong YW, Yang YY. Self-assembled oligopeptide nanostructures for co-delivery of drug and gene with synergistic therapeutic effect. Biomaterials 2009; 30(17): 3100-9. [http://dx.doi.org/10.1016/j.biomaterials.2009.03.006] [PMID: 19342093]

[135] Torchilin VP. Micellar nanocarriers: pharmaceutical perspectives. Pharm Res 2007; 24(1): 1-16. [http://dx.doi.org/10.1007/s11095-006-9132-0] [PMID: 17109211]

[136] Kim BS, Park SW, Hammond PT. Hydrogen-bonding layer-by-layer-assembled biodegradable polymeric micelles as drug delivery vehicles from surfaces. ACS Nano 2008; 2(2): 386-92. [http://dx.doi.org/10.1021/nn700408z] [PMID: 19206641]

[137] Daugherty AL, Mrsny RJ. Transcellular uptake mechanisms of the intestinal epithelial barrier Part one. Pharm Sci Technol Today 1999; 4(2): 144-51. [http://dx.doi.org/10.1016/S1461-5347(99)00142-X] [PMID: 10322371]

[138] Rajput G, Majmudar F, Patel J, Thakor R, Rajgor NB. Stomach specific mucoadhesive microsphere as a controlled drug delivery system. Syst Rev Pharm 2010; 1: 70-8.

[http://dx.doi.org/10.4103/0975-8453.59515]

[139] Wallmark B, Larsson H, Humble L. The relationship between gastric acid secretion and gastric H+,K+-ATPase activity. J Biol Chem 1985; 260(25): 13681-4. [http://dx.doi.org/10.1016/S0021-9258(17)38778-1] [PMID: 2997178]

[140] Wikberg M, Ulmius J, Ragnarsson G. Review article: Targeted drug delivery in treatment of intestinal diseases. Aliment Pharmacol Ther 1997; 11 (Suppl. 3): 109-15. [http://dx.doi.org/10.1111/j.1365-2036.1997.tb00815.x] [PMID: 9467985]

[141] Evans DF, Pye G, Bramley R, Clark AG, Dyson TJ, Hardcastle JD. Measurement of gastrointestinal pH profiles in normal ambulant human subjects. Gut 1988; 29(8): 1035-41. [http://dx.doi.org/10.1136/gut.29.8.1035] [PMID: 3410329]

[142] McConnell EL, Fadda HM, Basit AW. Gut instincts: explorations in intestinal physiology and drug delivery. Int J Pharm 2008; 364(2): 213-26. [http://dx.doi.org/10.1016/j.ijpharm.2008.05.012] [PMID: 18602774]

[143] Rijcken CJ, Soga O, Hennink WE, van Nostrum CF. Triggered destabilisation of polymeric micelles and vesicles by changing polymers polarity: an attractive tool for drug delivery. J Control Rel 2007; 120(3): 131-48. [http://dx.doi.org/10.1016/j.jconrel.2007.03.023] [PMID: 17582642]

[144] Wu CY, Benet LZ. Predicting drug disposition *via* application of BCS: transport/absorption/ elimination interplay and development of a biopharmaceutics drug disposition classification system. Pharm Res 2005; 22(1): 11-23. [http://dx.doi.org/10.1007/s11095-004-9004-4] [PMID: 15771225]

[145] Xu W, Ling P, Zhang T. Polymeric micelles, a promising drug delivery system to enhance bioavailability of poorly water-soluble drugs. J Drug Deliv 2013; 2013: 340315. [http://dx.doi.org/10.1155/2013/340315] [PMID: 23936656]

[146] Gelderblom H, Verweij J, Nooter K, Sparreboom A, Cremophor EL. Cremophor EL: the drawbacks and advantages of vehicle selection for drug formulation. Eur J Cancer 2001; 37(13): 1590-8. [http://dx.doi.org/10.1016/S0959-8049(01)00171-X] [PMID: 11527683]

[147] Savić R, Eisenberg A, Maysinger D. Block copolymer micelles as delivery vehicles of hydrophobic drugs: micelle-cell interactions. J Drug Target 2006; 14(6): 343-55. [http://dx.doi.org/10.1080/10611860600874538] [PMID: 17092835]

[148] Chiappetta DA, Hocht C, Taira C, Sosnik A. Oral pharmacokinetics of the anti-HIV efavirenz encapsulated within polymeric micelles. Biomaterials 2011; 32(9): 2379-87. [http://dx.doi.org/10.1016/j.biomaterials.2010.11.082] [PMID: 21186055]

[149] Gaucher G, Dufresne MH, Sant VP, Kang N, Maysinger D, Leroux JC. Block copolymer micelles: preparation, characterization and application in drug delivery. J Control Rel 2005; 109(1-3): 169-88. [http://dx.doi.org/10.1016/j.jconrel.2005.09.034] [PMID: 16289422]

[150] Burt HM, Zhang X, Toleikis P, Embree L, Hunter WL. Development of copolymers of poly(d,l-lactide) and methoxypolyethylene glycol as micellar carriers of paclitaxel. Coll Surf B Biointerfaces 1999; 16: 161-71. [http://dx.doi.org/10.1016/S0927-7765(99)00067-3]

[151] Movassaghian S, Merkel OM, Torchilin VP. Applications of polymer micelles for imaging and drug delivery. Wiley Interdiscip Rev Nanomed Nanobiotechnol 2015; 7(5): 691-707. [http://dx.doi.org/10.1002/wnan.1332] [PMID: 25683687]

[152] Gill KK, Nazzal S, Kaddoumi A. Paclitaxel loaded PEG(5000)-DSPE micelles as pulmonary delivery platform: formulation characterization, tissue distribution, plasma pharmacokinetics, and toxicological evaluation. Eur J Pharm Biopharm 2011; 79(2): 276-84. [http://dx.doi.org/10.1016/j.ejpb.2011.04.017] [PMID: 21575719]

[153] Hamaguchi T, Matsumura Y, Suzuki M, *et al.* NK105, a paclitaxel-incorporating micellar nanoparticle

formulation, can extend in vivo antitumour activity and reduce the neurotoxicity of paclitaxel. Br J Cancer 2005; 92(7): 1240-6.
[http://dx.doi.org/10.1038/sj.bjc.6602479] [PMID: 15785749]

[154] Gaucher G, Satturwar P, Jones M-C, Furtos A, Leroux JC. Polymeric micelles for oral drug delivery. Eur J Pharm Biopharm 2010; 76(2): 147-58.
[http://dx.doi.org/10.1016/j.ejpb.2010.06.007] [PMID: 20600891]

[155] Bromberg L. Polymeric micelles in oral chemotherapy. J Control Rel 2008; 128(2): 99-112.
[http://dx.doi.org/10.1016/j.jconrel.2008.01.018] [PMID: 18325619]

[156] Julien DC, Behnke S, Wang G, Murdoch GK, Hill RA. Utilization of monoclonal antibody-targeted nanomaterials in the treatment of cancer. MAbs 2011; 3(5): 467-78.
[http://dx.doi.org/10.4161/mabs.3.5.16089] [PMID: 21814040]

[157] Lukyanov AN, Torchilin VP. Micelles from lipid derivatives of water-soluble polymers as delivery systems for poorly soluble drugs. Adv Drug Deliv Rev 2004; 56(9): 1273-89.
[http://dx.doi.org/10.1016/j.addr.2003.12.004] [PMID: 15109769]

[158] Maeda H, Greish K, Fang J. The EPR effect and polymeric drugs: a paradigm shift for cancer chemotherapy in the 21st century. Adv Polym Sci 2006; 193: 103-21.
[http://dx.doi.org/10.1007/12_026]

[159] Matsumura Y. Poly (amino acid) micelle nanocarriers in preclinical and clinical studies. Adv Drug Deliv Rev 2008; 60(8): 899-914.
[http://dx.doi.org/10.1016/j.addr.2007.11.010] [PMID: 18406004]

[160] Vetvicka D, Hruby M, Hovorka O, *et al.* Biological evaluation of polymeric micelles with covalently bound doxorubicin. Bioconjug Chem 2009; 20(11): 2090-7.
[http://dx.doi.org/10.1021/bc900212k] [PMID: 19835372]

[161] Watanabe M, Kawano K, Yokoyama M, Opanasopit P, Okano T, Maitani Y. Preparation of camptothecin-loaded polymeric micelles and evaluation of their incorporation and circulation stability. Int J Pharm 2006; 308(1-2): 183-9.
[http://dx.doi.org/10.1016/j.ijpharm.2005.10.030] [PMID: 16324807]

[162] Zhang Y, Jiang M. New approaches to stimuli responsive polymeric micelles and hollow spheres. Front Chem 2006; 4: 364-8.

[163] Nishiyama N, Kataoka K. Current state, achievements, and future prospects of polymeric micelles as nanocarriers for drug and gene delivery. Pharmacol Ther 2006; 112(3): 630-48.
[http://dx.doi.org/10.1016/j.pharmthera.2006.05.006] [PMID: 16815554]

[164] Schmaljohann D. Thermo- and pH-responsive polymers in drug delivery. Adv Drug Deliv Rev 2006; 58(15): 1655-70.
[http://dx.doi.org/10.1016/j.addr.2006.09.020] [PMID: 17125884]

[165] Gu J, Cheng WP, Liu J, *et al.* pH-triggered reversible "stealth" polycationic micelles. Biomacromolecules 2008; 9(1): 255-62.
[http://dx.doi.org/10.1021/bm701084w] [PMID: 18095651]

[166] Torchilin VP. Targeted polymeric micelles for delivery of poorly soluble drugs. Cell Mol Life Sci 2004; 61(19-20): 2549-59.
[http://dx.doi.org/10.1007/s00018-004-4153-5] [PMID: 15526161]

[167] Liu B, Yang M, Li R, *et al.* The antitumor effect of novel docetaxel-loaded thermosensitive micelles. Eur J Pharm Biopharm 2008; 69(2): 527-34.
[http://dx.doi.org/10.1016/j.ejpb.2008.01.015] [PMID: 18359617]

[168] Xiong XB, Mahmud A, Uludağ H, Lavasanifar A. Conjugation of arginine-glycine-aspartic acid peptides to poly(ethylene oxide)-b-poly(epsilon-caprolactone) micelles for enhanced intracellular drug delivery to metastatic tumor cells. Biomacromolecules 2007; 8(3): 874-84.
[http://dx.doi.org/10.1021/bm060967g] [PMID: 17315946]

[169] Musacchio T, Laquintana V, Latrofa A, Trapani G, Torchilin VP. PEG-PE micelles loaded with paclitaxel and surface-modified by a PBR-ligand: synergistic anticancer effect. Mol Pharm 2009; 6(2): 468-79.
[http://dx.doi.org/10.1021/mp800158c] [PMID: 19718800]

[170] Torchilin VP, Lukyanov AN, Gao Z, Papahadjopoulos-Sternberg B. Immunomicelles: targeted pharmaceutical carriers for poorly soluble drugs. Proc Natl Acad Sci USA 2003; 100(10): 6039-44.
[http://dx.doi.org/10.1073/pnas.0931428100] [PMID: 12716967]

[171] Zeng YC, Li S, Liu C, *et al.* Soluplus micelles for improving the oral bioavailability of scopoletin and their hypouricemic effect *in vivo.* Acta Pharmacol Sin 2017; 38(3): 424-33.
[http://dx.doi.org/10.1038/aps.2016.126] [PMID: 28112183]

[172] Chen YC, Su CY, Jhan HJ, Ho HO, Sheu MT. Physical characterization and *in vivo* pharmacokinetic study of self-assembling amphotericin B-loaded lecithin-based mixed polymeric micelles. Int J Nanomed 2015; 10: 7265-74.
[PMID: 26664117]

[173] Han X, Wang Z, Wang M, *et al.* Liver-targeting self-assembled hyaluronic acid-glycyrrhetinic acid micelles enhance hepato-protective effect of silybin after oral administration. Drug Deliv 2016; 23(5): 1818-29.
[http://dx.doi.org/10.3109/10717544.2015.1108374] [PMID: 26556526]

[174] Zhang P, Xu Y, Zhu X, Huang Y. Goblet cell targeting nanoparticle containing drug-loaded micelle cores for oral delivery of insulin. Int J Pharm 2015; 496(2): 993-1005.
[http://dx.doi.org/10.1016/j.ijpharm.2015.10.078] [PMID: 26541299]

[175] Han L, Yang Q, Shen T, Qing J, Wang J. Lymphatic transport of orally administered probucol-loaded mPEG-DSPE micelles. Drug Deliv 2016; 23(6): 1955-61.
[PMID: 25816053]

[176] Choi YA, Yoon YH, Choi K, *et al.* Enhanced oral bioavailability of morin administered in mixed micelle formulation with PluronicF127 and Tween80 in rats. Biol Pharm Bull 2015; 38(2): 208-17.
[http://dx.doi.org/10.1248/bpb.b14-00508] [PMID: 25747979]

[177] Wang T, Wang N, Song H, *et al.* Preparation of an anhydrous reverse micelle delivery system to enhance oral bioavailability and anti-diabetic efficacy of berberine. Eur J Pharm Sci 2011; 44(1-2): 127-35.
[http://dx.doi.org/10.1016/j.ejps.2011.06.015] [PMID: 21742030]

[178] Binkhathlan Z, Hamdy DA, Brocks DR, Lavasanifar A. Development of a polymeric micellar formulation for valspodar and assessment of its pharmacokinetics in rat. Eur J Pharm Biopharm 2010; 75(2): 90-5.
[http://dx.doi.org/10.1016/j.ejpb.2010.03.010] [PMID: 20332028]

[179] Lee SC, Huh KM, Lee J, Cho YW, Galinsky RE, Park K. Hydrotropic polymeric micelles for enhanced paclitaxel solubility: *in vitro* and *in vivo* characterization. Biomacromolecules 2007; 8(1): 202-8.
[http://dx.doi.org/10.1021/bm060307b] [PMID: 17206808]

[180] Vaishya RD, Gokulgandhi M, Patel S, Minocha M, Mitra AK. Novel dexamethasone-loaded nanomicelles for the intermediate and posterior segment uveitis. AAPS PharmSciTech 2014; 15(5): 1238-51.
[http://dx.doi.org/10.1208/s12249-014-0100-4] [PMID: 24895075]

[181] Jaiswal M, Kumar M, Pathak K. Zero order delivery of itraconazole *via* polymeric micelles incorporated in situ ocular gel for the management of fungal keratitis. Coll Surf B Biointerfaces 2015; 130: 23-30.
[http://dx.doi.org/10.1016/j.colsurfb.2015.03.059] [PMID: 25889081]

[182] Kanoujia J, Kushwaha PS, Saraf SA. Evaluation of gatifloxacin pluronic micelles and development of

its formulation for ocular delivery. Drug Deliv Transl Res 2014; 4(4): 334-43.
[http://dx.doi.org/10.1007/s13346-014-0194-y] [PMID: 25787066]

[183] Duan Y, Cai X, Du H, Zhai G. Novel in situ gel systems based on P123/TPGS mixed micelles and gellan gum for ophthalmic delivery of curcumin. Coll Surf B Biointerfaces 2015; 128: 322-30.
[http://dx.doi.org/10.1016/j.colsurfb.2015.02.007] [PMID: 25707750]

[184] Famili A, Kahook MY, Park D. A combined micelle and poly(serinol hexamethylene urea)-co-poly-N-isopropylacrylamide) reverse thermal gel as an injectable ocular drug delivery system. Macromol Biosci 2014; 14(12): 1719-29.
[http://dx.doi.org/10.1002/mabi.201400250] [PMID: 25187427]

[185] Luschmann C, Tessmar J, Schoeberl S, Strauß O, Luschmann K, Goepferich A. Self-assembling colloidal system for the ocular administration of cyclosporine A. Cornea 2014; 33(1): 77-81.
[http://dx.doi.org/10.1097/ICO.0b013e3182a7f3bf] [PMID: 24162754]

[186] Lin HR, Chang PC. Novel pluronic-chitosan micelle as an ocular delivery system. J Biomed Mater Res B Appl Biomater 2013; 101(5): 689-99.
[http://dx.doi.org/10.1002/jbm.b.32871] [PMID: 23359519]

[187] Li X, Zhang Z, Li J, Sun S, Weng Y, Chen H. Diclofenac/biodegradable polymer micelles for ocular applications. Nanoscale 2012; 4(15): 4667-73.
[http://dx.doi.org/10.1039/c2nr30924f] [PMID: 22732776]

[188] Ribeiro A, Sosnik A, Chiappetta DA, Veiga F, Concheiro A, Alvarez-Lorenzo C. Single and mixed poloxamine micelles as nanocarriers for solubilization and sustained release of ethoxzolamide for topical glaucoma therapy. J R Soc Interface 2012; 9(74): 2059-69.
[http://dx.doi.org/10.1098/rsif.2012.0102] [PMID: 22491977]

[189] Civiale C, Licciardi M, Cavallaro G, Giammona G, Mazzone MG. Polyhydroxyethylaspartamide-based micelles for ocular drug delivery. Int J Pharm 2009; 378(1-2): 177-86.
[http://dx.doi.org/10.1016/j.ijpharm.2009.05.028] [PMID: 19465101]

[190] Parayath NN, Nehoff H, Norton SE, *et al.* Styrene maleic acid-encapsulated paclitaxel micelles: antitumor activity and toxicity studies following oral administration in a murine orthotopic colon cancer model. Int J Nanomed 2016; 11: 3979-91.
[http://dx.doi.org/10.2147/IJN.S110251] [PMID: 27574427]

[191] Ma ZG, Ma R, Xiao XL, *et al.* Azo polymeric micelles designed for colon-targeted dimethyl fumarate delivery for colon cancer therapy. Acta Biomater 2016; 44: 323-31.
[http://dx.doi.org/10.1016/j.actbio.2016.08.021] [PMID: 27544813]

[192] Zhu P, Zhao N, Sheng D, *et al.* Inhibition of growth and metastasis of colon cancer by delivering 5-Fluorouracil-loaded pluronic P85 copolymer micelles. Sci Rep 2016; 6: 20896.
[http://dx.doi.org/10.1038/srep20896] [PMID: 26864651]

[193] Chang T, Trench D, Putnam J, Stenzel MH, Lord MS. Curcumin-loading-dependent stability of PEGMEMA-based micelles affects endocytosis and exocytosis in colon carcinoma cells. Mol Pharm 2016; 13(3): 924-32.
[http://dx.doi.org/10.1021/acs.molpharmaceut.5b00820] [PMID: 26755445]

[194] Xu G, Shi H, Ren L, *et al.* Enhancing the anti-colon cancer activity of quercetin by self-assembled micelles. Int J Nanomed 2015; 10: 2051-63.
[PMID: 25844036]

[195] Chen Y, Wu Q, Song L, *et al.* Polymeric micelles encapsulating fisetin improve the therapeutic effect in colon cancer. ACS Appl Mater Interfaces 2015; 7(1): 534-42.
[http://dx.doi.org/10.1021/am5066893] [PMID: 25495760]

[196] Gao X, Wang B, Wei X, *et al.* Preparation, characterization and application of star-shaped PCL/PEG micelles for the delivery of doxorubicin in the treatment of colon cancer. Int J Nanomed 2013; 8: 971-82.

[http://dx.doi.org/10.2147/IJN.S39532] [PMID: 23493403]

[197] Li P, Zhou X, Qu D, *et al.* Preliminary study on fabrication, characterization and synergistic anti-lung cancer effects of self-assembled micelles of covalently conjugated celastrol-polyethylene glycol-ginsenoside Rh2. Drug Deliv 2017; 24(1): 834-45. [http://dx.doi.org/10.1080/10717544.2017.1326540] [PMID: 28532223]

[198] Zhu WT, Liu SY, Wu L, *et al.* Delivery of curcumin by directed self-assembled micelles enhances therapeutic treatment of non-small-cell lung cancer. Int J Nanomed 2017; 12: 2621-34. [http://dx.doi.org/10.2147/IJN.S128921] [PMID: 28435247]

[199] Yan H, Song J, Jia X, Zhang Z. Hyaluronic acid-modified didecyldimethylammonium bromide/ d-a-tocopheryl polyethylene glycol succinate mixed micelles for delivery of baohuoside I against non-small cell lung cancer: *in vitro* and *in vivo* evaluation. Drug Deliv 2017; 24(1): 30-9. [http://dx.doi.org/10.1080/10717544.2016.1228713] [PMID: 28155337]

[200] Zhang Y, Tong D, Che D, *et al.* Ascorbyl palmitate/d-α-tocopheryl polyethylene glycol 1000 succinate monoester mixed micelles for prolonged circulation and targeted delivery of compound K for antilung cancer therapy *in vitro* and *in vivo.* Int J Nanomed 2017; 12: 605-14. [http://dx.doi.org/10.2147/IJN.S119226] [PMID: 28144142]

[201] Wu H, Zhong Q, Zhong R, *et al.* Preparation and antitumor evaluation of self-assembling oleanolic acid-loaded Pluronic P105/d-α-tocopheryl polyethylene glycol succinate mixed micelles for non-small-cell lung cancer treatment. Int J Nanomed 2016; 11: 6337-52. [http://dx.doi.org/10.2147/IJN.S119839] [PMID: 27932881]

[202] Zhu JJ, Zhang XX, Miao YQ, *et al.* Delivery of acetylthevetin B, an antitumor cardiac glycoside, using polymeric micelles for enhanced therapeutic efficacy against lung cancer cells. Acta Pharmacol Sin 2017; 38(2): 290-300. [http://dx.doi.org/10.1038/aps.2016.113] [PMID: 27917871]

[203] Suzuki H, Ueno K, Mizumoto T, Seto Y, Sato H, Onoue S. Self-micellizing solid dispersion of cyclosporine A for pulmonary delivery: Physicochemical, pharmacokinetic and safety assessments. Eur J Pharm Sci 2017; 96: 107-14. [http://dx.doi.org/10.1016/j.ejps.2016.09.015] [PMID: 27634579]

[204] Russo A, Pellosi DS, Pagliara V, *et al.* Biotin-targeted Pluronic(®) P123/F127 mixed micelles delivering niclosamide: A repositioning strategy to treat drug-resistant lung cancer cells. Int J Pharm 2016; 511(1): 127-39. [http://dx.doi.org/10.1016/j.ijpharm.2016.06.118] [PMID: 27374195]

[205] Zhuang B, Du L, Xu H, *et al.* Self-assembled Micelle Loading Cabazitaxel for therapy of lung cancer. Int J Pharm 2016; 499(1-2): 146-55. [http://dx.doi.org/10.1016/j.ijpharm.2015.12.073] [PMID: 26762884]

[206] Cheng X, Qiu N, Yang J, *et al.* Preparation, characterization, and *in vivo* study of 7-Ethyl--4-Aminocamptothecin-Loaded Poly(Ethylene Glycol)2000-Poly(Lactic Acid) 2000 polymeric micelles against H460 human nonsmall cell lung carcinoma. J Pharm Sci 2015; 104(11): 3934-42. [http://dx.doi.org/10.1002/jps.24613] [PMID: 26306026]

[207] Gupta N, Ibrahim HM, Ahsan F. Peptide-micelle hybrids containing fasudil for targeted delivery to the pulmonary arteries and arterioles to treat pulmonary arterial hypertension. J Pharm Sci 2014; 103(11): 3743-53. [http://dx.doi.org/10.1002/jps.24193] [PMID: 25266507]

[208] Lapteva M, Santer V, Mondon K, *et al.* Targeted cutaneous delivery of ciclosporin A using micellar nanocarriers and the possible role of inter-cluster regions as molecular transport pathways. J Control Rel 2014; 196: 9-18. [http://dx.doi.org/10.1016/j.jconrel.2014.09.021] [PMID: 25278258]

[209] Bachhav YG, Mondon K, Kalia YN, Gurny R, Möller M. Novel micelle formulations to increase cutaneous bioavailability of azole antifungals. J Control Rel 2011; 153(2): 126-32.

[http://dx.doi.org/10.1016/j.jconrel.2011.03.003] [PMID: 21397643]

[210] Deng P, Teng F, Zhou F, *et al.* Y-shaped methoxy poly (ethylene glycol)-block-poly (epsilon-caprolactone)-based micelles for skin delivery of ketoconazole: *in vitro* study and *in vivo* evaluation. Mater Sci Eng C 2017; 78: 296-304.
[http://dx.doi.org/10.1016/j.msec.2017.04.089] [PMID: 28575987]

[211] Sharma G, Kamboj S, Thakur K, Negi P, Raza K, Katare OP. Delivery of thermoresponsive-tailored mixed micellar nanogel of lidocaine and prilocaine with improved dermatokinetic profile and therapeutic efficacy in topical anaesthesia. AAPS PharmSciTech 2017; 18(3): 790-802.
[http://dx.doi.org/10.1208/s12249-016-0561-8] [PMID: 27317572]

[212] Cholkar K, Gilger BC, Mitra AK. Topical delivery of aqueous micellar resolvin E1 analog (RX-10045). Int J Pharm 2016; 498(1-2): 326-34.
[http://dx.doi.org/10.1016/j.ijpharm.2015.12.037] [PMID: 26706439]

[213] Seino H, Arai Y, Nagao N, Ozawa N, Hamada K. Efficient percutaneous delivery of the antimelanogenic agent glabridin using cationic amphiphilic chitosan micelles. PLoS One 2016; 11(10): e0164061.
[http://dx.doi.org/10.1371/journal.pone.0164061] [PMID: 27695112]

[214] Liang K, Xu K, Bessarab D, Obaje J, Xu C. Arbutin encapsulated micelles improved transdermal delivery and suppression of cellular melanin production. BMC Res Notes 2016; 9: 254.
[http://dx.doi.org/10.1186/s13104-016-2047-x] [PMID: 27129306]

[215] Wang Y, Zhao B, Wang S, *et al.* Formulation and evaluation of novel glycyrrhizic acid micelles for transdermal delivery of podophyllotoxin. Drug Deliv 2016; 23(5): 1623-35.
[http://dx.doi.org/10.3109/10717544.2015.1135489] [PMID: 26786787]

[216] Lapteva M, Mondon K, Möller M, Gurny R, Kalia YN. Polymeric micelle nanocarriers for the cutaneous delivery of tacrolimus: a targeted approach for the treatment of psoriasis. Mol Pharm 2014; 11(9): 2989-3001.
[http://dx.doi.org/10.1021/mp400639e] [PMID: 25057896]

[217] Agrawal V, Gupta V, Ramteke S, Trivedi P. Preparation and evaluation of tubular micelles of pluronic lecithin organogel for transdermal delivery of sumatriptan. AAPS PharmSciTech 2010; 11(4): 1718-25.
[http://dx.doi.org/10.1208/s12249-010-9540-7] [PMID: 21128126]

[218] Hassan AS, Soliman GM, El-Mahdy MM, El-Gindy GE. Solubilization and enhancement of *ex vivo* vaginal delivery of progesterone using solid dispersions, inclusion complexes and micellar solubilization. Curr Drug Deliv 2018; 15: 110-21.
[PMID: 28322164]

[219] Wen D, Peng Y, Lin F, Singh RK, Mahato RI. Micellar delivery of miR-34a modulator rubone and paclitaxel in resistant prostate cancer. Cancer Res 2017; 77(12): 3244-54.
[http://dx.doi.org/10.1158/0008-5472.CAN-16-2355] [PMID: 28428276]

[220] Chen L, Ji F, Bao Y, *et al.* Biocompatible cationic pullulan-g-desoxycholic acid-g-PEI micelles used to co-deliver drug and gene for cancer therapy. Mater Sci Eng C 2017; 70(Pt 1): 418-29.
[http://dx.doi.org/10.1016/j.msec.2016.09.019] [PMID: 27770912]

[221] Zhao M, Li M, Zhang Z, Gong T, Sun X. Induction of HIV-1 gag specific immune responses by cationic micelles mediated delivery of gag mRNA. Drug Deliv 2016; 23(7): 2596-607.
[http://dx.doi.org/10.3109/10717544.2015.1038856] [PMID: 26024387]

[222] Wu Y, Zhang Y, Zhang W, Sun C, Wu J, Tang J. Reversing of multidrug resistance breast cancer by co-delivery of P-gp siRNA and doxorubicin *via* folic acid-modified core-shell nanomicelles. Coll Surf B Biointerfaces 2016; 138: 60-9.
[http://dx.doi.org/10.1016/j.colsurfb.2015.11.041] [PMID: 26655793]

[223] Mao X, Liu J, Gong Z, *et al.* iRGD-conjugated DSPE-PEG2000 nanomicelles for targeted delivery of

salinomycin for treatment of both liver cancer cells and cancer stem cells. Nanomed (Lond) 2015; 10(17): 2677-95.
[http://dx.doi.org/10.2217/nnm.15.106] [PMID: 26355733]

[224] Kanazawa T, Sugawara K, Tanaka K, Horiuchi S, Takashima Y, Okada H. Suppression of tumor growth by systemic delivery of anti-VEGF siRNA with cell-penetrating peptide-modified MPEG-PCL nanomicelles. Eur J Pharm Biopharm 2012; 81(3): 470-7.
[http://dx.doi.org/10.1016/j.ejpb.2012.04.021] [PMID: 22579732]

[225] Huang HY, Kuo WT, Chou MJ, Huang YY. Co-delivery of anti-vascular endothelial growth factor siRNA and doxorubicin by multifunctional polymeric micelle for tumor growth suppression. J Biomed Mater Res A 2011; 97(3): 330-8.
[http://dx.doi.org/10.1002/jbm.a.33055] [PMID: 21465641]

[226] Harada-Shiba M, Takamisawa I, Miyata K, *et al.* Intratracheal gene transfer of adrenomedullin using polyplex nanomicelles attenuates monocrotaline-induced pulmonary hypertension in rats. Mol Ther 2009; 17(7): 1180-6.
[http://dx.doi.org/10.1038/mt.2009.63] [PMID: 19337232]

[227] Zeng Q, Li H, Jiang H, *et al.* Tailoring polymeric hybrid micelles with lymph node targeting ability to improve the potency of cancer vaccines. Biomaterials 2017; 122: 105-13.
[http://dx.doi.org/10.1016/j.biomaterials.2017.01.010] [PMID: 28110170]

[228] Layek B, Lipp L, Singh J. APC targeted micelle for enhanced intradermal delivery of hepatitis B DNA vaccine. J Control Rel 2015; 207: 143-53.
[http://dx.doi.org/10.1016/j.jconrel.2015.04.014] [PMID: 25886704]

[229] Lu Y, Miao L, Wang Y, *et al.* Curcumin Micelles remodel tumor microenvironment and enhance vaccine activity in an advanced melanoma model. Mol Ther 2016; 24(2): 364-74.
[http://dx.doi.org/10.1038/mt.2015.165] [PMID: 26334519]

[230] Luo Z, Shi S, Jin L, *et al.* Cationic micelle based vaccine induced potent humoral immune response through enhancing antigen uptake and formation of germinal center. Coll Surf B Biointerfaces 2015; 135: 556-64.
[http://dx.doi.org/10.1016/j.colsurfb.2015.07.079] [PMID: 26311662]

[231] Trent A, Ulery BD, Black MJ, *et al.* Peptide amphiphile micelles self-adjuvant group A streptococcal vaccination. AAPS J 2015; 17(2): 380-8.
[http://dx.doi.org/10.1208/s12248-014-9707-3] [PMID: 25527256]

[232] Miki K, Nagaoka K, Harada M, *et al.* Combination therapy with dendritic cell vaccine and IL-2 encapsulating polymeric micelles enhances intra-tumoral accumulation of antigen-specific CTLs. Int Immunopharmacol 2014; 23(2): 499-504.
[http://dx.doi.org/10.1016/j.intimp.2014.09.025] [PMID: 25284343]

[233] Noh YW, Hong JH, Shim SM, *et al.* Polymer nanomicelles for efficient mucus delivery and antigen-specific high mucosal immunity. Angew Chem Int Ed Engl 2013; 52(30): 7684-9.
[http://dx.doi.org/10.1002/anie.201302881] [PMID: 23765547]

[234] Estevan M, Irache JM, Grilló MJ, Blasco JM, Gamazo C. Encapsulation of antigenic extracts of Salmonella enterica serovar. Abortusovis into polymeric systems and efficacy as vaccines in mice. Vet Microbiol 2006; 118(1-2): 124-32.
[http://dx.doi.org/10.1016/j.vetmic.2006.07.004] [PMID: 16914273]

[235] Shi-Xiao J, Shi-Ying J, Lv QY, Shan-Shan F, Jin H, Yuan HL. Mixed nanomicelles loaded with thymopeptides-sodium deoxycholate/phospholipid improve drug absorption. Chin J Nat Med 2014; 12(1): 65-70.
[http://dx.doi.org/10.1016/S1875-5364(14)60011-X] [PMID: 24484599]

[236] Torchilin VP. Polymeric contrast agents for medical imaging. Curr Pharm Biotechnol 2000; 1(2): 183-215.
[http://dx.doi.org/10.2174/1389201003378960] [PMID: 11467336]

[237] Nishiyama N, Kataoka K. Current state, achievements, and future prospects of polymeric micelles as nanocarriers for drug and gene delivery. Pharmacol Ther 2006; 112(3): 630-48. [http://dx.doi.org/10.1016/j.pharmthera.2006.05.006] [PMID: 16815554]

[238] Hu X, Jing X. Biodegradable amphiphilic polymer-drug conjugate micelles. Expert Opin Drug Deliv 2009; 6(10): 1079-90. [http://dx.doi.org/10.1517/17425240903158917] [PMID: 19645633]

[239] Yang H, Mao H, Wan Z, *et al.* Micelles assembled with carbocyanine dyes for theranostic near-infrared fluorescent cancer imaging and photothermal therapy. Biomaterials 2013; 34(36): 9124-33. [http://dx.doi.org/10.1016/j.biomaterials.2013.08.022] [PMID: 24008037]

[240] Torchilin VP, Frank-Kamenetsky MD, Wolf GL. CT visualization of blood pool in rats by using long-circulating, iodine-containing micelles. Acad Radiol 1999; 6(1): 61-5. [http://dx.doi.org/10.1016/S1076-6332(99)80063-4] [PMID: 9891154]

[241] Patil RR, Yu J, Banerjee SR, *et al.* Probing *in vivo* trafficking of polymer/DNA micellar nanoparticles using SPECT/CT imaging. Mol Ther 2011; 19(9): 1626-35. [http://dx.doi.org/10.1038/mt.2011.128] [PMID: 21750533]

[242] Trubetskoy VS, Torchilin VP. Polyethyleneglycol based micelles as carriers of therapeutic and diagnostic agents. STP Pharma Sci 1996; 6: 79.

[243] Schmiedl UP, Nelson JA, Teng L, Starr F, Malek R, Ho RJ. Magnetic resonance imaging of the hepatobiliary system: intestinal absorption studies of manganese mesoporphyrin. Acad Radiol 1995; 2(11): 994-1001. [http://dx.doi.org/10.1016/S1076-6332(05)80702-0] [PMID: 9419672]

[244] Lauffer RB. Magnetic resonance contrast media: principles and progress. Magn Reson Q 1990; 6(2): 65-84. [PMID: 2202424]

[245] Grant CWM, Karlik S, Florio E. A liposomal MRI contrast agent: phosphatidylethanolamine-DTPA. Magn Reson Med 1989; 11(2): 236-43. [http://dx.doi.org/10.1002/mrm.1910110211] [PMID: 2779414]

[246] Kabalka GW, Buonocore E, Hubner K, Davis M, Huang L. Gadolinium-labeled liposomes containing paramagnetic amphipathic agents: targeted MRI contrast agents for the liver. Magn Reson Med 1988; 8(1): 89-95. [http://dx.doi.org/10.1002/mrm.1910080111] [PMID: 3173073]

[247] Unger E, Fritz T, Wu G, *et al.* J Liposomal MR contrast agents. Liposome Res 1994; 4: 811. [http://dx.doi.org/10.3109/08982109409018601]

[248] Enoch HG, Strittmatter P. Formation and properties of 1000-A-diameter, single-bilayer phospholipid vesicles. Proc Natl Acad Sci USA 1979; 76(1): 145-9. [http://dx.doi.org/10.1073/pnas.76.1.145] [PMID: 34148]

[249] Davis SS, Illum L, Moghimi SM, *et al.* Microspheres for targeting drugs to specific body sites. J Control Rel 1993; 24: 157. [http://dx.doi.org/10.1016/0168-3659(93)90175-5]

[250] Trubetskoy VS, Frank-Kamenetsky MD, Whiteman KR, Wolf GL, Torchilin VP. Stable polymeric micelles: lymphangiographic contrast media for gamma scintigraphy and magnetic resonance imaging. Acad Radiol 1996; 3(3): 232-8. [http://dx.doi.org/10.1016/S1076-6332(96)80448-X] [PMID: 8796670]

[251] Torchilin VP. Polymeric micelles in diagnostic imaging. Coll Surf B Biointerfaces 1999; 16: 305-19. [http://dx.doi.org/10.1016/S0927-7765(99)00081-8]

[252] Kim SC, Kim DW, Shim YH, *et al.* *In vivo* evaluation of polymeric micellar paclitaxel formulation: toxicity and efficacy. J Control Rel 2001; 72(1-3): 191-202.

[http://dx.doi.org/10.1016/S0168-3659(01)00275-9] [PMID: 11389998]

[253] Kim TY, Kim DW, Chung JY, *et al.* Phase I and pharmacokinetic study of Genexol-PM, a cremophor-free, polymeric micelle-formulated paclitaxel, in patients with advanced malignancies. Clin Cancer Res 2004; 10(11): 3708-16.
[http://dx.doi.org/10.1158/1078-0432.CCR-03-0655] [PMID: 15173077]

[254] Madaan A, Singh P, Awasthi A, *et al.* Efficiency and mechanism of intracellular paclitaxel delivery by novel nanopolymer-based tumor-targeted delivery system, Nanoxel(TM). Clin Transl Oncol 2013; 15(1): 26-32.
[http://dx.doi.org/10.1007/s12094-012-0883-2] [PMID: 22855169]

[255] Hamaguchi T, Kato K, Yasui H, *et al.* A phase I and pharmacokinetic study of NK105, a paclitaxel-incorporating micellar nanoparticle formulation. Br J Cancer 2007; 97(2): 170-6.
[http://dx.doi.org/10.1038/sj.bjc.6603855] [PMID: 17595665]

[256] van Gaal EV, Crommelin DJ. Polymeric Micelles. Non-Biological Complex Drugs. Springer 2015; pp. 11-76.
[http://dx.doi.org/10.1007/978-3-319-16241-6_2]

[257] Tsuji A, Hamaguchi T, Yamaguchi K, *et al.* A phase II study of NK012, a polymeric micelle formulation of SN-38, in colorectal cancer patients who had received prior oxaliplatin-based regimen. Cancer Chemother Pharmacol 2018; 82(6): 1021-9.
[http://dx.doi.org/10.1200/jco.2015.33.15_suppl.3527]

[258] Vergote I, Brize A, Lisyanskaya AS, Lichinitser M. Randomized phase III study comparing paclical-carboplatin with paclitaxel-carboplatin in patients with recurrent platinumsensitiveepithelial ovarian cancer. J Clin Oncol 2015; 33: 5517.
[http://dx.doi.org/10.1200/jco.2015.33.15_suppl.5517]

[259] Pitto-Barry A, Barry NP. Pluronic® block-copolymers inmedicine: From chemical and biological versatility to rationalization and clinical advances. Polym Chem 2014; 5: 3291.
[http://dx.doi.org/10.1039/C4PY00039K]

[260] Endo K, Ueno T, Kondo S, *et al.* Tumor-targeted chemotherapy with the nanopolymer-based drug NC-6004 for oral squamous cell carcinoma. Cancer Sci 2013; 104(3): 369-74.
[http://dx.doi.org/10.1111/cas.12079] [PMID: 23216802]

[261] Ghamande S, Lin C-C, Cho DC, *et al.* A phase I study of the novel DNA topoisomerase-1inhibitor, TLC388 (Lipotecan®), administered intravenouslyto patients with advanced solid tumors. Mol Cancer Ther 2011; 10 (Suppl. 1): 89.

[262] Ueno T, Endo K, Hori K, *et al.* Assessment of antitumor activity and acute peripheral neuropathy of 1,2-diaminocyclohexane platinum (II)-incorporating micelles (NC-4016). Int J Nanomed 2014; 9: 3005-12.
[http://dx.doi.org/10.2147/IJN.S60564] [PMID: 24971011]

[263] Takahashi A, Yamamoto Y, Yasunaga M, *et al.* NC-6300, an epirubicin-incorporating micelle, extends the antitumor effect and reduces the cardiotoxicity of epirubicin. Cancer Sci 2013; 104(7): 920-5.
[http://dx.doi.org/10.1111/cas.12153] [PMID: 23495762]

[264] Owen SC, Chan DP, Shoichet MS. Polymeric micellestability. Nano Today 2012; 7: 53.
[http://dx.doi.org/10.1016/j.nantod.2012.01.002]

CHAPTER 6

Potential and Promises of Carbon Nanotubes in Drug Delivery

Nagarani Thotakura, Vipin Kumar and **Kaisar Raza***

Department of Pharmacy, School of Chemical Sciences and Pharmacy, Central University of Rajasthan, Bandarsindri, Dist. Ajmer, Rajasthan, India

Abstract: For almost the last three decades, carbon nanotubes (CNTs) are under continuous scientific investigation due to the inimitable properties and immense applications in diverse fields, especially in drug delivery. CNTs are the rolled graphene sheets with sp^2 hybridization. Various synthetic approaches like arc discharge method, laser ablation methods, chemical vapor deposition, and pyrolysis are employed to produce CNTs. Despite applications in biomedical sensors, artificial implants, preservatives, solar cells, and transistors, CNTs have also been explored to deliver anti-cancer agents, anti-malarials, anti-HIV agents, and anti-bacterial drugs. CNTs can even deliver nucleic acids and genes into cells of the target site. The present chapter explains the discovery of CNTs, their types, synthetic methods, various reactions possible with the usage of CNTs, applications of CNTs, safety profile, and challenges involved.

Keywords: Carbon nanocarriers, Gene therapy, Nanotechnology, Nucleic acids, Pharmacokinetics, Safety profile, Synthetic approaches.

INTRODUCTION

Most researchers contribute to nanotechnology and other nanocomposites in the present era, which help to understand biological systems and disease diagnosis and treatment [1]. Nanoparticulate drug delivery systems using biodegradable polymers help overcome certain problems like unwanted toxicity, less efficiency, dose dumping, solubility, and other metabolic issues [2].

Various nanocarriers used in novel drug delivery systems are generally made up of either inorganic or organic matter [3]. Along with various nanocarriers like liposomes, niosomes, ethosomes, polymeric micelles, mixed micelles, solid lipid nanoparticles, and nanolipoidal carriers, carbon-based nanocarriers like

* **Corresponding author Kaisar Raza:** Department of Pharmacy, School of Chemical Sciences and Pharmacy, Central University of Rajasthan, Bandarsindri, Dist.Ajmer, Rajasthan, India 305 817; E-mail: drkaisar@curaj.ac.in

Surendra Nimesh, Nidhi Gupta and Ramesh Chandra (Eds.)

graphenes, fullerenes, and carbon nanotubes are being widely used in the drug delivery process [4 - 8].

Until 1980, a common perception in science was that the chemistry of carbon was well known. According to the discovery of buckminsterfullerene in 1985 by Rick Smalley and coworkers (as recognized by the Nobel prize in Chemistry awarded to Kroto, Curl and Smalley in 1996) [9], the allotropes of carbon increased to three, namely, graphite, diamond, and fullerenes [10]. During the carbon-carbon composites workshop in 1990, Rick Smalley proposed the existence of cylindrical fullerenes, which he named buckytubes. He anticipated that these tubes might be due to the elongation of C_{60} fullerene [11]. In 1991, during the oral presentation at the fullerene workshop on symmetry in Philadelphia, Dresselhaus proposed that carbon nanotubes are capped at both ends by fullerenes [12]. But finally, in 1991, when Ijima proved the existence of carbon nanotubes (CNTs) with an image of multiwalled carbon nanotubes (MWCNTs), obtained through transmission electron microscope [13]. Two years after observing MWCNTs, both the Bethune and Ijima groups identified single-walled carbon nanotubes (SWCNTs) [14, 15].

Even though the official discovery of carbon nanotubes is credited to Ijima, these nanosystems were observed thirty years back, when Bacon was doing carbon research to determine carbon fibers' properties at Union Carbide 1956. He reported that he found carbon nanotubes while heating graphite at high temperatures [16]. In the 1970s, Endo synthesized these nanotubes by pyrolysis of benzene and ferrocene at 1000°C and imaged using high-resolution transmission electron microscopy (HRTEM) [17].

Iijima proved that CNTs are concentrically rolled graphene sheets, but with many chiralities and helicities. Initially, he observed only MWCNTs, but later on, he reported SWCNTs' existence and structural illustrations. The structure of SWCNTs depends on how the graphene sheet is rolled. The shape of the SWCNT was based on vector C, a circumference of the tube surface connecting two carbon atoms. It can be determined by using the below formula [14].

$$\mathbf{C = na_1 + ma_2} \quad \mathbf{(1)}$$

Where, a_1 an a_2 are two basis vectors of graphene, whereas m and n are two indexes that determine the chiral angle 'θ' (Equation 2).

$$\boldsymbol{\theta} = \mathbf{Tan^{-1}[\sqrt{3}(n/(2m + n))]} \quad \mathbf{(2)}$$

Based on this chiral angle, SWCNTs are divided into three types (Fig. **1**). These are armchair (n=m, θ=30°), zig-zag (m=0, n > 0, θ=0°) and chiral (0 <|m|< n, 0 < θ < 30°). In general, it is observed that arm chaired SWCNTs are metallic with no bandgap. Zig-zag and chiral SWCNTs show semi-metallic and semi-conductive properties with a definite bandgap. The bandgap of both zig-zag and chiral SWCNTs is equal to an inverse of the tube diameter. The Equation used for determining the diameter is shown as Equation (3).

$$\mathbf{d = \sqrt{3}[a_{c-c}(m^2 + mn\ + n^2)^{1/2}/\pi} \qquad \mathbf{(3)}$$

Where $\mathbf{a}_{\mathbf{c\text{-}c}}$ is C-C bond length and m and n are indexes as discussed earlier in Equation (1)

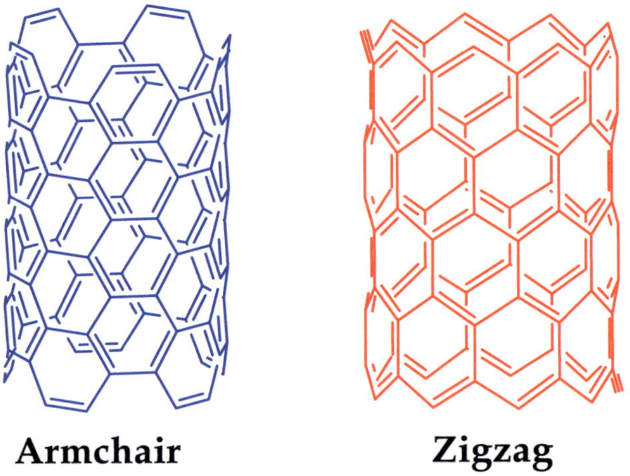

Fig. (1). Examples of armchair and zig-zag SWCNTs identified by integers (m,n).

Based on the number of graphene rolls, CNTs are classified into three types. They are single-walled CNTs (single graphene roll), double-walled CNTs (two CNTs, one inside the other), and multiwalled CNTs (multilayers of graphene rolls with an interlayer distance of 0.34 nm) [18, 19]. Both SWCNTs and MWCNTs are compared in Table **1** regarding various aspects.

Table 1. Comparison between SWCNTs and MWCNTs.

S.No	SWCNTs	MWCNTs
1	A single layer of graphene. It can be easily twisted.	Multiple layers of graphene. It cannot be easily twisted.
2	The catalyst is required for synthesis.	It can be produced without a catalyst.

(Table 1) cont.....

S.No	SWCNTs	MWCNTs
3	Bulk synthesis is difficult, as it requires proper control overgrowth and atmospheric conditions.	Bulk synthesis is relatively easy.
4	Purity is poor. A chance of defect is more during functionalization.	Purity is high. The chance of defect is less, but once occurred, it is difficult to improve.
5	Less accumulation in the body.	More accumulation in the body.
6	Characterization and evaluation are easy.	It has a very complex structure.

METHODS INVOLVED IN SYNTHESIS OF CNTS

The synthetic process of CNTs involves specific conditions, which include the use of a hot furnace with hydrocarbon gas [20], a high-powered laser [21], and two arcing graphite rods [13]. In overview, the basic requirements for the synthesis of CNTs include a carbon source, active catalyst, and adequate energy. Various synthetic approaches of CNTs have been discussed in the subsequent sections.

Arc Discharge Method

Arc discharge method is the first method used for the synthesis of all three types of CNTs. A thirty years old process of preparing carbon whiskers, reported by Roger Bacon and Kratschmer-Huffman for producing fullerenes, is analogous to this method. In this method, for synthesis, high temperatures above 1700°C are used, leading to structural defects. Using this method, the synthesis of MWCNTs is straightforward vis-à-vis the other two types of CNTs. For the synthesis of the SWCNTs transition, a metal catalyst is used [22].

Initially, between the two graphite electrodes (5-20 mm diameter), an arc is fashioned at a gap of 1 mm. Inert gases such as helium (He) and argon (Ar) at a pressure of 100 to 1000 torr are employed [15, 23]. Nanotubes, along with fullerenes and carbon soot, are deposited on the cathode. During the synthesis process, Iijima used Fe: C as anode and methane: argon for the inert environment. Bethune produced CNTs by changing the Fe: C anode to Co: C anode and carried out the whole experiment in a helium environment [24]. The arc discharge chamber is filled with a suitable inert gas mixture. Two thin electrodes are installed in the chamber's center and the cathode with a shallow dip to hold a small piece of a catalyst. Now the power of 200 A at 20 V is passed between the two electrodes [25]. Nowadays, several variations of arcs can be used for the arc discharge chamber. The maximum number of scientists are currently using Ar: He environment to carry out the reaction process. The diameter of CNTs can be easily controlled by changing the ratio of Ar: He. The higher the percentage of Ar, the smaller is the diameter of synthesized tubes [26].

Laser Ablation Method

In 1996, SWCNTs were first produced on a large scale by the Smalley group at Rice University [27]. CNTs produced by this method depends on specific parameters like flow and pressure of the buffer gas, chamber chemical composition and pressure, the distance between a target and the substrates, temperature control, laser properties, structural and chemical properties of target material [28]. The mechanism and principle of synthesis are similar to that of the arc discharge method. This method's energy source is the hitting of graphite pellet containing catalysts (cobalt or nickel) with a laser [29]. In this method, the target material, graphite and cobalt/nickel, is placed in a quartz tube furnace at 1200°C with an inert atmosphere produced by argon or helium and subjected to vaporization with a laser pulse. Both pulsed or continuous waved laser can be used for this process [21].

The metal catalyst plays a vital role in the formation of SWCNTs with a large number of by-products. Both synthesized nanotubes and derivatives are collected through the condensation process. Generally, these by-products include nano-onions (concentric fullerene spheres), amorphous and graphitic carbon, and fullerenes (C_{60}, C_{70}) [30]. The diameter of SWCNTs produced by this method ranges from1.0 nm to 1.6 nm. As the technique helps in creating a good yield of CNTs, scientists are interested in using and exploring the new changes. The latest developments are recently reported in this field: the continuous-wave laser-powder method and ultra-fast pulses from a free-electron laser method [31].

Chemical Vapor Deposition (CVD)

The above discussed two methods are capable of producing a good yield of CNTs, which are highly unpurified. Therefore, by the continuous evaluation of researchers, CVD's process came into the limelight to control by-products' formation. In 1993, Endo *et al.* first used CVD to produce MWCNTs [32]. Later on, in 1996, Smalley and their group used Co-based CVD for the synthesis of SWCNTs [33]. This method helps to produce nanotubes in bulk quantities and synthesize individual SWCNTs aligned on SiO_2 for electronic applications [34]. This method even helps in the synthesis of vertical MWCNTs, which are used in high-performance field emitters. It is generally observed that there is an increase in the purity and the yield of SWCNTs compared to other currently available methods.

The basic principle involved in the synthesis of CNTs using this method is similar to that of the arc-discharge process *i.e.*, excitation of carbon atoms due to the presence of metallic catalysts. This is a type of thermal dehydrogenation reaction that helps break down a hydrocarbon molecule into carbon and hydrogen in the presence of metal catalysts like cobalt, nickel, and iron. On the other hand, it is analogous to the synthesis of hydrocarbon fuels (octane) by catalytic cleavage, which is very successful in large-scale throughput. To attain all the aims and objectives, the tubes were first punctured into silicon and metal particles implanted at the bottom. A hydrocarbon like acetylene is subjected to heating and decomposed on the substrate [35]. The carbon particles that decompose, come in contact with the metal particles and form a tunnel's shape. This helps in growing a long and properly aligned CNT. Along with the metal catalysts, the support of MgO or Al_2O_3 is also used to increase the surface area for a higher rate of reaction between the catalytic metal particles and pure carbon [36].

There are various CVD types like thermal or microwave plasma CVD, plasma-enhanced oxygen assisted CVD, radiofrequency CVD, water-assisted CVD, hot filament CVD and catalytic chemical vapor deposition (CCVD [37 - 39]). Nowadays, CCVD is the standard technique used for the production of CNTs. Through this method, CNTs of size ranging from 0.4 nm to 5 nm can be synthesized, depending on various factors like catalyst and feedstock. The final yield can be high (> 99%) and devoid of amorphous carbon and other by-products [40].

Flame Pyrolysis

This is the new method used to synthesize CNTs on a large scale under simple experimental conditions and less equipment. This method was first used by Liu *et al.*, which reported high yield and low by-products [41]. They used a V-type pyrolysis flame and carbon monoxide as the source. Heating was provided by the combustion of acetylene or air, which is pretreated with gas (H_2/ He). The catalyst used was pentacarbonyl iron, and premixed gas acts as diluent and protective. The observed diameter of synthesized CNTs ranged between 10-20 nm. The length of tubes also increased and was found to be in dozens of microns [42]. The effect of various parameters like premixed gases, the amount of carbon source, temperature, and sampling time on the growth process of CNTs has been well studied [43].

COMMON CHEMICAL REACTIONS OF CNTS

Due to extraordinary characteristic properties, CNTs are considered promising candidates in drug delivery, biosensors, nanotechnology-related applications, and various other purposes. However, the main limitation of using the CNTs is the low solubility in various solvent systems used regularly. After multiple reactions, CNTs are only in such a condition that they quickly form dispersions through sonication, resulting in precipitation, as soon as the sonication is ceased [44]. In various instances, CNTs react with a wide range of compounds, resulting in supramolecular complexes' formation. These complexes help in the production of novel and innovative nanocarriers and devices. CNTs can undergo numerous chemical reactions, increasing solubility in organic, inorganic, and biological systems. These procedures can be grouped into three categories like (a) covalent attachment of CNTs with the functional groups, (b) noncovalent bonding of chemical moieties on the skeleton of the CNT molecules, and filling the inner cavity of CNTs as shown in Fig. (**2**) [45].

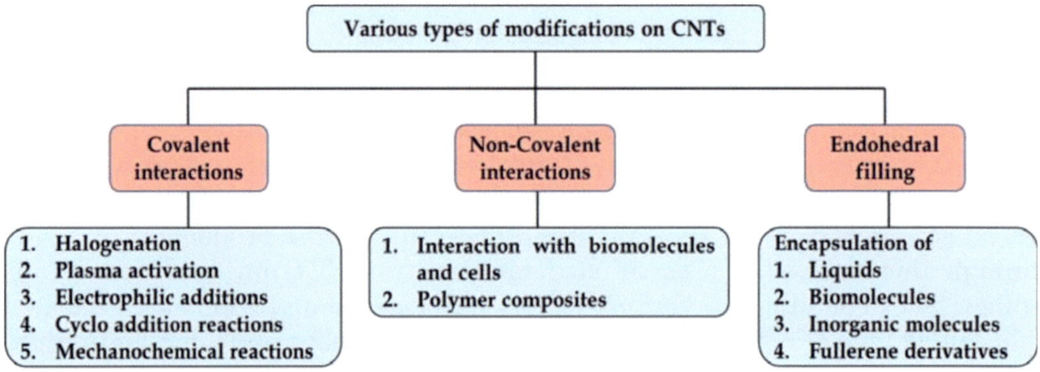

Fig. (2). Diagrammatic representation of various types of reactions using CNTs.

Covalent Attachment of Functional Groups

Some of the covalent approaches used for the functionalization of CNTs are discussed in subsequent sections:

Sidewall Halogenation of CNTs

The CNTs produced through the arc-discharge method have been subjected to fluorination at 600°C using elemental fluorine [46]. After several studies, it was observed that better fluorination results were observed when the temperature used was in the range of 150-400°C. The use of higher temperatures increases the propensity for decomposition in the graphitic network of tubes. Through this method, moderate solubility (1mg/mL) was observed in alcoholic solvents [47]. Fluorine can be detached from the tubes either by heat annealing or hydrazine solution in 2-propanol. Pristine CNTs are also obtained by defunctionalization using electron beam irradiation during microscopic observation [48]. This reaction offers opportunities for further functionalization of CNTs, like incorporating alkyl groups through the Grignard reaction [49]. Even diols and diamines react readily with the halogenated CNTs *via* nucleophilic substitution, resulting in the formation of amino alkylated CNTs that are freely soluble in dilute acids and water. These aminated CNTs can be functionalized further to form dicarboxylic chlorides. Halogenation of CNTs (bromination and chlorination) can also be achieved through electrochemical processes [50].

Hydrogenation

CNTs are reported to undergo hydrogenation in a mixture of methanol and ammonia, using lithium metal as a catalyst *via* Birch reduction [51]. The hydrogenated CNTs have been reported to be stable till 400ºC. Scientists observed the appearance of disorders and furrows all over the surface of CNTs after this process of functionalization using Transmission Electron Microscopy (TEM) [52]. CNTs can also be subjected to hydrogenation through proton bombardment or by supplying atomic hydrogen using glow discharge [53].

Ozonolysis

CNTs on ozonolysis at a temperature of -78ºC result in primary CNT-ozonides [54]. Naïve CNTs are treated with sodium borohydride or hydrogen peroxide to synthesize aldehyde/ketone, carboxylic acid/ester, and alcoholic functional groups over the CNT surface. Through this process, the walls of CNTs are decorated by the active functional groups, which increase the aqueous solubility and the chemical reactivity of tubes [55]. Banerjee *et al.* observed that the diameter of the CNTs determines the nature of their chemical reactivity. If the diameter of CNTs is smaller, then there will be high strain energy between the carbon atoms resulting in high rehybridization energy [56].

Mechanochemical Reactions

CNTs on exposure to ball-milling produce small-sized nanotubes with various functional groups like hydroxyl, amine, thiols, and mercaptans on the surface. In this process of mechanochemical reactions, the CNTs were initially mixed with the potassium hydroxide using a ball mill [57]. After this, a small amount of ethanol is added and subjected to chemical treatment. After completing this process, hydroxylated CNTs are produced with a high water solubility (up to 4 mg/mL). Using the same approach, even C_{60} fullerenes can also be attached to the surface of CNTs [58, 59].

Plasma Activation

To avoid the excessive exposure of CNTs to chemicals, a green technique known as the radio frequency glow-discharge plasma activation method is developed [60]. When CNTs are subjected to treatment with aldehyde-plasma, then Schiff base linkages are observed through the immobilization of amino dextran chains.

The final product obtained is highly hydrophilic due to carbohydrate groups' presence on the CNT surface [61].

Nucleophilic Additions

Basiuk *et al.* recently tried to perform solvent-free amination on the MWCNTs caps using octadecyl amine. They suggested that only pentagon rings of the graphitic network get involved in these type of addition reactions. They also stated that hexagonal graphite network rings avoid such chemical reactions due to their stability [62]. Chen and group synthesized CNTs with alkyl and the carboxylic acid groups by treating naïve CNTs with carbon dioxide and secondary butyl lithium. They observed that the functionalized CNTs have a size range of 100-200 nm and can easily be dispersed in water (0.5 mg/mL) [63]. Georgekilas *et al.,* performed the alkylation on SWCNTs, using layered smectite minerals as catalysts. They reported the intercalation of nanotubes between the layers of clay after alkyl modification. They also said that the spacing between the clay layers was increased by 2.5 nm showing the fractional exfoliation of an inorganic component due to the alkyl-modified nanotubes [64].

Noncovalent Interactions

Generally, CNT exists in bundles tied very strongly, so there is difficulty dispersing them homogenously in most solvent systems [65]. The most commonly used method to unwind these strong bundles is noncovalent functionalization on nanotubes' surface with reactants like polynuclear aromatic compounds, polymers, biomolecules, and surfactants [66]. This method is desirable in current day research due to its ability to attach the functional groups without damaging the graphitic network on the tubular surface. These interactions are commonly under the control of thermodynamics and are based on the weakest bonding forces like Π-Π stacking or Vander Waals forces [67]. It has been observed that these interactions between functional groups and nanotubes help in the proper replacement of CNTs on the nanoparticles [68].

Endohedral Filling

Through various studies, scientists observed that CNTs possess the capability to load materials of specific nature in their inner cavities. This process helps in storing liquid fuels as well as in the production of nanowires [69]. Any liquid having a surface tension below 180 mN can easily penetrate the inner cavity of CNTs at atmospheric pressure [70]. High pressure should be applied for the tubes' forceful entry in the case of fluids with high surface tension. Several attempts have been made to fill the CNTs during their growth process by sublimating metal-containing compounds [71].

CNTs play a crucial role in nanofluidics. In the designing of biomedical devices, nanofluidics is considered a powerful technology. In this nanofluidics, all the transportation process is carried through both natural as well as forced convection. Initially, water is incorporated into the inner cavities of the tubes. It has been observed that water as a solvent system can moisten the inner walls of CNTs, which helps in solution chemistry within the CNTs [72]. Molecular dynamic simulations are used to study water's behavioral properties, which is encapsulated in the tubes. Modeling is employed to understand the effect of encapsulation on the structure of hydrogen bonding. The obtained results showcase a substantial decrease in the number of hydrogen bonds than that of bulk water. In the case of narrow CNTs, it has been found that there is a destruction in the network of hydrogen bonding.

In some cases, it has been observed that these water molecules exist as long linear chains [73]. On the contrary, wide nanotubes can pack the water molecules as a stacked column of cyclic hexamers. Heating of water incorporated CNTs through electrical irradiation results in shrinkage due to solvent evaporation. The other crucial parameter which affects the confinement of liquids is pressure. Through certain studies, it is established that on an application of 50 to 500 Mpa of force, the phase transition occurs, and water converts into ice inside the CNTs [48]. Finally, scientists concluded that CNTs could be used as a channel for water [74, 75].

Other than liquids, even biomolecules within the size range of the diameter of internal cavities of CNTs can be entrapped easily. Various proteins and enzymes, along with DNA, can also enter into the free cavity space of CNTs. The enzyme known as lactamase was encapsulated into the inner cavity of the CNTs and tested for the activity with the free form of lactamase for the hydrolysis of Penicillin. It is found that the immobilized enzyme can be catalytically active without any conformational change [76]. Through molecular dynamic studies, it was found that DNA oligonucleotides made up of eight base pairs can be easily entrapped into the CNTs in aqueous media. These trapped DNA molecules can be observed using a fluorescent microscope [77]. These dynamic interactions between CNTs and biomolecules are based on the van der Waals and hydrophobic forces. Even RNA molecules (single-stranded) can also be trapped in the CNTs having a width of 1.5 nm, respectively. Inspite of having enormous energy and entropy values, they observed that RNA remains hydrophobically entrapped in the CNTs [78].

APPLICATIONS OF CNTS

Due to various characteristic properties of CNTs, they have several applications like the formation of waterproof fabrics, can be used like steel and concrete due to its strength. Its thermal and electrical properties can be used in the construction of sensors and electrical circuits. It can also be used in food packing technology [79]. Some of the applications of CNTs are explained in detail below.

Artificial Implants

Generally, a body cannot accept the organ implants which are placed into the biological system. This may lead to infection and post-administration pain. But as the nanotubes and nano horns are nanometric sized, they can efficiently deliver amino acids and proteins without any rejection. They can also act as an alternative to joints during replacement surgery without any reaction. They can be even used as the alternative to the bone when CNTs gets filled with calcium due to their tensile strength [80].

Preservatives

CNTs and nano horns themselves possess an oxidative nature. This helps in storing various formulations made up of water and other microbial attractive solvent systems. They also can prevent the oxidation of drugs. Due to this anti-oxidant property, they are also used in the production of anti-aging creams. It is also used along with zinc oxide in sunscreen formulation to stop skin oxidation [81].

Diagnostic Tool

CNTs, when filled with specific proteins and enzymes, show fluorescent properties. They can be used as biosensors within the body due to certain biomolecules [82] . CNTs act as nanocapsules and can be filled with materials with magnetic properties and enzymes which are radio-labeled. These nanocapsules can be used as biosensors. Nanotubes, nanomotors, and robots help study the cell structure and functioning within the biological system [83, 84].

As a Catalyst

As the surface area of CNTs is high, large amounts of catalyst can be incorporated and used in the reaction process. Recovery of the catalyst is easy by encapsulating them within the CNTs. They can be activated and reused for the next time. Through this method, the frequency of adding the catalyst can be reduced along with the amount of catalyst [85].

Formation of Batteries

Almost all electronic devices need rechargeable batteries made up of lithium ions. Inside these batteries, lithium ions from one electrode to another one and produce energy. The two electrodes are made up of graphite and metal oxide, respectively [81]. North Carolina scientists used the electrode made up of SWCNTs instead of graphite, which increased the storage almost two times. It was found that electrodes made up of CNTs are thinner and lighter by ten times than that of carbon electrodes. Even the conductivity of the batteries is increased by a thousand times [86]. In the case of electric vehicles, the weight of the vehicle affects the requirement of the battery power directly. CNTs are also used as supercapacitors providing energy with a density of 30 K Watts/Kg. These supercapacitors help in the reduction of time of charging by electrical devices like phones and laptops. Using CNT infused paper, ultra-thin flexible batteries have been made. In this battery, the paper dipped in the ionic solvent acts as an electrolyte. As the human body fluids like sweat, blood, and urine can also be used as the electrolyte in implants made up of CNTs. The main advantage of these ultra-thin batteries is they can be rolled, folded up, or cut down with any loss of efficiency [87].

Solar Cells

CNTs are also used in the manufacturing of solar cells. Scientists from Georgia Tech Research Institute used iron-coated silicon wafers and synthesized CNTs towers, 100 μm in length. In the square centimeter of an area, there are around forty thousand towers [48]. Around millions of CNTs should be aligned together to form the tower. The ability of absorption of light by CNTs was almost ten thousand times more than the normal cells [88]. The excess of light absorbed gets reflected from the sides of the tower. The cells' efficiency gets 100% increased significantly when the sun is over the head, *i.e.*, at 90°. These cells automatically rotate in the direction of the sun. This property can be utilized in the space, which prevents mechanical energy for the rotation cells to absorb sunlight [89].

Transistors

For the formation of integrated circuits and various modern devices like digital switches, transistors are critical. Different configurations of CNTs results in the formation of defects. These defects help in the formation of transistors. The controllers made up of nanotubes are having the electron size [90]. These switches can be observed only at cryogenic temperatures but with normal eyes. Inside the button made up of CNTs, the molecules capable of passing the electric current can be placed very quickly. Once the molecule is placed inside the switch, it controls electrons' movement, *i.e.*, current [91]. The gate used for this method is in the size of one nanometer, or in other words, it can also be three times smaller than the size of a silicon chip. In 2001, some scientists proved that room temperature is sufficient for these nanotubic transistors' work. Even IBM also performed specific experiments to produce these transistors. Henceforth, these CNTs can be a better option in the field of electrical circuits [92].

CNTS AS DELIVERY VEHICLES

There are two common ways to functionalize the CNTs. One is the well-established method, *i.e.*, oxidation using strong acids, which leads to structural damage and the generation of carboxylic acid groups on the surface. These carboxylic acid groups help in making the CNTs water-dispersible but not wholly soluble. The other way of functionalization is functional groups' addition to the surface or tips of the nanotubes. CNTs can be biocompatible only when they are soluble in physiological systems. Functionalized CNTs can be used to deliver various molecules like proteins, peptides, drug molecules, and nucleic acids.

Peptide Delivery

Scientists initially thought that CNTs could be used to deliver bi-active peptides, which helps the immune system activate and produce antibodies [93].To evaluate this, CNTs are subjected to amination, and the B-cell epitope of foot and mouth disease virus (FMDV) is attached covalently by using a bifunctional linker. The peptides bound to CNTs rearrange themselves to form the required secondary structure, which can be easily recognized by various antibodies (both monoclonal and polyclonal). Synthesized CNT conjugated peptide systems were subject to multiple *in-vivo* immunogenic studies. It was found that the production of antibodies (immunization) is high in mice, which are injected with the CNT-FMDV peptide conjugate system when compared to that of the free peptide. Antibodies produced in mice are due to the peptides but not because of CNTs, as they are peptide-specific antibodies. It was observed that the obtained antibodies have the ability to neutralize the virus. Even CNTs are used as the potential carriers in delivering vaccines, which is proved through the interactions with complement. Complement is produced in biological systems, which can kill the pathogens [94]. They are the protein series, which helps to recognize, opsonize, clear, and destroy various foreign particles, necrotic or apoptotic cells, and pathogens. Salvador-Morales and group proved that CNTs could activate complementarity in both classical and alternate ways through adsorption of selective proteins [95]. As these compliments are produced in response to the antigens, it can indirectly enhance antibodies' formation due to CNT-peptide conjugated systems' immune response.

Cellular Uptake of CNTs

The major key factor for functionalizing CNTs is its ability to penetrate the cell membranes. If CNTs are conjugated with fluorescent compounds, they can easily intake by the cells. The detection process in the fibroblast's cytoplasm and nucleoplasm becomes more straightforward with types of equipment like confocal laser scanning and epifluorescence microscopes. The cellular uptake mechanism of these derivatized CNTs is independent of specific mechanisms like endocytosis and passive transport. During the incubation of cells with endocytosis inhibitors, there is no particular influence on functionalized CNTs to penetrate cells. There is no effect of temperature on the cellular uptake of functionalized CNTs. This is observed by incubating the cells with CNTs at both lower and higher temperatures. Dai and his colleagues synthesized oxidized CNTs and attached biotin or fluorescein, which helps in the formation of fluorescent streptavidin and avidin-biotin complex after uptake by cells [95]. As there is the formation of fluorescent complexes, nanotubes can be easily identified in the cell. With contrary to the above statement, in this study, CNT conjugate entered into cells, especially endosomes, through the uptake mechanism known as endocytosis. Transmission electron microscopy (TEM) can also be used to observe the moment of CNTs within the cell. As CNTs are visualized perpendicular to the cell's cytoplasm, the possible mechanism of uptake is the same as microneedles. Microneedles can penetrate and diffuse the cell's bilipid layer, resulting in the leakage of the cellular fluids and, finally, cell death.

Delivery of Nucleic Acids

Till now, various cationic systems are used for the cellular delivery of nucleic acids. CNTs, which are derivatized with cationic functional groups like amine, can form complex structures with nucleic acids through electrostatic interactions. The main aim of these systems is to observe a significant enhancement in gene expression, as plasmid DNA only can enter the cells and the nucleus. Like cationic nanocarriers like liposomes, nanoparticles, and microparticles made up of cationic polymers, the cationic nature of the macromolecular functionalized CNTs can also help in the condensation process of plasmid DNA [96]. The potential of CNTs in gene transfer can be explored by adsorbing plasmid DNA pCMV-Bgal expressing b-galactosidase on the surface of the cationic functionalized CNTs. Both SWCNTs and MWCNTs can make stable complexes that can be characterized using electrophoresis, fluorescence dye exclusion, SEM, and TEM. Through gene transfer experiments and the formation of stable complexes, it was clear that functionalized CNTs affect the b-galactosidase expression. Gene expression levels were increased by five to ten times when compared to the DNA alone. When the external surface CNTs were functionalized covalently using polyethyleneimine (PEI), then the efficiency of the DNA delivery increased. PEI-derivatized CNTsis used in the delivery of plasmid DNA and tested in various cell types, but it was found that the expressed level of luciferase is the same as that of cells injected with PEI alone. Scientists used the same method for the condensation of oligodeoxynucleotide (ODN) chains and improving their activity to stimulate the immune system [96]. CNTs are also used for the delivery of RNA polymers into cells. CNTs use the method of non-specific binding for the condensation of RNA. Through various reports, it was clear that CNTs can be used in the delivery of nucleic acids. These studies open the scope of CNTs' usage in vaccination, gene, and immunization therapy.

Drug Delivery Using CNTs

There are continuous efforts for the effective delivery of drugs with reduced side effects and target specificity. Generally, a new drug delivery system was developed to improve the therapeutic and pharmacological profile of a drug. As the functionalized CNTs can penetrate the cells, it can offer a better drug delivery route [97]. These derivatized CNTs are used in the delivery process of various anti-cancer, anti-HIV, anti-malarial and anti-bacterial drugs. Another advantage of using these nanocarriers in the drug delivery process is their ability to carry more than one drug at a particular time, increasing the targeting capacity or producing a synergetic effect. In some instances, this mainly helps the delivery of anti-tumor drugs and antibiotics [98]. To achieve this multi-drug delivery with a single nanocarrier, there is the necessity of multi functionalized CNTs. In some specific cases, along with drug molecules, certain fluorescent dyes are also conjugated or adsorbed to study parameters like cellular uptake and conjugate availability in the cytoplasm and nucleoplasm of the cell [93].

MWCNTs were conjugated with an antibiotic named Amphotericin-B and fluorescein. On linking antibiotics to CNTs, they can quickly enter into the mammalian cells with lesser adverse effects than the pure antibiotics. It was also observed that CNT conjugated with Amphotericin-B have high anti-fungal activity on a wide range of fungal species like Cryptococcus neoformans, Candida albicans, and Candida parapsilosis [99]. A group of scientists functionalized CNTs with carborane cages and used as the carrier to deliver boron neutron capture therapy. As these functionalized CNTs are water-soluble, they can be used as the best source for the treatment of various tumors. It was found that already some of the tissues contain small concentrations of carborane. On intravenous administration of carborane functionalized CNTs, the carborane concentration increases at the tumor sites [94]. As like CNTs, there is another group of carbon nanocarriers called single-walled carbon Nano horns used for the delivery of drugs to the target site. These are spherical and made up of an aggregation of nanotubes [100]. Murakami and the group synthesized dexamethasone loaded nanotubes and nano horns. They performed drug release studies and binding efficiency studies and concluded that on oxidation of these nanohorns, a high amount of drug could be adsorbed without hampering the biological veracity. This was confirmed by observing the increase in the response of glucocorticoid in the bone marrow of mice. They also reported the production of alkaline phosphatase in the osteoblasts of the mice. Through all these results, it can be concluded that functionalized nanohorns can be used as the novel carrier for the delivery of various drugs [101]. It was reported that these CNT conjugated systems were less cytotoxic, which further require pharmacokinetic evaluation to know multiple parameters like biodistribution, bioavailability, metabolism, and elimination from the biological system. From these studies, it can be concluded that the functionalization of CNTs helps deliver various drug molecules [102].

PHARMACOKINETICS AND BIODISTRIBUTION STUDIES OF CNTS

Pharmacokinetics (absorption, distribution, metabolism, and elimination) of CNTs, *i.e.*, the fate of CNTs in the human body, helps produce various useful biomedical products. Till now, the biodistribution of CNTs has not been understood completely. After performing several well-established studies, scientists cannot determine the biodistribution of these nanotubes in the body because of its contradictory and complex results. For the determination of toxicity and accumulation of CNTs in tissues route of administration is a vital factor. To increase the use of CNTs in the medical field, its ADME profile should be thoroughly investigated. According to medical terms, various pharmacological factors like half-life, elimination, biodistribution, distribution volume, and clearance help make a compound successful. In this book chapter, we will discuss the pharmacokinetics of CNTs after both intravenous and oral administration. These routes are the most commonly employed for delivering CNT-based medical formulations, which are under development. Table **2** shows the findings after administering the CNTs intravenously, whereas Table **3** shows the administration of CNTs through the oral route.

Table 2. Brief report of *in-vivo* studies of CNTs and their derivatives through intravenous route by various research groups.

CNT Type	Model	Dose	Observation Items	Findings	References
^{13}C-enriched naïve SWCNTs	600 μg/mouse	Male KM mice	The TEM analysis of extracted tissues and the determination of ^{13}C content in the blood samples were collected at regular periods for 28 days.	The main accumulates in the liver along with the lungs and spleen. The retention time of CNTs in these organs is 28 days. There is no sign of acute toxicity.	Yang *et al.*

(Table 2) cont.....

CNT Type	Model	Dose	Observation Items	Findings	References
Naïve SWCNTs	20 μg/kg	Rabbits	UV-Visible, near IR, determines concentration nanotubes in serum at pre-determined time points and examination of tissues collected from various organs under fluorescent microscopy.	The only site of accumulation is the liver. Half-life is observed as one h without any acute toxic effects.	Cherukuri *et al.*
[^{14}C- taurine]-MWCNT	10 μg/mouse	Male KM mice	Observation of liver, lungs, and feces under TEM and determination of ^{14}C activity in blood samples collected from organs. Biological index examination.	The liver is the major accumulation site (80%), and most of the MWCNTs are entrapped in Kupffer cells, also known as hepatic macrophages. Retention time is 28 days. Small amounts are accumulated in the spleen and lungs too. A low level of acute toxicity is observed.	Deng *et al.*
^{111}In DTPA-SWCNT	60 μg (for biodistribution studies) and 400 μg (for *in-vivo* studies)	BALB/c mice	Determining blood clearance rate, TEM analysis of urine samples, and activity of ^{111}In in various organs.	Showed high affinity towards kidneys, skin, bone, muscle, and blood. A half-life of 3.5 h. Excreted through urine in intact form. No uptake by spleen and liver. No acute toxicity is observed.	Singh *et al.*
DOTA-SWCNTs labeled With ^{86}Y and ^{111}In	12 mg/mouse	Mice	Checking the whole-body distribution and clearance of ^{86}Y at 24 h after injecting the sample.	Distribution is observed in the liver, spleen, bones, and kidneys. Clearance is slow and occurred throughout the body. ^{86}Y is cleared from the blood within three h.	Mc Devitt

(Table 2) cont.....

CNT Type	Model	Dose	Observation Items	Findings	References
Pristine MWCNT	200 μg in 200 μL of mouse serum	BALB/c mice	Histological examination of liver, lungs, kidney, and spleen after 24 h of injection.	Tissue accumulation is low with increased elimination.	Lacerda *et al.*
PEG-MWCNT &Non-PEG-MWCNT	10 and 60 mg/kg	KM mice	Histopathological observation. Hepatotoxicity studies, AST/ALT, creatinine, and total bilirubin levels in the body. Determination of oxidative stress, TNF-α factor.	Both types induce necrosis, mitochondrial destruction of the liver, and inflammation. The latter showed more toxicity when compared to the PEG-MWCNTs.	Zang *et al.*
Taurine-MWCNT	60 or 100 mg of body weight	KM mice	Histopathological as well as the TEM examination of spleen and activity of malondialdehyde (MDA), superoxide dismutase (SOD) and reduced glutathione (GSH).	There no effect on the activity of SOD, MDA, and GSH after two months. No damage to the tissues in the spleen. There is a color change in accumulated CNTS (red to a white pulp).	Deng *et al.*

Table 3. Brief report of *in-vivo* studies of CNTs and their derivatives administer through oral route by various research groups.

CNT Type	Dose	Model	Observation Items	Findings	References
SWCNTs	64 μg/kg	Female Fisher 344 rats	Checked for the oxidative damage of DNA in mucosa, lungs, liver, and colon (8-oxodG levels measurement).	There is an increase in the levels of 8-oxide in both lungs and liver showing direct genotoxicity.	Folkmann *et al.*
MWCNT	60 and 100 mg/kg	Swiss Albino rats	Determination of levels of catalase (CAT) and superoxide dismutase (SOD). Histopathological examination of the liver.	Change in the levels of SOD and CAT. Necrosis, macrophage damage, and pathological change in the liver.	Awasthi *et al.*

(Table 3) cont.....

CNT Type	Dose	Model	Observation Items	Findings	References
[^{14}C- taurine]-MWCNT	10 μg/mouse	Male KM mice	Acute toxicity in the liver (measured LDH, TBIL, TBA, ALP, ALT). Histological examination of liver.	Radioactivity detectable in stomach, small, and large intestine, feces. CNTs not absorbed in the intestinal tracts, cannot enter the blood circulation. No acute toxicity.	Deng *et al.*
Ultra-short and full SWCNTs	100 mg/kg of body weight	Swiss Albino Rat	Pathological examination (*e.g.*, animal growth, animal behavior).	No abnormalities observed in pathological examination with all CNT materials used.	Kolosnjaj-Tabi *et al.*
MWCNT	500 mg/kg daily	CBA mice	Histopathological analysis of different sections of the GI tract and of the urothelium of the renal nephron.	Hypertrophy of lymphoid structures in the gastrointestinal mucosa, hypertrophy of renal capsule, and parenchyma, no pronounced toxic effects.	Golokhvast *et al.*
Hydroxylated CNTs	10 and 100 mg/kg	Pregnant CD1 mice	Histological and TEM assessment of fetal tissue.	Fetal morphological and skeletal abnormalities. No observed effects.	Philbrook *et al.*

ABSORPTION

Suppose a formulation is administered through an intravenous route. In that case, the drug's absorption rate is considered 100 because the full dose (amount of drug delivered to the body) is circulated in the blood. But if the same is administered through oral rate, then the drug should get dissolved and cross the gastrointestinal (GI) barrier to mix with the blood circulation and reach the target site. Till now, no reports are showing that CNTs can cross the GI barrier and deliver the bioavailability. In their studies, Deng *et al.* mentioned that the absorption of water-soluble MWCNTs through the intestinal tract is negligible [103]. Folkmann *et al.* reported that they observed genotoxicity in both lungs and liver by administering the single dose of SWCNTs dispersed in corn oil and a saline solution through the intragastric route. SWCNTs dispersed in corn oil cannot be considered for delivering through the oral route because the absorption of corn oil is the toughest than the intragastric route's adverse effects. The rate of absorption of CNTs through the GI barrier is dependent on the dose [104]. Philbrook *et al.*

reported that there was fetal toxicity during the gestation period observed by oral administration of CNTs of low concentrations (10 mg/kg). At higher concentrations (100 mg/kg), there was no toxicity due to aggregates' formation. These aggregates cannot cross the intestinal barrier [105].

Distribution

As the living organisms contain large amounts of the carbon-based network in the body, it is challenging to determine the CNTs through the biological test. Various isotopic labeling methods are used to track some rare impurities (like using cobalt for bay tubes) and biodistribution studies' performance. Specific proteins like opsonins get adsorbed onto the surface of CNTs resulting in easy recognition and capturing by the reticuloendothelial system (RES). After administering intravenously, it was observed that the distribution of CNTs in various organs and high amounts are accumulated mainly in the lungs, spleen, and liver. Out of all, the organ liver is found to be the major site of accumulation. Deng *et al.* found almost 80% of CNT accumulation in the liver within the 5 minutes of administering water-soluble [^{14}C-taurine]-MWCNTs intravenously [103]. They reported that the retention time of CNTs in the liver is 28 days, and after three months' duration from injecting, it was reduced to 20%. A group of scientists (Yang *et al.*) found that CNTs can even get accumulated in the brain and lungs, liver, and spleen. The time of retention in the brain is similar to that of the liver, *i.e.*, 28 days. This indicates that the CNTs possess the ability to cross the penetrating barrier known as BBB (blood-brain barrier) [106].

Metabolism

There is no much information about the metabolism of CNTs to date. It is reported that CNTs are known for their persistent nature in biological systems. This may be due to the structure and chemical composition of CNTs. After administering these CNTs in the biological systems, they carry the ability to accumulate and form granulomas in macrophages. Contrary to this statement, Kagan *et al.* reported that SWCNTs are subjected to biodegradation in the human body by the neutrophilic enzyme known as myeloperoxidase. This report proved that if the concentration of CNTs is not enough to initiate the process of accumulation within the biological systems, they can be the best option for novel drug delivery systems [107].

Elimination

The major disadvantage of using the CNTs in drug delivery and water insolubility is its rapid clearance from the body. The half-life of CNTs is reported as one h by Cherukuri *et al.* They also showcased that there is the immediate replacement of surfactant used for making the dispersion of CNTs with the blood proteins [108]. By these studies, scientists concluded that the elimination of CNTs depends on the formation of the serum protein layer on to the surface of the CNTs. To overcome this problem, the nanotube surface can be functionalized with various biocompatible molecules like PEG, PLGA, and glycine, which reduces the displacement of proteins, which helps indirectly increase the stability of CNTs within the biological systems. Various reports are showing that the liver and bile acts as the major elimination organs. Along with these organs, some reports indicate that there is elimination through kidneys also. The major factors influencing the elimination profile of CNTs are their length, type of derivatization, and the ability to disperse in a solvent system.

SAFETY CONCERNS AND CHALLENGES

Inspite of the huge increase in the production and use of CNTs, there is no such report that showcases the complete profile of its interactions within the human body. Other than the surface chemistry, coating of nanotubes, manipulating the surface charge, stiffness, length, and width, no other properties of CNTs are completely understood by the researchers. Even though various organs and tissues of the human body are exposed to the CNTs, the nasal route is found to be the most common route of exposure by inhalation. Hence it can be concluded that the lungs act as a key organ in promoting the toxic effects. This also leads to the development of various pulmonary disorders like chronic obstructive pulmonary disorder, caused due to air pollution. Maynard *et al.* prepared the CNTs for inhalation and checked for their toxic nature. They prepared the aerosolized CNTs using both laboratory and field settings [109]. In laboratory settings, they used diverse technologies, namely, two-component fluid bed technology. In field settings, they used SWCNTs produced either by laser ablation or by the HiPCO ablation method(high-pressure carbon monoxide). They concluded that the aerosols produced using a laboratory setting could release fine particles into the atmosphere. They found that these CNTs exist in lumps or clusters in a field study but not as fibers. These lumps cannot be inhaled as their size is not in the range of particles entering the biological system through inhalation. But during the synthesis of CNTs using the HiPCO ablation method, it was observed that there was 0.2 to 6 mg of CNTs depositing on the workers' gloves. This raised another debate that CNTs can be toxic, not only by inhalation. After this study, scientists started exploring the adverse effects of CNTs on the Human skin surface. Using different *in-vitro* and *in-vivo* methods, researchers started checking for the toxic effects of CNTs on various human body organs along with the skin. Based on the various biochemical applications of CNTs, a huge amount of literature is found with both known and unknown interactions of biological systems and CNTs. Ryman-Rasmussen *et al.* used a murine allergic asthma model to determine toxicity produced by inhalation of CNTs. After continuous exposure for 14 days, they observed that CNTs could produce airway fibrosis in both CNT and ovalbumin exposed mice [110]. In their next report, the same group mentioned that CNTs could enter the sub-pleura of non-pathogenic C7L6 mice on a single exposure. They also reported that they observed the sub-pleural fibrosis one day and regularly exposed up to 6 weeks from the first day. These findings mentioned above may be interesting and attract the researcher in understanding the toxic nature of the CNTs. The most important point to be noticed in all these reports is that they used the doses in extremely high amounts, which are not possible in day-to-day life. Nanoparticulate aerosols like CNTs, graphene, and fullerene are highly expensive, and the person handling them should have sound knowledge about the format and rate of deposition. For specific equipment is used for the synthesis of particulate dispersions. To date, there is no complete investigation of the pulmonary effect of CNTs. So, these studies are still in the craze for new inventions. Another study by Takagi *et al.*, which is similar to the research of Ryman-Rasmussen *et al.* Takagi and group injected CNTs intra-peritoneally into p53 mice at a dose of 3 mg/mL. They found that these CNTs show similar effects to asbestos fibers and induce mesothelioma in the pleural cavity [111]. But after reporting this data, they also faced a lot of criticism as the dose used is high than the normal animal dose. Other than these two reports, another group Poland *et al.*, also studied the effect of CNTs on the biological systems. This can be considered important information for determining the toxic effects produced by CNTs on the human body. Poland *et al.* injected 100 mg/mL (50 g per mouse) of a variety of CNTs (like long, short, stiffed, and bundled) through an intraperitoneal route and observed the formation of granuloma, which is based on the physical characteristics of CNTs [112]. CNTs, which are long and stiff, show similar characteristics to long asbestos fibers and cause granulomatous inflammation on the diaphragm's surface as the sample is injected intraperitoneally. Administration of short or bundled CNTs didn't show any such toxic effects. These findings are similar to that of the findings found in the case of asbestos fibers. Now the same "fiber paradigm" used for asbestos came into the limelight. The three principal rules of the paradigm are (i). Fiber should be non-bio soluble and bio-persistent; (ii). Fiber should have a thickness of < 5 nm); and (iii). The length of fiber should be 15 mm. Another Brown and group also reported Poland *et al.*'s points that long CNTs' biological activity is similar to asbestos fibers. Brown *et al.* mentioned that CNTs interact with the macrophages in *in-vitro* conditions resulting in frustrated phagocytosis. They also found that there is the tangling of fibers and amplification in the production of reactive oxygen species (ROS)both in cultured cells and the cell-free system. The straight fibers are intended to produce ROS slowly but have extended effects in animals. In addition to all these reports, Poland *et al.* also discussed the presence of metal contaminants and their role in promoting the toxicity profile of CNTs inside the biological systems as already mentioned by Kagan *et al.* Finally, after various studies, they reported that the toxicity profile of CNTs is not influenced by the presence of heavy metals (used during the process of synthesis) as contaminants. The major factors affecting the use of these metals are the type and form of CNTs synthesized. This indicates that there is a need for high-quality production along with complete characterization before performing biological studies. The similar biological activity of asbestos fibers and CNTs helps to provoke the oxidative stress by the production of ROS or by fitting into the oxidative stress paradigm. This property of CNTs helps in increasing the potential use for the treatment of various diseases.

There are very high-quality research publications over the past ten years. Until now, no study shows the complete interaction between the biological system and CNTs. The available data is not sufficient for use as a reference for future research and cannot be related to any form of policy. It is well-known that many concepts regarding CNT interactions within the body should be investigated. Even the minute things should be understood clearly before their production and application for biological purposes. Some of the significant aspects include (i). relation of CNTs with both innate and adaptive immune systems; (ii). The factors that affect human health by exposure to CNTs. As the usages of CNTs are extended to various fields like medical, electronic, and making of car tires, it is highly impossible to prevent the inhalation of CNTs in a human's life cycle. It is even more difficult to avoid exposure if the particular human being is involved in the synthesis, isolating, and processing of CNTs or during high-wear processes like cutting and sanding, as in asbestos fibers. As mentioned earlier, there is no need to make the aerosol formulations using CNTs as they can be quickly entrenched into a polymer matrix which is highly non-toxic and biodegradable. The release profile and % amount of release from these polymer matrices are not revealed. In conclusion, alternate routes like oral, intravenous, intraperitoneal, or dermal can be used to deliver drugs other than the nasal route.

The functionalization of CNTs helps in the reduction of their toxicity profile. As per the literature survey from the last 20 years, it was concluded that despite various disadvantages and side effects, the usage of CNTs is increasing in day-t--day life. To overcome the different undesirable and damaging properties of CNTs, multiple routes are followed. The most widely used route is the derivatization of CNTs for various functional groups. These derivatized CNTs play a prominent role in developing pharmaceuticals due to their use in imaging, detection, and treatment of several diseases. Still, now this is considered as the most unexplored area for the delivery of drug molecules. A vast number of tactics are used to deliver numerous bioactive molecules, drugs, and contrasting agents through CNTs either by covalent, noncovalent bonding, or endohedral filling.

These functionalized CNTs can be used for the inhibition of enzymes. On the Dendron functionalization of CNTs, there is the efficient and successful delivery of siRNA. An RGD-functionalized CNTs act as a photoacoustic molecular imaging agent in the detection of tumors. Liu *et al.* reported that CNTs could be used as a drug delivery vehicle after performing *in-vivo* studies using BALB/c mice. They conjugated paclitaxel with PEG-SWCNTs and injected them into the mice [113]. After three weeks from the administration time, they observed a significant reduction in the tumor size compared to that of pure taxanes. They found that intravenous administration of PTX-PEG-SWCNTs consecutively every six days does not produce any toxicity in different organs. These all the findings are in support of the statement, *i.e.*, "The simple carbon tubes can be progressed into an effective tool for drug delivery." Thus, CNTs can be used in the development process of several biomedical tools. Even though nanotubes have several advantages, specific challenges and queries should be solved to make CNTs the most successful carbon nanocarrier in the field of biomedicine. Chronic effect determination of CNTs in biological systems and their interactions of CNTs with the immune system are the essential factors to be studied in detail.

CONCLUSIONS

Till 1980, scientists thought carbon chemistry was well-known and established. They considered that this is the only element responsible for the rich organic as well as biochemistry. It can adapt to specific typical inorganic structures like graphite and diamond. In 1985, after discovering fullerenes, the scientists were amused that they did not have a complete carbon profile. They again started the research thinking about the hidden secrets of this element. Kroto, Curl, and Smalley were awarded the Nobel prize in 1996 for their contribution to chemistry. The second surprise after the discovery of fullerene is carbon nanotubes, which are cylindrical structures. There are various methods used for the synthesis of CNTs. Some of them include the laser ablation method, CVD, and pyrolysis.

Even though there are several well-known methods for synthesizing nanotubes, the exact mechanism is not identified. It was found that the new techniques can produce CNTs with high purity, high yield, and low defects. This can be applied in several biomedical, electrical, nanotechnology, and nanoscience fields due to its potential applications. To overcome the major disadvantage of CNTs known as water solubility, CNTs are subjected to functionalization using various techniques like covalent interactions, non-covalent-interactions, and endohedral filling. CNTs on functionalization shows potential applications. They can be used in biosensors, implants, transistors, and as preservatives. They can also be used to deliver various drugs, genes, immunoprotective agents, and nucleic acids. It was also reported that functionalized CNTs have a low toxicity profile with high efficiency. In conclusion, there are several aspects of CNTs that have to study in detail to extend its use in pharmacy and medicine.

CONSENT FOR PUBLICATION

Not applicable.

CONFLICT OF INTEREST

The author/editor has no conflicts of interest, financial or otherwise, to declare.

ACKNOWLEDGEMENTS

Declered None.

REFERENCES

[1] Nikalje AP. Nanotechnology and its applications in medicine. Med Chem (Los Angeles) 2015; 5: 2. [http://dx.doi.org/10.4172/2161-0444.1000247]

[2] Jabir NR, Tabrez S, Ashraf GM, Shakil S, Damanhouri GA, Kamal MA. Nanotechnology-based approaches in anticancer research. Int J Nanomed 2012; 7: 4391-408. [http://dx.doi.org/10.2147/IJN.S33838] [PMID: 22927757]

[3] Kulkarni RP. Nano-Bio-Genesis: tracing the rise of nanotechnology and nanobiotechnology as 'big science'. J Biomed Discov Collab 2007; 2: 3. [http://dx.doi.org/10.1186/1747-5333-2-3] [PMID: 17629932]

[4] Madhwi KR, Kumar R, Kumar P, *et al. In vivo* pharmacokinetic studies and intracellular delivery of methotrexate by means of glycine-tethered PLGA-based polymeric micelles. Int J Pharm 2017; 519(1-2): 138-44. [http://dx.doi.org/10.1016/j.ijpharm.2017.01.021] [PMID: 28093326]

[5] Singh A, Thotakura N, Kumar R, *et al.* PLGA-soya lecithin based micelles for enhanced delivery of methotrexate: Cellular uptake, cytotoxic and pharmacokinetic evidences. Int J Biol Macromol 2017; 95: 750-6. [http://dx.doi.org/10.1016/j.ijbiomac.2016.11.111] [PMID: 27919818]

[6] Misra C, Kumar M, Sharma G, *et al.* Glycinated fullerenes for tamoxifen intracellular delivery with improved anticancer activity and pharmacokinetics. Nanomed (Lond) 2017; 12(9): 1011-23. [http://dx.doi.org/10.2217/nnm-2016-0432] [PMID: 28440713]

[7] Raza K, Kumar D, Kiran C, *et al.* Conjugation of docetaxel with multiwalled carbon nanotubes and codelivery with piperine: Implications on pharmacokinetic profile and anti-cancer activity. Mol Pharm 2016; 13(7): 2423-32. [http://dx.doi.org/10.1021/acs.molpharmaceut.6b00183] [PMID: 27182646]

[8] Raza K, Thotakura N, Kumar P, *et al.* C60-fullerenes for delivery of docetaxel to breast cancer cells: A promising approach for enhanced efficacy and better pharmacokinetic profile. Int J Pharm 2015; 495(1): 551-9. [http://dx.doi.org/10.1016/j.ijpharm.2015.09.016] [PMID: 26383841]

[9] Acquah SFA, Penkova AV, Markelov DA, Semisalova AS, Leonhardt BE, Magi JM. Review—The beautiful molecule: 30 years of C_{60} and its derivatives. ECS J Solid State Sci Technol 2017; 6: M3155-62. [http://dx.doi.org/10.1149/2.0271706jss]

[10] Haley MM, Tykwinski RR, Rik R. Carbon-rich compounds : from molecules to materials 2006. [http://dx.doi.org/10.1002/ange.200685429]

[11] Popov V. Carbon nanotubes: properties and application. Mater Sci Eng Rep 2004; 43: 61-102. [http://dx.doi.org/10.1016/j.mser.2003.10.001]

[12] O'Connell M, Michael J. Carbon nanotubes : properties and applications. 2006. [http://dx.doi.org/10.1201/9781315222127]

[13] Iijima S. Helical microtubules of graphitic carbon. Nat 1991; 354: 56-8. [http://dx.doi.org/10.1038/354056a0]

[14] Iijima S, Ichihashi T. Single-shell carbon nanotubes of 1-nm diameter. Nat 1993; 363: 603-5. [http://dx.doi.org/10.1038/363603a0]

[15] Bethune DS, Kiang CH, de Vries MS, Gorman G, Savoy R, Vazquez J, *et al.* Cobalt-catalysed growth of carbon nanotubes with single-atomic-layer walls. Nat 1993; 363: 605-7. [http://dx.doi.org/10.1038/363605a0]

[16] Bacon R. Growth, Structure, and Properties of Graphite Whiskers. J Appl Phys 1960; 31: 283-90. [http://dx.doi.org/10.1063/1.1735559]

[17] Oberlin A, Endo M, Koyama T. Filamentous growth of carbon through benzene decomposition. J Cryst Growth 1976; 32: 335-49. [http://dx.doi.org/10.1016/0022-0248(76)90115-9]

[18] Abrahamson J, Wiles PG, Rhoades BL. Structure of carbon fibres found on carbon arc anodes. Carbon N Y 1999; 37: 1873-4. [http://dx.doi.org/10.1016/S0008-6223(99)00199-2]

[19] Liu Z, Tabakman S, Welsher K, Dai H. Carbon nanotubes in biology and medicine: *In vitro* and *in vivo* detection, imaging and drug delivery. Nano Res 2009; 2(2): 85-120. [http://dx.doi.org/10.1007/s12274-009-9009-8] [PMID: 20174481]

[20] Dai H, Rinzler AG, Nikolaev P, Thess A, Colbert DT, Smalley RE. Single-wall nanotubes produced by metal-catalyzed disproportionation of carbon monoxide. Chem Phys Lett 1996; 260: 471-5. [http://dx.doi.org/10.1016/0009-2614(96)00862-7]

[21] Thess A, Lee R, Nikolaev P, *et al.* Crystalline ropes of metallic carbon nanotubes. Sci 1996; 273. [http://dx.doi.org/10.1126/science.273.5274.483]

[22] Waldorff EI, Waas AM, Friedmann PP, Keidar M. Characterization of carbon nanotubes produced by arc discharge: Effect of the background pressure. J Appl Phys 2004; 95: 2749-54. [http://dx.doi.org/10.1063/1.1642737]

[23] Farhat S, Lamy de La Chapelle M, Loiseau A, *et al.* Diameter control of single-walled carbon nanotubes using argon–helium mixture gases. J Chem Phys 2001; 115: 6752-9. [http://dx.doi.org/10.1063/1.1390526]

[24] Szabó A, Perri C, Csató A, Giordano G, Vuono DB, Nagy J. Synthesis methods of carbon nanotubes and related materials. Materials (Basel) 2010; 3: 3092-140. [http://dx.doi.org/10.3390/ma3053092]

[25] Prasek J, Drbohlavova J, Chomoucka J, *et al.* Methods for carbon nanotubes synthesis—review. J Mater Chem 2011; 21: 15872. [http://dx.doi.org/10.1039/c1jm12254a]

[26] Eatemadi A, Daraee H, Karimkhanloo H, *et al.* Carbon nanotubes: properties, synthesis, purification, and medical applications. Nanoscale Res Lett 2014; 9(1): 393. [http://dx.doi.org/10.1186/1556-276X-9-393] [PMID: 25170330]

[27] de la Chapelle ML, Stéphan C, Nguyen TP, *et al.* Raman characterization of singlewalled carbon nanotubes and PMMA-nanotubes composites. Synth Met 1999; 103: 2510-2.

[http://dx.doi.org/10.1016/S0379-6779(98)01080-7]

[28] Chrzanowska J, Hoffman J, Małolepszy A, *et al.* Synthesis of carbon nanotubes by the laser ablation method: Effect of laser wavelength. Phys Status Solidi 2015; 252: 1860-7. [http://dx.doi.org/10.1002/pssb.201451614]

[29] Das R, Shahnavaz Z, Ali ME, Islam MM, Abd Hamid SB. Can we optimize arc discharge and laser ablation for well-controlled carbon nanotube synthesis? Nanoscale Res Lett 2016; 11(1): 510. [http://dx.doi.org/10.1186/s11671-016-1730-0] [PMID: 27864819]

[30] Bystrzejewski M, Rümmeli MH, Lange H, *et al.* Single-walled carbon nanotubes synthesis: a direct comparison of laser ablation and carbon arc routes. J Nanosci Nanotechnol 2008; 8(11): 6178-86. [http://dx.doi.org/10.1166/jnn.2008.SW05] [PMID: 19198361]

[31] Zhang Q, Huang J-Q, Zhao M-Q, Qian W-Z, Wei F. Carbon nanotube mass production: principles and processes. ChemSusChem 2011; 4(7): 864-89. [http://dx.doi.org/10.1002/cssc.201100177] [PMID: 21732544]

[32] Endo M, Takeuchi K, Igarashi S, Kobori K, Shiraishi M, Kroto HW. The production and structure of pyrolytic carbon nanotubes (PCNTs). J Phys Chem Solids 1993; 54: 1841-8. [http://dx.doi.org/10.1016/0022-3697(93)90297-5]

[33] Zhou O, Fleming RM, Murphy DW, Chen CH, Haddon RC, Ramirez AP, *et al.* Defects in carbon nanostructures. Sci (80-) 1994; 263: 1744-7. [http://dx.doi.org/10.1126/science.263.5154.1744]

[34] Height MJ, Howard JB, Tester JW, Vander Sande JB. Flame synthesis of single-walled carbon nanotubes. Carbon N Y 2004; 42: 2295-307. [http://dx.doi.org/10.1016/j.carbon.2004.05.010]

[35] Shah KA, Tali BA. Synthesis of carbon nanotubes by catalytic chemical vapour deposition: A review on carbon sources, catalysts and substrates. Mater Sci Semicond Process 2016; 41: 67-82. [http://dx.doi.org/10.1016/j.mssp.2015.08.013]

[36] Che G, Lakshmi BB. C. R. Martin and, Fisher* ER, Ruoff RS. Chemical vapor deposition based synthesis of carbon nanotubes and nanofibers using a template method. Chem Mater 1998; 10(1): 260-7. [http://dx.doi.org/10.1021/cm970412f]

[37] Bronikowski MJ, Willis PA, Colbert DT, Smith KA, Smalley RE. Gas-phase production of carbon single-walled nanotubes from carbon monoxide *via* the HiPco process: A parametric study. J Vac Sci Technol A Vacuum, Surfaces. Film 2001; 19: 1800-5. [http://dx.doi.org/10.1116/1.1380721]

[38] Nikolaev P, Bronikowski MJ, Bradley RK, *et al.* Gas-phase catalytic growth of single-walled carbon nanotubes from carbon monoxide. Chem Phys Lett 1999; 313: 91-7. [http://dx.doi.org/10.1016/S0009-2614(99)01029-5]

[39] O'Connell MJ, Bachilo SM, Huffman CB, *et al.* Band gap fluorescence from individual single-walled carbon nanotubes. Sci (80-) 2002; 593-6. [http://dx.doi.org/10.1126/science.1072631]

[40] Che G, Lakshmi BB, Martin CR, Fisher ER, Ruoff RS. Chemical vapor deposition based synthesis of carbon nanotubes and nanofibers ssing a template method. Chem Mater 1998; 10(1): 260-7. [http://dx.doi.org/10.1021/cm970412f]

[41] Howard J. Flame synthesis of single-walled carbon nanotubes. Carbon 2004; 42(11): 2295-307. [http://dx.doi.org/10.1016/j.carbon.2004.05.010]

[42] Shaijumon MM, Ramaprabhu S. Synthesis of carbon nanotubes by pyrolysis of acetylene using alloy hydride materials as catalysts and their hydrogen adsorption studies. Chem Phys Lett 2003; 374(5-6): 513-20. [http://dx.doi.org/10.1016/S0009-2614(03)00741-3]

[43] Moothi K, Simate GS, Falcon R, Iyuke SE, Meyyappan M. Carbon nanotube synthesis using coal pyrolysis. Langmuir 2015; 31(34): 9464-72.
[http://dx.doi.org/10.1021/acs.langmuir.5b01894] [PMID: 26262467]

[44] Miners SA, Rance GA, Khlobystov AN, *et al.* Chemical reactions confined within carbon nanotubes. Chem Soc Rev 2016; 45(17): 4727-46.
[http://dx.doi.org/10.1039/C6CS00090H] [PMID: 27301444]

[45] Tasis D, Tagmatarchis N, Bianco A, Prato M. Chemistry of carbon nanotubes. Chem Rev 2006; 106(3): 1105-36.
[http://dx.doi.org/10.1021/cr050569o] [PMID: 16522018]

[46] Abdelkader VK, Domingo-García M, Gutiérrez-Valero MD, *et al.* Sidewall chlorination of carbon nanotubes by iodine trichloride. J Phys Chem C 2014; 118: 2641-9.
[http://dx.doi.org/10.1021/jp411935g]

[47] Hirsch A, Vostrowsky O. Functionalization of single-walled carbon nanotubes. Angew Chem Int Ed Engl 2002; 41(11): 1853-9.
[http://dx.doi.org/10.1002/1521-3773(20020603)41:11<1853::AID-ANIE1853>3.0.CO;2-N] [PMID: 19750614]

[48] Koga K. How does water freeze inside carbon nanotubes? Physica A 2002; 341: 462-9.
[http://dx.doi.org/10.1016/S0378-4371(02)01074-9]

[49] Janas D, Boncel S, Koziol KKK. Electrothermal halogenation of carbon nanotube films. Carbon N Y 2014; 73: 259-66.
[http://dx.doi.org/10.1016/j.carbon.2014.02.062]

[50] Bulusheva LG, Okotrub AV, Flahaut E, *et al.* Bromination of double-walled carbon nanotubes. Chem Mater 2012; 24: 2708-15.
[http://dx.doi.org/10.1021/cm3006309]

[51] Pekker* S, Salvetat J-P, Jakab E, Bonard J-M, Forró L. Hydrogenation of carbon nanotubes and graphite in liquid ammonia. J Phys Chem B 2001; 105(33): 7938-43.
[http://dx.doi.org/10.1021/jp010642o]

[52] Talyzin AV, Luzan S, Anoshkin IV, *et al.* Hydrogenation, purification, and unzipping of carbon nanotubes by reaction with molecular hydrogen: road to graphane nanoribbons. ACS Nano 2011; 5(6): 5132-40.
[http://dx.doi.org/10.1021/nn201224k] [PMID: 21504190]

[53] Tang X, Zhao Y, Jiao Q, Cao Y. Hydrogenation of multiwalled carbon nanotubes in Ethylenediamine. Fullerenes, Nanotubes and Carbon Nanostructures 2010; 18(1): 14-23.
[http://dx.doi.org/10.1080/15363830903291697]

[54] Yeo ESY, Mathys GI, Brack N, Thostenson ET, Rider AN. Functionalization and dispersion of carbon nanomaterials using an environmentally friendly ultrasonicated ozonolysis process. J Vis Exp 2017; (123):
[http://dx.doi.org/10.3791/55614] [PMID: 28605378]

[55] Chen Z, Ziegler KJ, Shaver J, Hauge RH, Smalley RE, Smalley RE. Cutting of single-walled carbon nanotubes by ozonolysis. J Phys Chem B 2006; 110(24): 11624-7.
[http://dx.doi.org/10.1021/jp057494c] [PMID: 16800455]

[56] Rajendran Royan NR, Sulong AB, Sahari J, Suherman H. Effect of acid- and ultraviolet/Ozonolysis-treated MWCNTs on the electrical and mechanical properties of epoxy nanocomposites as bipolar plate applications. J Nanomater 2013; 2013: 1-8.
[http://dx.doi.org/10.1155/2013/717459]

[57] Pan H, Liu L, Guo Z-X, *et al.* Carbon nanotubols from mechanochemical reaction. Nano Lett 2002; 3(1): 29-32.
[http://dx.doi.org/10.1021/nl025856b]

[58] Chen L, Xie H, Li Y, Yu W. Surface chemical modification of multiwalled carbon nanotubes by a wet-mechanochemical reaction. J Nanomater 2008; 2008: 1-5. [http://dx.doi.org/10.1155/2008/783981]

[59] Chen L, Xie H, Li Y, Yu W. Nanofluids containing carbon nanotubes treated by mechanochemical reaction. Thermochim Acta 2008; 477: 21-4. [http://dx.doi.org/10.1016/j.tca.2008.08.001]

[60] Ritts AC, Yu Q, Li H, *et al.* Plasma treated Multi-Walled Carbon Nanotubes (MWCNTs) for epoxy nanocomposites. Polymers (Basel) 2011; 3: 2142-55. [http://dx.doi.org/10.3390/polym3042142]

[61] Shenggao W, Tao W, Yanqiong L, Xiujian Z, Jianjun H, Jianhua W. Plasma activation of integrated carbon nanotube electrodes for electrochemical detection of catechol. Plasma Sci Technol 2007; 9: 194-7. [http://dx.doi.org/10.1088/1009-0630/9/2/16]

[62] Elena V. Interaction of oxidized single-walled carbon nanotubes with vaporous aliphatic amines. J Phys Chem B 2002; 106(7): 1588-97. [http://dx.doi.org/10.1021/jp0120110]

[63] Chen S, Shen W, Wu G, Chen D, Jiang M. A new approach to the functionalization of single-walled carbon nanotubes with both alkyl and carboxyl groups. Chem Phys Lett 2005; 402(4–6): 312-7. [http://dx.doi.org/10.1016/j.cplett.2004.12.035]

[64] Graupner Ralf, Abraham Jürgen, Wunderlich David, Vencelová Andrea, Lauffer Peter, Röhrl Jonas, *et al.* Nucleophilic−Alkylation−Reoxidation: A Functionalization Sequence for Single-Wall Carbon Nanotubes. J Am Chem Soc 2006; 128(20): 6683-9. [http://dx.doi.org/10.1021/ja0607281]

[65] Chen RJ, Bangsaruntip S, Drouvalakis KA, *et al.* Noncovalent functionalization of carbon nanotubes for highly specific electronic biosensors. Proc Natl Acad Sci USA 2003; 100(9): 4984-9. [http://dx.doi.org/10.1073/pnas.0837064100] [PMID: 12697899]

[66] Saifuddin N, Raziah AZ, Junizah AR. Carbon Nanotubes: A review on structure and their interaction with proteins. J Chem 2013; 2013: 1-18. [http://dx.doi.org/10.1155/2013/676815]

[67] Zhao Y-L, Stoddart JF. Noncovalent functionalization of single-walled carbon nanotubes. Acc Chem Res 2009; 42(8): 1161-71. [http://dx.doi.org/10.1021/ar900056z] [PMID: 19462997]

[68] Di Crescenzo A, Ettorre V, Fontana A. Non-covalent and reversible functionalization of carbon nanotubes. Beilstein J Nanotechnol 2014; 5: 1675-90. [http://dx.doi.org/10.3762/bjnano.5.178] [PMID: 25383279]

[69] Byong MKim, Shizhi Qian, Bau HH. Filling carbon nanotubes with particles. Nano Lett 2005; 5(5): 873-8. [http://dx.doi.org/10.1021/nl050278v]

[70] Campo J, Piao Y, Lam S, *et al.* Enhancing single-wall carbon nanotube properties through controlled endohedral filling. Nanoscale Horiz 2016; 1(4): 317-24. [http://dx.doi.org/10.1039/C6NH00062B] [PMID: 32260652]

[71] Campo J, Piao Y, Lam S, Streit JK, Walker ARH, Fagan JA. Alkane filled single-wall carbon nanotubes: Endohedral volume control for improved nanotube properties. Meet Abstr 2016; 677-7.

[72] Chaban V. Filling carbon nanotubes with liquid acetonitrile. Chem Phys Lett 2010; 496: 50-5. [http://dx.doi.org/10.1016/j.cplett.2010.07.003]

[73] Mattia D, Gogotsi Y. Review: static and dynamic behavior of liquids inside carbon nanotubes. Microfluid Nanofluidics 2008; 5: 289-305.

[http://dx.doi.org/10.1007/s10404-008-0293-5]

[74] Agrawal KV, Shimizu S, Drahushuk LW, Kilcoyne D, Strano MS. Observation of extreme phase transition temperatures of water confined inside isolated carbon nanotubes. Nat Nanotechnol 2017; 12(3): 267-73.
[http://dx.doi.org/10.1038/nnano.2016.254] [PMID: 27893731]

[75] Kipper AC, da Silva LB. Water filling of carbon nanotubes membranes: Porosity and temperature effects. Chem Phys Lett 2012; 552: 84-7.
[http://dx.doi.org/10.1016/j.cplett.2012.09.038]

[76] Davis JJ, Green MLH, Allen O, *et al.* The immobilisation of proteins in carbon nanotubes. Inorg Chim Acta 1998; 272: 261-6.
[http://dx.doi.org/10.1016/S0020-1693(97)05926-4]

[77] Gao Huajian, Kong Yong, Cui D, Ozkan CS. Spontaneous Insertion of DNA Oligonucleotides into Carbon Nanotubes. Nano Letters 2003; 3(4): 471-3.
[http://dx.doi.org/10.1021/nl025967a]

[78] Kang Yu, Wang Qi, Liu Ying-Chun, Wu Tao. Dynamic mechanism of collagen-like peptide encapsulated into carbon Nanotubes. J Phys Chem B 2008; 112(15): 4801-7.
[http://dx.doi.org/10.1021/jp711392g]

[79] Robertson J. Realistic applications of CNTs. Mater Today 2004; 7: 46-52.
[http://dx.doi.org/10.1016/S1369-7021(04)00448-1]

[80] He H, Pham-Huy LA, Dramou P, Xiao D, Zuo P, Pham-Huy C. Carbon nanotubes: applications in pharmacy and medicine. BioMed Res Int 2013; 2013: 578290.
[http://dx.doi.org/10.1155/2013/578290] [PMID: 24195076]

[81] Abdalla S, Al-Marzouki F, Al-Ghamdi AA, Abdel-Daiem A. Different technical applications of carbon nanotubes. Nanoscale Res Lett 2015; 10(1): 358.
[http://dx.doi.org/10.1186/s11671-015-1056-3] [PMID: 26377211]

[82] Tîlmaciu C-M, Morris MC. Carbon nanotube biosensors. Front Chem 2015; 3: 59.
[http://dx.doi.org/10.3389/fchem.2015.00059] [PMID: 26579509]

[83] He P, Xu Y, Fang Y. Applications of carbon nanotubes in Electrochemical DNA Biosensors. Mikrochim Acta 2006; 152: 175-86.
[http://dx.doi.org/10.1007/s00604-005-0445-1]

[84] Wang W, Zhu Y, Liao S, Li J. Carbon nanotubes reinforced composites for biomedical applications. BioMed Res Int 2014; 2014: 518609.
[http://dx.doi.org/10.1155/2014/518609] [PMID: 24707488]

[85] Yan Y, Miao J, Yang Z, *et al.* Carbon nanotube catalysts: recent advances in synthesis, characterization and applications. Chem Soc Rev 2015; 44(10): 3295-346.
[http://dx.doi.org/10.1039/C4CS00492B] [PMID: 25855947]

[86] Sehrawat P, Julien C, Islam SS. Carbon nanotubes in Li-ion batteries: A review. Mater Sci Eng B 2016; 213: 12-40.
[http://dx.doi.org/10.1016/j.mseb.2016.06.013]

[87] Liu X-M. Carbon nanotube (CNT)-based composites as electrode material for rechargeable Li-ion batteries: A review. Compos Sci Technol 2012; 72: 121-44.
[http://dx.doi.org/10.1016/j.compscitech.2011.11.019]

[88] Habisreutinger SN, Nicholas RJ, Snaith HJ. Carbon nanotubes in perovskite solar cells. Adv Energy Mater 2017; 7: 1601839.
[http://dx.doi.org/10.1002/aenm.201601839]

[89] Arici E, Karazhanov S. Carbon nanotubes for organic/inorganic hybrid solar cells. Mater Sci Semicond Process 2016; 41: 137-49.

[http://dx.doi.org/10.1016/j.mssp.2015.07.086]

[90] Wu Y, Lin X, Zhang M. Carbon nanotubes for thin film transistor: Fabrication, properties, and applications. J Nanomater 2013; 2013: 1-16.
[http://dx.doi.org/10.1155/2013/627215]

[91] Sun D-M, Liu C, Ren W-C, Cheng H-M. A review of carbon nanotube- and graphene-based flexible thin-film transistors. Small 2013; 9(8): 1188-205.
[http://dx.doi.org/10.1002/smll.201203154] [PMID: 23519953]

[92] Ruhil S, Sehgal J, Rohilla K. Carbon nanotubes field effect transistor: A review. Int J Sci Res ISSN (Online Index Copernicus Value Impact Factor) 2013; 14: 2319-7064.

[93] Zhang W, Zhang Z, Zhang Y, *et al.* The application of carbon nanotubes in target drug delivery systems for cancer therapies. Nanoscale Res Lett 2011; 6: 555.
[http://dx.doi.org/10.1186/1556-276X-6-555] [PMID: 21995320]

[94] Klumpp C, Kostarelos K, Prato M, Bianco A. Functionalized carbon nanotubes as emerging nanovectors for the delivery of therapeutics. Biochim Biophys Acta 2006; 1758(3): 404-12.
[http://dx.doi.org/10.1016/j.bbamem.2005.10.008] [PMID: 16307724]

[95] Pondman KM, Salvador-Morales C, Paudyal B, Sim RB, Kishore U. Interactions of the innate immune system with carbon nanotubes. Nanoscale Horiz 2017; 2(4): 174-86.
[http://dx.doi.org/10.1039/C6NH00227G] [PMID: 32260639]

[96] Battigelli A. Cationic Carbon Nanotubes for Nucleic Acids Delivery 2012. http://hdl.handle.net/10077/7416

[97] Chen G. Nanotube-based controlled drug delivery. Pharm Anal Acta 2012; 3: 9.
[http://dx.doi.org/10.4172/2153-2435.1000e136]

[98] Foldvari M, Bagonluri M. Carbon nanotubes as functional excipients for nanomedicines: II. Drug delivery and biocompatibility issues. Nanomed (Lond) 2008; 4(3): 183-200.
[http://dx.doi.org/10.1016/j.nano.2008.04.003] [PMID: 18550450]

[99] Wu W, Wieckowski S, Pastorin G, *et al.* Targeted delivery of amphotericin B to cells by using functionalized carbon nanotubes. Angew Chem Int Ed Engl 2005; 44(39): 6358-62.
[http://dx.doi.org/10.1002/anie.200501613] [PMID: 16138384]

[100] Ajima K, Masako Yudasaka, Tatsuya Murakami, *et al.* Carbon nanohorns as anticancer drug carriers. Mol Pharm 2005; 2(6): 475-80.
[http://dx.doi.org/10.1021/mp0500566]

[101] Murakami T, Kumiko Ajima, Jin Miyawaki, Masako Yudasaka, Sumio Iijima, Kiyotaka Shiba. Drug-loaded carbon nanohorns: Adsorption and release of dexamethasone *in vitro*. Mol Pharma 2004; 1(6): 399-405.
[http://dx.doi.org/10.1021/mp049928e]

[102] Zhou S, Hashida Y, Kawakami S, *et al.* Preparation of immunostimulatory single-walled carbon nanotube/CpG DNA complexes and evaluation of their potential in cancer immunotherapy. Int J Pharm 2014; 471(1-2): 214-23.
[http://dx.doi.org/10.1016/j.ijpharm.2014.05.037] [PMID: 24861942]

[103] Deng X, Jia G, Wang H, Sun H, Wang X, Yang S, *et al.* Translocation and fate of multiwalled carbon nanotubes *in vivo*. Carbon N Y 2007; 45: 1419-24.
[http://dx.doi.org/10.1016/j.carbon.2007.03.035]

[104] Folkmann JK, Risom L, Jacobsen NR, Wallin H, Loft S, Møller P. Oxidatively damaged DNA in rats exposed by oral gavage to C60 fullerenes and single-walled carbon nanotubes. Environ Health Perspect 2009; 117(5): 703-8.
[http://dx.doi.org/10.1289/ehp.11922] [PMID: 19479010]

[105] Philbrook NA, Walker VK, Afrooz ARMN, Saleh NB, Winn LM. Investigating the effects of

functionalized carbon nanotubes on reproduction and development in Drosophila melanogaster and CD-1 mice. Reprod Toxicol 2011; 32(4): 442-8.
[http://dx.doi.org/10.1016/j.reprotox.2011.09.002] [PMID: 21963887]

[106] Yang Z, Zhang Y, Yang Y, *et al.* Pharmacological and toxicological target organelles and safe use of single-walled carbon nanotubes as drug carriers in treating Alzheimer disease. Nanomed (Lond) 2010; 6(3): 427-41.
[http://dx.doi.org/10.1016/j.nano.2009.11.007] [PMID: 20056170]

[107] Kagan VE, Konduru NV, Feng W, *et al.* Carbon nanotubes degraded by neutrophil myeloperoxidase induce less pulmonary inflammation. Nat Nanotechnol 2010; 5(5): 354-9.
[http://dx.doi.org/10.1038/nnano.2010.44] [PMID: 20364135]

[108] Cherukuri P, Gannon CJ, Leeuw TK, *et al.* Mammalian pharmacokinetics of carbon nanotubes using intrinsic near-infrared fluorescence. Proc Natl Acad Sci USA 2006; 103(50): 18882-6.
[http://dx.doi.org/10.1073/pnas.0609265103] [PMID: 17135351]

[109] Maynard AD. Are we ready for spray-on carbon nanotubes? Nat Nanotechnol 2016; 11(6): 490-1.
[http://dx.doi.org/10.1038/nnano.2016.99] [PMID: 27272137]

[110] Ryman-Rasmussen JP, Tewksbury EW, Moss OR, Cesta MF, Wong BA, Bonner JC. Inhaled multiwalled carbon nanotubes potentiate airway fibrosis in murine allergic asthma. Am J Respir Cell Mol Biol 2009; 40(3): 349-58.
[http://dx.doi.org/10.1165/rcmb.2008-0276OC] [PMID: 18787175]

[111] Takagi A, Hirose A, Futakuchi M, Tsuda H, Kanno J. Dose-dependent mesothelioma induction by intraperitoneal administration of multi-wall carbon nanotubes in p53 heterozygous mice. Cancer Sci 2012; 103(8): 1440-4.
[http://dx.doi.org/10.1111/j.1349-7006.2012.02318.x] [PMID: 22537085]

[112] Poland CA, Duffin R, Kinloch I, *et al.* Carbon nanotubes introduced into the abdominal cavity of mice show asbestos-like pathogenicity in a pilot study. Nat Nanotechnol 2008; 3(7): 423-8.
[http://dx.doi.org/10.1038/nnano.2008.111] [PMID: 18654567]

[113] Liu Y, Zhang J, Cheng Y, Jiang SP. Effect of carbon nanotubes on direct electron transfer and electrocatalytic activity of immobilized glucose oxidase. ACS Omega 2018; 3(1): 667-76.
[http://dx.doi.org/10.1021/acsomega.7b01633] [PMID: 30023785]

CHAPTER 7

Magnetic Nanoparticles for Advanced Drug Delivery System

Arindam Saha* and **P. Sujatha Devi***

CSIR-Central Glass and Ceramic Research Institute, Jadavpur, Kolkata-700032, India

Abstract: Magnetic nanoparticles have attracted tremendous attention in drug delivery and other biomedical applications since the last two decades. Many magnetic nanoparticle based formulations are already available in the market and even many more are under different stages of development. The incredible progress in this area is due to well-established synthesis procedures of size and shape-controlled magnetic nanoparticles as well as their well-recognized surface functionalization approaches. Advanced drug delivery systems consider the shape, colloidal stability, density of surface anchoring groups on nanoparticle surface as well as responsive and sustained release of drug molecules along with integrated imaging modality.

Keywords: Magnetic nanoparticles, Responsive release, Responsive release, Release kinetics, Surface functionalization, Surface coating.

INTRODUCTION

Nanoparticles, by definition, are inorganic or organic particles in size range of 1-100 nm, although generally, particles between 1-1000 nm are considered as nanoparticles [1, 2]. Nanoparticles exhibit different properties compared to their bulk counterparts having same composition, owing to their high surface to volume ratio [3] and quantum confinement effects [4]. As for example, cadmium selenide (CdSe) nanoparticles (<5 nm) have shown tunable emission properties and have been extensively used as light emitting diode (LED), whereas bulk CdSe does not have any emission property [5, 6]. Nanoparticles exhibit exciting magnetic and electronic properties depending on their size and shape which can be precisely

* **Corresponding author Arindam Saha:** CSIR-Central Glass and Ceramic Research Institute, Jadavpur, Kolkata-700032, India; E-mail: arindamchemiitkgp@gmail.com
P. Sujatha Devi: CSIR-Central Glass and Ceramic Research Institute, Jadavpur, Kolkata-700032, India; E-mail: psujathadevi@cgcri.res.in

Surendra Nimesh, Nidhi Gupta and Ramesh Chandra (Eds.)

tuned based on their applications. Currently, nanoparticles are being extensively used in electronic devices, energy applications, healthcare products,catalytic applications, environmental applications and many others [7, 8].

Two major synthetic pathways exist for the top-down approach to nanoparticles (Fig. **1**), where the bulk materials are converted into nanomaterials by mechanical grinding, lithography or micro-patterning. This method is costly and requires sophisticated technology but results in low yield. The most widely used method is the bottom-up approach where nanoparticles are prepared from atoms under a controlled environment (Fig. **1**) [9 - 12]. Various methods of synthesis like sol-gel, thermal decomposition, solvothermal, co-precipitation, *etc.*, fall in this category.

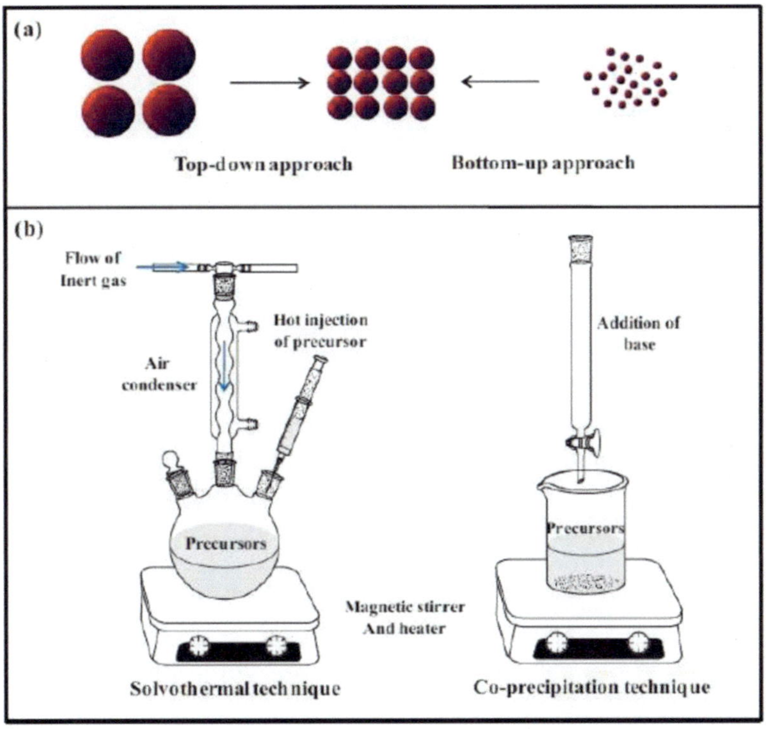

Fig. (1). (**a**) Schematic showing the top-down and bottom-up approach for nanoparticle synthesis. (**b**) Schematic representation of solvothermal and co-precipitation techniques for nanoparticle synthesis.

There are three major classes of inorganic nanoparticles - magnetic, plasmonic and semiconductor, which serve different purposes based on their unique properties. Magnetic nanoparticles include Fe_2O_3, Fe_3O_4, Gd_2O_3, NiO, Co_3O_4, *etc.*, whereas plasmonic nanoparticles involve Au, Ag and Cu nanoparticles. Semiconductor nanoparticles are of different types based on their constituent atoms and bandgap properties. Well-known semiconductor nanoparticles include TiO_2, ZnO, SnO_2, CdSe, CdS, InP, ZnSe, and various doped and alloyed modifications of them.

The most promising application of nanoparticles has been found in healthcare products and in biomedical fields with several ongoing researches focusing on this area [13 - 15]. Magnetic nanoparticles are most suitable and widely used to develop various biomedical devices and theranostic applications [15 - 18]. Various applications of magnetic nanoparticles in the biomedical field involve therapy (drug delivery, hyperthermia) [19 - 21], magnetic resonance imaging (MRI) [22], separation (protein, cells) [23] and sensor development [24]. Magnetic nanoparticles possess several advantages like size and shape-dependent tunable magnetic properties, biocompatibility, widely available synthesis and functionalization techniques, cost-effective and uncomplicated administration in targeted site. A recent focus is to apply the properties of magnetic nanoparticles for drug delivery applications leading to a major breakthrough in this field. Perhaps the most promising breakthrough anticipated would be cancer therapy in a non-invasive way for an early-stage diagnosis. The need for more advanced theranostic approaches is emerging. Conventional treatments like surgery, chemotherapy or radiation therapy have limited success due to lack of selectivity of tumor tissue, difficult accessibility of tumor tissues, chances of spreading the tumor throughout the body, risk in surgery of sophisticated organs and complicated side effects [25, 26]. Immunotherapies, introduced recently, have been found to be good for small tumors in the early stage. Late-stage cancer treatment is the major challenge with very low success rates in recent times. Thus, the demand for targeted and specific combinatorial therapy, where more than one therapeutic approach can be used in the targeted organ, is in high demand for effective cancer treatment. The major advantage of nanoparticle-based drug delivery systems is that it can act at the cellular level, that is, the drug carrier can pass through the cell membrane *via* endocytosis. Properly designed nanosystem can even target the nucleus and deliver drugs to specific organs, which can substantially reduce the toxic effect on normal cells and enhance the effective drug concentration at the target site.

Various drug delivery systems have been explored and available in the market and many more are under development [27, 28]. Among the conventional drug delivery systems, most widely used are micellar systems [29], polymeric systems [30], liposomes [31], dendrimer-based drug delivery systems [32], solid lipid nanoparticles, *etc.* [33]. The major disadvantages of these systems are large size, poor colloidal stability, lack of specificity and lack of selective delivery approach. An effective drug delivery system must fulfill the following criteria [34]: (i) small hydrodynamic diameter (typically between 1-100 nm), (ii) proper shape, (iii) superior colloidal stability for a long period which enhances the blood circulation time by preventing aggregation, (iv) proper surface functionalization to obtain surface anchoring groups for maximum drug loading (v) PEGylation to avoid non-specific interaction, (vi) targeting group to specifically target the tumor site and (vii) responsive linkage with the drug molecules which helps to release the drug under specific conditions. Keeping such requirements in mind, magnetic nanoparticle-based drug delivery vehicles have been explored that may be able to deliver vehicles to the site non-invasively after administration.

Magnetic nanoparticle-based drug delivery can be achieved by passive targeting or active targeting and sometimes both occur simultaneously [35, 36]. Passive targeting is based on the extravasation of the delivery vehicles in the leaky blood vessels of the tumor tissues during circulation in the blood. Enhanced permeation and retention (EPR) effect dictates the maximum accumulation of nanoparticles by this approach [35]. In active targeting drug delivery, vehicles are functionalized with specific targeting ligands which target the overexpressed receptors on the cancer cell membrane. Ligand receptor interaction initiates endocytosis mechanism which can be clathrin-mediated or caveolae-mediated. These endocytosed nanoparticles deliver drug molecules within the cancer cells under specific stimuli activation. This is highly specific which helps in precise control of the drug delivery.

Here, we focussed on magnetic drug delivery systems with an emphasis on the types of magnetic nanoparticles, synthesis techniques available for magnetic nanoparticles, surface coating and functionalization techniques followed by drug loading chemistry and responsive delivery systems. We will also highlight the effect of the shape and size of the nanoparticles in drug delivery.

TYPES OF MAGNETIC NANOPARTICLES

The broad classification of magnetic nanoparticles based on the composition includes four different types, *viz.* metals, magnetic alloys, magnetic oxides and ferrites.

Metal Nanoparticles

Although metal nanoparticles possess higher magnetization compared to their oxide counterparts, they are not suitable for practical applications due to their strong oxidation tendency. Among the metal nanoparticles, Fe, Co and Ni are the most commonly used ones. Fe is ferromagnetic in nature with high saturation magnetization at room temperature (218 emu/gm) and high Curie temperature (1043 K). The major challenges in the preparation of Fe nanoparticles are phase purity and monodispersity with respect to shape and size [37 - 39]. It also possesses low magnetocrystalline anisotropy which directly affects the superparamagnetic behaviour. There are two different phases of Fe nanoparticles: α-Fe having body centered cubic (bcc) crystal structure and γ-Fe having face centered cubic (fcc) chemical structure. γ-Fe normally exists at a higher temperature, which is higher than Curie temperature and exhibits antiferromagnetic property in some alloys [40]. Ferromagnetic Co and Ni possess comparatively low saturation magnetization. Co has a high Curie temperature (1388 K), whereas Ni exhibits a low Curie temperature (627 K) [41 - 44]. All the metal nanoparticles are oxidation prone and form an oxide shell if not properly surface coated. The metal-metal oxide core-shell materials exhibit interesting magnetic properties. Another group of interesting materials includes rare earth elements which show ferromagnetic properties, among which Gd, Eu, Tb and Dy nanoparticles are recently under investigation. They exhibit complex magnetic behaviour.

Magnetic Alloys

Magnetic alloys, explored recently, have found applications in storage device, catalysis and in biomedical field. Among them, Fe-Co, Fe-Ni, Fe-Pt and Co-Pt are widely studied [45 - 48]. Fe-Co is a soft magnetic alloy with bcc structure having high saturation magnetization, high Curie temperature and low magnetocrystalline anisotropy. Saturation magnetization is maximum at ~40 atom% of Co and maximum coercive force is attained at ~80 atom% of Co. Fe-Ni magnetic alloys, in comparison, have low saturation magnetization and low Curie temperature. But, the major advantage is that they exhibit superparamagnetic behaviour in wide size ranges (~10-80 nm) and temperature ranges. Fe-Pt alloy has been studied most extensively due to its high magnetocrystalline anisotropy and coercivity at room temperature which qualify it for application in next-generation storage devices. Various synthetic methods have been explored to prepare shape and size-dependent Fe-Pt nanoalloys. Superparamagnetic Fe-Pt alloy with ~3.5 nm size has also been studied. Generally Fe-Pt alloy has face-centered tetragonal (fct) structure which is also known as L10 phase. The chemical synthetic route leads to a superparamagnetic fcc structure, which converts into L10 phase after thermal annealing. FePt-Fe_3Pt nanocomposite with complex magnetic behaviour has also been prepared by varying the annealing temperature. Co-Pt is another important magnetic alloy that has gained increasing importance in the scientific community due to its chemical stability with high magneto-crystalline anisotropy.

Magnetic Oxide Nanoparticles

Magnetic iron oxide nanoparticles are most extensively studied due to their wide range of biomedical applications in multimodal imaging, drug delivery, protein enrichment and sensor development. This is because of their controlled and well-documented size/shape dependent synthetic techniques, biocompatibility, controlled surface properties, chemical stability and suitable magnetic properties. These oxides are available in various forms of structure and composition. Apart from biomedical applications, magnetic iron oxides have found important applications in storage media, catalyst, gas sensors, environmental applications, *etc.* Other than iron, both nickel and cobalt oxides are also well established. Gadolinium oxides have recently become popular due to their paramagnetic properties which are useful as contrast agents in magnetic resonance imaging (MRI). Well developed synthetic methods are available to synthesize iron oxide nanoparticles of different shapes and sizes. Superparamagnetic iron oxide nanoparticles (SPION) have shown more advantages over other nanoparticles with respect to drug delivery and imaging applications. Superparamagnetism arises on the nanoscale due to thermal fluctuation of the magnetization vector in the absence of external magnetic field. Typically the duration between two fluctuations is known as Neel relaxation time and if the timescale of measurement is larger than Neel relaxation time, the oxide material behaves as paramagnetic material. This phenomenon is known as superparamagnetism which is crucial for magnetic drug delivery applications.

Fe_2O_3 nanoparticles, where both the iron atoms are in a ferric state, exist mostly as hematite (α-Fe_2O_3), which has rhombohedral crystalline phase and maghemite (γ-Fe_2O_3) having a cubic structure [49 - 51]. In hematite, all the Fe^{3+} ions remain in octahedral position whereas maghemite exhibits AB_2O_4 type of spinel structure. Hematite exhibits antiferromagnetic behaviour below 950K and it has a bandgap of 2.2 eV. The Maghemite is ferrimagnetic in nature and mostly shows superparamagnetic behaviour at room temperature, and is widely explored for biomedical applications since its shape and size dependent synthesis is well developed.

Fe_3O_4 consists of Fe^{2+} and Fe^{3+} ions in inverse spinel structure with cubic phase. It is ferromagnetic in nature below 858K. Synthesis of various sizes of Fe_3O_4 nanoparticles ranging from 2 nm to 200 nm and of various shapes have been well developed [52]. It has shown extensive applications in the biomedical field in drug delivery and sensor applications. FeO exhibits a cubic phase and is metastable and antiferromagnetic in nature [53]. The oxyhydroxide FeOOH has also gained substantial attention in recent times. Various phases of FeOOH have been synthesized and studied extensively [54]. The α-FeOOH or Goethite is antiferromagnetic in nature whereas β-FeOOH is paramagnetic at room temperature. Other forms, γ-FeOOH and δ-FeOOH, are comparatively less commonly occurring.

Among the other magnetic oxides, CoO is antiferromagnetic in nature, whereas Co_3O_4 exhibits ferromagnetic properties below 33K. Co_3O_4 has a cubic spinel structure and many synthetic methods have been developed for its shape and size dependent synthesis [55]. On the other hand, NiO is antiferromagnetic in nature but exhibits superparamagnetism on the nanoscale [56]. NiO also possesses electrical properties.

Ferrite Nanoparticle

Ferrites are iron (III) oxides chemically combined with one or more metallic oxides. Ferrites are generally ferromagnetic in nature. Different types of ferrites have been studied with respect to their nanocrystalline properties and synthesis techniques. Technically there are two types of ferrites depending on the extent of coercivity, hard ferrite and soft ferrite. Hard ferrites are having high coercivity (*e.g.* strontium ferrite, barium ferrite, cobalt ferrite, *etc.*) whereas soft ferrites have low coercivity (*e.g.* manganese zinc ferrite, nickel zinc ferrite, *etc.*). Nano ferrites have recently been studied widely due to their magnetic as well as optical properties. Among the ferrites, $ZnFe_2O_4$, $MnFe_2O_4$, $CoFe_2O_4$ and $NiFe_2O_4$ are most extensively studied [57 - 59]. These nano ferrite materials have been synthesized in different shapes and sizes, and their application potential in biomedical fields is being explored. These nano ferrites, especially manganese and zinc ferrite, often exhibit fluorescence due to surface defects.

SYNTHESIS OF MAGNETIC NANOPARTICLES

In this section, the synthesis of magnetic oxide nanoparticles and ferrites that are the most popular and compatible materials for biomedical applications will be presented. There are three major synthesis techniques generally followed (i) Physical methods which include gas phase deposition and ion beam lithography. These methods produce atomically pure nanoparticle phase, although this method is not suitable for large scale synthesis or to obtain shape/size controlled nanoparticles. (ii) Wet chemical methods which include sol-gel technique, co-precipitation, solvothermal, flow injection method, electrochemical, sonochemical, oxidation, aerosol or vapour phase method, *etc.* These methods have been most extensively used and can control the shape and size of the nanoparticles precisely, although several post-synthetic purification techniques must be followed. (iii) Microbial methods which are versatile, environment-friendly, economic and have moderate control over shape, size and the properties of the nanoparticles.

Electron Beam Lithography

In this technique, iron oxide nanoparticles can be directly prepared from iron particles under the exposure of electron beam [60]. In this method, sometimes specific polymer masks are used to obtain the uniform size distribution of nanoparticles. Although the method is seldom used, it produces a pure crystalline phase.

Gas Phase Deposition

Chemical vapour deposition method is suitable to synthesize iron oxide sheets. Volatile molecular precursors are normally heated to vapour and deposited on a solid substrate like alumina. Gold acts as a catalyst in this process. Various Fe-molecular precursors have been used; Bis(hexafluoroacetylacetonato)(N,N,N',N'-tetramethylethylenediamine) iron(II), $Fe[(C_2H_5)_2NCS_2]_3$, $[Fe(OBu^t)_3]_2$, *etc.* are common examples of molecular precursors [61].

Sol-Gel Technique

Sol-gel technique is a widely accepted wet chemical technique where a molecular precursor solution undergoes hydroxylation, followed by condensation to obtain 'sol' of nanosized particles. The 'sol' is then converted to 'gel' by solvent evaporation or by chemical means which then results in nanoparticles of desired shape and size [62, 63]. Often the final product is annealed at high temperature to obtain well-defined crystallinity, shape, size and size distribution. Solvent used is mostly water and hydroxylation can be catalyzed by both acid and base. There are various parameters like kinetics of growth reaction, condensation and hydrolysis which determine the shape and size of the as-synthesized oxide nanoparticles. These parameters can be controlled by adjusting the reaction pH, temperature, precursor addition sequence or by varying the precursor ratio. The stability of the as-prepared magnetic oxide nanoparticles depends on the surfactant used with metal salt precursors. Tetraethyl orthosilicate (TEOS) is commonly used during the sol-gel technique which provides a thin silica layer on the oxide nanoparticles, thus preventing aggregation. In addition, sodium benzene sulfonate, dextran, cetrimonium bromide (CTAB), sodium dodecyl sulphate (SDS), sodium dodecyl benzene sulfonate (SDBS), polyethylene glycol (PEG) and other polyols are commonly employed as surfactants during synthesis. These surfactants not only prevent aggregation of nanoparticles, but control the kinetics of the growth process and in turn control the shape, size and size distribution of the as-synthesized particles. The major advantage of this method is well-controlled synthesis and large scale production. But, low crystallinity, poor surface properties and synthesis of ultra-small SPIONS are the major drawbacks of this technique.

Co-Precipitation Method

Co-precipitation is perhaps the most simplest and reliable method for the magnetic iron oxide synthesis (Fig. **1**). This method is suitable due to well-separated nucleation and growth stage, and thus more control over the shape and size of the nanoparticles, although size distribution has been compromised [54, 64]. Moreover, a gram scale yield in this method is advantageous from the synthetic point of view. In this method, mixed iron oxide salts of both Fe^{2+} and Fe^{3+} are taken in water and hydroxide formation is initiated by adding ammonium hydroxide (NH_4OH) or ammonia (NH_3) solution or sodium hydroxide (NaOH) under sonication or heating condition. Heating or sonication leads to oxide formation from the hydroxide thus precipitated. Final product is separated by decantation, centrifugation or magnetic separation and dried at 100°C to remove water. The control over shape and size depends on the reaction temperature, pH of the solution, addition sequence of the base, nature of the base, ratio of the precursors, *etc.* Crystal phase also depends on the above factors. In a typical example, drop-wise addition of NH_4OH to ferric nitrate solution leads to γ-Fe_2O_3, whereas a one time addition results in α-FeOOH. In some cases, the as-synthesized iron oxide powder is further annealed for better crystallinity. One of the major issues of this process is the colloidal stability of the oxide nanoparticles. Electrostatic stability can be achieved by adding an excess base to the nanoparticle solution or tri-octyl phosphine oxide (TOPO), which has been used widely. Steric stability can be achieved by adding long chain fatty acids during synthesis (*e.g.* oleic acid, oleyl amine, *etc.*). To increase the stability, different types of block copolymers (pluorinc polymers), poly-vinyl alcohol (PVA), PEG, tween solutions and dextran have been extensively studied. Hydrolysis can also be initiated in non-aqueous solvents like ethylene glycol, where the precursors are mostly iron carbonyls or iron complexes.

Micro-Emulsion Method

Micro-emulsion method or two-phase method provides much more control over size and size-distribution, although the shape is not well controlled in this method [65]. Moreover, the yield in this method is comparatively low. The method is easy and can be performed for various nanoparticles with less agglomeration tendency. This method consists of the formation of water in oil (w/o) microemulsion by sonication or vigorous stirring with suitable surfactant. This surfactant stabilized nanocavities or water droplets with ~10 nm diameter provide a confinement effect to nanoparticle nucleation and growth. After nanoparticle formation in the cavities, emulsions are lysed by ethanol or acetone to obtain the magnetic nanoparticles. Since this technique results in impurities, surfactant or co-surfactant selection and optimization of their ratio play a crucial role. Cationic, anionic and non-ionic surfactants are widely used in this method. The reverse technique that is oil in water emulsion is also employed where the precursor is metal-oleates or stearates.

Thermal Decomposition Method

This technique is most widely used in recent times due to its several advantages like precise control of size and shape, very narrow size distribution, well crystallinity, well-separated nucleation and growth, chemical and colloidal stability, *etc.* (Fig. **1**). Size ranges between 2 nm to 200 nm can be easily obtained by this technique. Moreover, this method is applicable to almost all types of nanoparticles, *viz.* magnetic, plasmonic, semiconductor or doped nanoparticles. Thermal decomposition is commonly known as solvothermal and hydrothermal if the solvent is water. Solvothermal method for iron oxide nanoparticles is well known and has been used to prepare both Fe_3O_4 and γ-Fe_2O_3 of different shapes and sizes with extremely narrow size distribution. Common precursors are Fe (III) oleate or stearate, which is dissolved in non-coordinating solvent octadecene (ODE). At a temperature of 300-320°C, the iron complex decomposes to result in Fe_3O_4 nanoparticles. In this method, oleic acid, oleyl amine, TOPO or octyl amine are commonly used as steric, stabilising agent or capping ligand [66]. Shape and size can be controlled by changing the precursor to capping ligand ratio, heating rate, heating time or sometimes cooling rate. As for example, in a 1:1 mixture of Fe (III) oleate and oleic acid, heating at 300°C for 30 minutes results in spherical Fe_3O_4 nanoparticles, whereas 3-hour heating leads to cubic nanoparticles. To obtain γ-Fe_2O_3, mild organic oxidising morpholine-N-oxide is used. These reactions are normally performed in a Schleck line setup under an inert atmosphere. Solvothermal method can also be performed with other solvents under autoclave condition. A common non-aqueous solvent other than ODE is ethylene glycol. In hydrothermal process, metal salts dissolved in water with appropriate reactants are placed in reactors or autoclave [67]. The pressure can go as high as 2000 psi and the temperature can be raised to 200°C. At high temperature and pressure, metal salts are hydrolysed by water and then dehydrated. Due to the low solubility of the metal oxides thus formed, they generate supersaturating solutions. Precursor concentration, residence time, heating rate and temperature affect the particle size and shape.

Polyol Method

Polyol method employs PEG of various molecular weights, PVP or other diols/polyols as solvents [68]. Due to the ranges of the boiling point of these solvents, different operating temperatures can be achieved in this process. Normally the solvent mixed with the metal salt is refluxed for the desired time period. The major advantage is that solvents play a critical role in stabilising the nanoparticles and thus prevent aggregation. No external reagent or surfactant is required. Shape and size can be controlled in this method, although small particle (<50 nm) synthesis is not reliable by this method.

Sonochemical Method

In this method, metal precursors are mixed with a suitable solvent, mostly water, followed by sonication using a high-energy probe sonicator [69]. Metal salts are generally organometallic precursors like metal acetate or metal acetylacetonate. High sonication energy hydrolyses the particles and forms metallic hydroxide, which further decomposes to metallic oxide nanoparticles. Metal carbonyls are often used as a precursor and toluene is used as a suitable solvent. Sonication power, time and precursor concentration control the shape and size of the nanoparticles. Although the method is convenient, it lacks narrow size distribution and particle stability. To stabilize the as-synthesized nanoparticles, PEG, PVP, *etc.* are used as surfactants in an aqueous solvent, whereas octyl amine, oleyl amine, oleic acid are commonly used organic solvents.

Vapour Phase or Aerosol Method

In this method, metal precursors mostly metal carbonyls and suitable reducing agents are vaporized and released in the form of spray by a tiny whole onto some solid substrate [70]. It is also known as spray pyrolysis. The particle size depends on the size of the spray droplets. Sometimes laser light is used to vaporize the precursors and is known as laser pyrolysis. Silica or alumina is generally used as solid support. In this method, 2D sheet of oxide nanoparticles is formed with broad size distribution.

Electrochemical Method

Electrochemical methods have been employed to synthesize small-sized (~5-10 nm) Fe_3O_4 and γ-Fe_2O_3 nanoparticles with uniform size. Fe has been used as anode and Pt as cathode [71]. Tetraoctyl ammonium bromide (TOAB) or N,N-dimethyl formamide (DMF) acts as supporting electrolyte. The size and shape of as-prepared nanoparticles are controlled by changing the current density and time of residence. In this method, surfactants are often used as stabilizing agents. Large scale synthesis or rigorous control over shape and size is not possible in this method.

Flow Injection Method

This method depends on a continuous laminar flow of reactant mixtures inside a capillary reactor [72]. This method is generally used to prepare small size oxide nanoparticles, although their size and shape are not well controlled. The advantage is its high reproducibility due to homogeneous mixing.

Super-Critical Fluid Method

Super critical fluid method (SCF) has become popular for nanoparticle synthesis due to its green chemistry approach that can avoid the usage of hazardous chemicals and organic solvents. Basically, SCFs are single-phase substances that can exist above critical temperature and pressure [73]. Commonly used SCFs are trifluoromethane, chlorodifluoromethane, diethyl ether, CO_2, acetone, water, *etc.* Among them, CO_2 and water are most commonly used due to their environment-friendly nature. Both Fe_3O_4 and γ-Fe_2O_3 have been synthesized in super critical CO_2. In this method, organic long chain fatty acids act as capping ligands to prevent aggregation. Size can be controlled by varying the molar ratio of the metal salt to the capping ligand [73].

Oxidation Method

This method relies on partial oxidation of Fe (II) hydroxide to small size Fe_3O_4 nanoparticles [74]. The Fe_3O_4 nanoparticles further get oxidized at a higher temperature and form γ-Fe_2O_3. This method has poor control over shape and size distribution.

Microbial Method

This method has received attention of the scientific community due to its reliable approach, *i.e.* low temperature, high yield, high reproducibility and low cost [75]. Moreover, the method reduces the use of toxic chemical substances. The major disadvantage, however, is control over shape/size and it is time-consuming. This method relies on Fe (III) salt reducing bacteria like Thermoanaerobacter species (*i.e.*, Thermo-anaero-bacterethanolicus strain TOR 39) and Shewanella species (*e.g.*, Shewanellaloihica strain PV-4). They reduce Fe (III) salt to prepare Fe_3O_4 nanoparticles under anaerobic condition. It requires a low temperature (~65°C) and 3/4 weeks to complete the process.

SURFACE MODIFICATION OF MAGNETIC NANOPARTICLES

Surface modification of as-synthesized nanoparticles plays a pivotal role in biomedical applications [76 - 79]. The surface modification reduces interparticle aggregation, provides appropriate surface charge, offers surface active functional groups which acts as "chemical hand" to hold target molecules, drugs *etc.*, provide colloidal stability for longer period and also facilitate the incorporation of multimodal imaging modality. Although there are various techniques available for surface modification both in solution and in solid phase, we will focus mostly on the wet chemical routes for surface modification. Surface modification can be subdivided into two categories: (i) Surface coating which prevents interparticle aggregation and provides colloidal stability of the nanoparticles and (ii) Surface functionalization which provides "chemical hand" to the nanoparticle surface which is further used to hold various small molecules, drug, *etc.* It also includes covalent and noncovalent binding of these molecules to the nanoparticle. Often, surface coating and functionalization are performed in one step.

Surface Coating

Surface coating can be carried out using (a) organic materials which involve polymer coating or surfactant coating and liposomes or biomolecule coating and (b) inorganic materials which involve silica shell or inert metal/metal oxide coatings. These surface coatings prevent agglomeration by way of electrostatic stabilization or steric stabilization. In electrostatic stabilization, nanoparticle surface is coated with cationic/anionic small molecules or surfactants which repeal the like-charged particles and thus avoid interparticle aggregation. In steric stabilization, nanoparticles are coated with polymers or long chain fatty acids which provide substantial space between two nanoparticles. Although steric stabilization is more appreciated for biomedical applications, it enhances the hydrodynamic diameter of the nanoparticles substantially. Steric stabilization can provide either cationic or anionic or zwitterionic surface charge.

Organic Material

In recent years, considerable effort has been given to optimize the surface coating of the nanoparticles to prevent agglomeration without compromising the properties of the nanoparticles. Generally, these organic coating materials are added during the nanoparticle synthesis and coating is developed *in-situ*. This is most common in aqueous phase synthesis. In the case of synthesis in an organic medium, surface coating is done post-synthesis of the nanoparticles. The main objective of the coating material is to preserve the magnetic properties of the magnetic nanoparticles and to provide surface anchor groups. This type of coating can be core-shell type, matrix type and shell-core-shell type where magnetic nanoparticles are always considered as core (Fig. **2**). In matrix type coating, certain polymers are used to form the homogeneous matrix where several magnetic nanoparticles are embedded in it. In core-shell type of coating, core and shell are either covalently linked to each other or shell material adsorbed on the surface of the core nanoparticles. Polymers are most widely chosen as shell or matrix materials due to their suitable properties and biocompatibility without compromising the particle properties. *In situ* polymerization, layer-by-layer assembly, simple mixing, self-assembly are a few common techniques widely accepted. The organic materials can be subdivided into-small molecules or surfactants and polymers. Molecules bearing terminal carboxylate, phosphate or hydroxide groups are most suitable for surface modification of iron oxide nanoparticles.

Small Molecules or Surfactants

Surface modification with small molecules in suitable solvents is most convenient and straightforward. It offers sufficient surface repulsion to prevent aggregation and surface anchoring groups for further modification. Such small molecules may be hydrophobic, hydrophilic or amphiphilic. Hydrophobic molecules involve long chain (10-18 carbon) fatty acids, amines or thiols, alkyl phenols, *etc.* [66, 80, 81]. These molecules provide steric repulsion between the nanoparticles and thus prevent aggregation (Fig. **2**). They are most suitably used *in situ* during nanoparticle synthesis. In solvothermal synthesis, with ODE solvent at high temperature, these small molecules are widely used as capping ligands and facilitate to obtain monodisperse nanoparticles. Among these long chain fatty acids, oleic acid, which has a C18 tail with a double bond in the middle offers much more flexibility to control shape and size. On the other hand, stearic acid also contains C18 tail but is not suitable to control the shape of nanoparticles, although it can efficiently stabilize nanoparticles. Oleyl amine, octanoic acid, octadecane thiol are a few other examples of long chain small molecules. Small molecules with carboxylate terminal group are most suitable for iron oxide nanoparticles due to the affinity between oxide surface and carboxylate groups. Other types of small hydrophobic molecules found suitable for iron oxide nanoparticles are alkyl phosphates and phosphonates like dodecyl phosphonate, hexadecyl phosphonate, dihexadecyl phosphate, *etc.* Normally these molecules form bilayer structure on nanoparticle surface and the primary layer is attached to the nanoparticles *via* phosphate or phosphonate groups. These small molecules are more biocompatible than carboxylate bearing groups. Generally, they are used during post synthesis surface modification stage in a reverse micelle with the hydrophobic nanoparticles. One major disadvantage of these molecules is that they render the nanoparticle surface hydrophobic which require further surface treatment for biomedical applications.

Hydrophilic small molecules possess the advantage of one-step synthesis of hydrophilic nanoparticles with biocompatibility and biodegradability feature along with minimized cytotoxicity. Small molecules like amino acids [82], citric acid [83], dopamine [84], vitamins [85], cyclodextrin [86], and carbohydrates [87] are directly added to the nanoparticle solution during synthesis under optimized condition. These small molecules are attached to the magnetic nanoparticle surface *via* non-covalent approach and provide water solubility. Sol-gel technique and co-precipitation methods are generally used for this type of surface modification. Another common and most extensively used method for surface coating is the ligand exchange method. Here, hydrophobic nanoparticles are converted into hydrophilic ones in one step. For the ligand exchange method, various small molecules and surfactants have been explored. Among the small molecules, dopamine, citric acid, tetra methyl ammonium hydroxide (TMAH), mercapto propionic acid (MPA) are commonly used to modify iron oxide nanoparticle surface. Among the surfactants, CTAB, tetramethylammonium-1--aminoundecanoate, mercapto undecanoic acid have been employed (Fig. **2**). Ligand exchange method is simple and easy to execute, although the reproducibility of ligand density on the nanoparticle surface and controlling the kinetics of ligand exchange is still under research. Ligand exchange is most commonly encountered in reverse micelle, which solubilises both hydrophobic nanoparticles and hydrophilic ligands and initiates the ligand exchange method in a single phase. Zwitterionic molecules like amino acids and amino sulfonate molecules are of special interest due to their almost neutral surface charge and superior colloidal stability [88]. These molecules are of particular interest in biology due to their neutral surface which prevents non-specific interactions with the cell walls.

Polymers

Although small molecules are convenient for surface coating, recently polymer-coated nanoparticles have gained immense importance due to their robust nature, improved colloidal stability, various anchoring groups, controlling the concentration of anchoring groups, *etc.* As for example, polymer coating can incorporate fluorescent nanoparticles/molecules or plasmonic nanoparticles in the same system with magnetic nanoparticles. This provides multiple imaging capability to track them *in-vitro* and *in-vivo*. On the other hand, different polymers have different responsive features like poly-N-isopropyl acryalamide (p-NIPAM) shows thermo responsive behaviour and poly acrylic acid (PAA) exhibits pH-responsive behaviour. This stimuli responsive switching between swelling and collapsing of polymer has several advantages in drug delivery. Commonly used polymers can be subdivided into the natural polymer and synthetic polymer. Among the natural polymers, dextran [89], starch [90], gelatin [91], chitosan [92] are widely used since they are biocompatible and biodegradable. For coating with natural polymer, as-synthesized nanoparticles are treated with the desired amount of polymer in water for a definite time period under optimized conditions. In many instances, acid or base can be used as a catalyst. All the natural polymers contain few hundreds to few thousands of surface hydroxyl groups which can efficiently anchor with the magnetic nanoparticle surface. Polymer coating increases the hydrodynamic diameter of the nanoparticles from a few ten to hundreds of nanometres. This can be controlled by controlling the molecular weight of the polymer or by controlling the coating time. Higher molecular weight polymers often lead to precipitation and poor colloidal stability.

Table 1. List of some important polymers and their roles.

Name	Structure	Role
Polyethylene glycol (PEG)	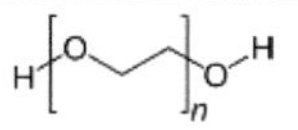	Enhances biocompatibility, EPR effect
Polyvinyl alcohol (PVA)	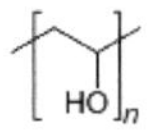	Surface coating
Polyacrylic acid (PAA)	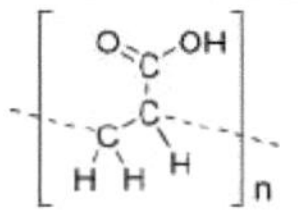	Provide surface carboxyl group, pH responsive coating
Poly(N-isopropylacrylamide) (PNIPAM)	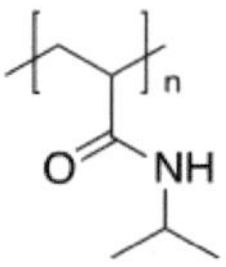	Temperature responsive coating
Poly lactic-co-glycolic acid (PLGA)	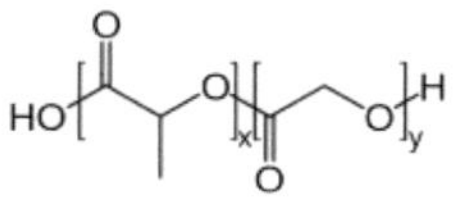	Biodegradable polymer
Poly lactic acid (PLA)	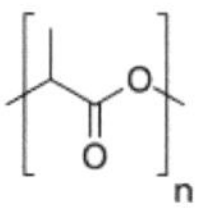	Biodegradable polymer
Polymethyl methacrylate (PMMA)	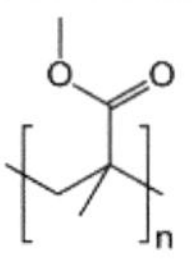	Surface coating
Dextran	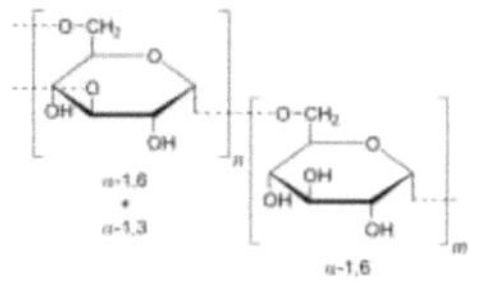	Biocompatible coating

Among the synthetic polymers used for magnetic nanoparticle coating, PEG is the most widely used and most versatile for biomedical applications [79, 93]. PEGs of various molecular weights and various end group functionality are commonly available. PEG coating offers superior colloidal stability, high biocompatibility, enhanced blood circulation time and minimum non-specific interactions. PEG molecules with terminal amine, acid, thiol or other small molecules have been used to provide anchoring groups to the nanoparticle surface for preferential drug loading. Other than PEG, poly (vinyl alcohol) (PVA) [94], poly lactic acid (PLA) [95], Alginate [96], polymethyl methacrylate (PMMA) [97], PAA [98], poly lactic-co-glycolic acid (PLGA) [99], p-NIPAM [100], polycaprolactone (PCL) [101] are widely employed for polymer coating of magnetic nanoparticles (Table **1**) . Often, block copolymers like PEG-PLGA, PLA-PLGA, PEG-PCL, PEG-PPO-PEG, PEG-PLGA-PLA *etc.* have been used to tailor the surface properties and to enhance the biocompatibility.

Polymer coating can be done by *in-situ* polymerization, coating of nanoparticles by pre-synthesized polymer or by layer by layer (LBL) assembly. *In-situ* polymerization is most commonly employed for magnetic nanoparticles. Radical assisted polymerization is the most convenient route to prepare this coating. In this polymerization, desired monomers and nanoparticles are dissolved in suitable solvents or in suitable reverse micelle media. Crosslinkers are used to make the coating more robust. The polymerization is initiated by adding a suitable initiator and it continues for the desired time period under an inert atmosphere. Polyacrylate coating, poly histidine coating, poly dopamine coating, poly imidazole coating are commonly used for magnetic nanoparticles. In polyacrylate coating, acrylate monomers are chosen first depending on the desired surface properties [79, 102]. As for example, if amine surface functionality is required, N-(3-aminopropyl) methacrylate is chosen as monomer, and for acid surface functionality, acrylic acid is the suitable monomer. To incorporate PEG functionality, PEG methacrylate is used. The monomers and nanoparticles are mixed vigorously in reverse micelle made by igepal and cyclohexane. Methylene bis-acryalamide (MBA) or phosphate crosslinkers are used in this technique. Under an inert atmosphere and in the presence of ammonium persulphate (APS) initiator, the polymerization is allowed to continue for 30 min to 1 hour depending on the desired thickness of the coating layer. Longer polymerization leads to gel-like polymer coated nanoparticles. In poly histidine coating, histidine was mixed with nanoparticles in reverse micelle in the presence of 2,4,6-tris (bromomethyl) mesitylene under basic conditions. The polymerization was continued for a desired time period, typically 12 hours. Alternatively, poly histidine methacrylate can be directly used for the polymerization process. Poly dopamine coating is very common in the case of magnetic nanoparticles due to its high affinity towards cis diol of dopamine. Under mild basic condition, these polymer-coated nanoparticles can be achieved. Poly imidazole coating can be seen as an extension of poly histidine coating where any imidazole derivative can be used as a monomer [103]. This type of in-situ polymerization has much more control over the shell thickness, number of reactive surface functional groups, surface charge, *etc.* As for example, in polyacrylate coating, a number of reactive surface amine groups can be controlled by using competitive monomers (amine and acid monomer or amine and sulphate monomer) in desired ratio during polymerization. Moreover, stimuli responsive behaviour can be controlled by incorporating NIPAM (thermo responsive) or acrylic acid (pH responsive) as the monomer along with other monomers. In many cases, *in-situ* polymer synthesis requires harsh chemical conditions and difficult techniques and that often compromises with the nanoparticle properties.

In these cases, polymers are directly assembled on the nanoparticle surface or nanoparticles are directly formed on the polymer matrix. PEG, PLGA, PLA, PCL and block copolymers fall in this category and directly employed in the nanoparticle solution under suitable conditions. Another technique commonly employed is LBL, where cationic and anionic polymers are sequentially added to the nanoparticle surface. Many biological polymers like albumin, proteins or carbohydrates have also been employed for nanoparticle coating.

Inorganic Materials

Nanoparticle coating with inorganic materials is well known and has been utilized since many years. This type of coating is convenient, easy to perform, cheap and can provide colloidal stability to the nanoparticles. Inorganic material coated iron oxide nanoparticles are less prone to oxidation and possess enhanced potential in various applications such as catalysis, separation, enrichment, *etc.* Silica-based inorganic coating is well studied and widely accepted amongst the scientific community. Carbon-based coating has also received attention recently due to its low cost, easy processing and biocompatibility. Noble metal nanoparticles have also been used to coat magnetic nanoparticles in a core-shell or shell-core-shell model, which prevents the magnetic nanoparticles from adverse reaction conditions.

Silica Coating

Silica coating is most common and most widely studied for magnetic nanoparticles [104, 105]. It possesses several advantages like superior hydrophilicity, biocompatibility, controlled surface modification, controlled thickness, *etc.* Furthermore, silica coating can be made porous by using proper surfactants and optimized conditions. Porous silica coating has several advantages in controlled and sustained drug loading and delivery. Typically silica shell can be controlled between 5-200 nm in common synthetic procedures. Although it possesses several advantages, it suffers from poor colloidal stability, especially in acidic and basic conditions, and magnetic properties are often compromised in the case of thick shell.

Sol-gel process and Stober method are most commonly employed for silica coating (Fig. **2**) [106]. Tetraethyl orthosilicate (TEOS) is the most widely used monomer for silica coating. TEOS to water ratio controls the thickness of the silica shell. TEOS can provide a protective silica layer on the nanoparticle surface, although TEOS does not provide any active surface functional groups. To provide surface functionality, (3-aminopropyl) triethoxysilane (APTES) or 3-mercaptopropylteimethoxysilane (MPTES) are mostly employed along with TEOS, which provide amine or sulfhydryl groups on the nanoparticle surface. This silica coating proceeds with the formation of Si-O bond on the nanoparticle surface. Silica coated nanoparticles are often employed to prepare composite nanoparticles with different imaging modality incorporated in them. As for example, silane modified dye molecules are often mixed with other silane monomer, which provides fluorescent property to magnetic nanoparticles. Often ligand exchange technique is employed for composite preparation between two nanoparticles *via* silica coating.

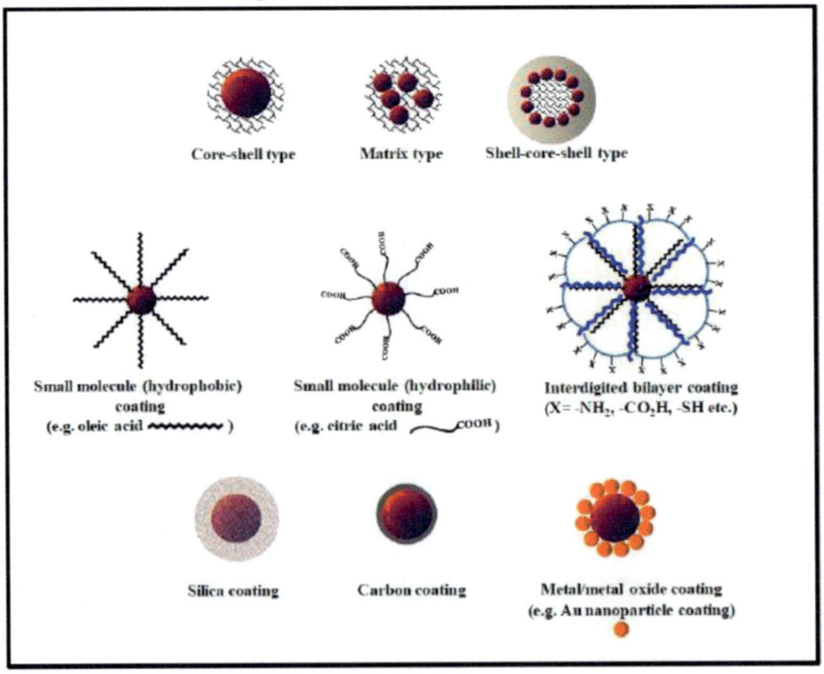

Fig. (2). Various coating approaches available for nanoparticles.

Carbon Coating

This coating approach is relatively new and recently explored with various carbon-containing nanostructures like carbon particles, carbon sheet, grapheme, *etc.* [107]. This coating provides a thin blanket of carbon on the magnetic nanoparticle surface (Fig. **2**). Often this carbon blanket is partially oxidized to provide surface functional groups which improve the colloidal stability of the nanoparticles as well as provide surface anchoring groups. The major advantages of this coating are, thin shell, biocompatibility and inexpensive methodology. Generally, magnetic nanoparticles are mixed with carbonaceous materials, as for example, glucose followed by vigorous stirring and heating in an autoclave. For graphene wrapping, nanoparticles are either synthesized *in situ* in a bed of graphene or post-synthetic nanoparticles are mixed with reduced graphene oxide under optimized conditions.

Metal and Metal Oxide Coating

Magnetic nanoparticles coated with metal or metal oxides are more common for catalysis applications, although recently, biomedical applications of such systems have also been explored in a large number [108, 109]. Noble metal coated magnetic iron oxide nanoparticles are abundantly found, for *e.g.* Au@γ-Fe_2O_3, Ag@γ-Fe_2O_3, Pd@Fe_3O_4, Ag@Au-Fe_3O_4 *etc.*, (Fig. **2**). These nanocomposites are the core-shell type where magnetic iron oxide nanoparticles act as core and noble metal nanoparticles act as shell. These core-shell nanocomposites are often prepared by the solvothermal method where heterojunction between the nanoparticles is formed. Seeded growth methods are also followed, especially for Au@γ-Fe_2O_3 nanoparticles where Au seeds are deposited on the iron oxide nanoparticle surface followed by growth of Au nanoparticles from the seeds under appropriate conditions. Often magnetic nanoparticles and noble metal nanoparticles are separately synthesized followed by composite preparation. This can be readily achieved by polyacrylate coating or ligand exchange technique. For ligand exchange technique, one nanoparticle surface is suitably modified with functional groups (-NH_2 or -CO_2H) and that is used to exchange the functional groups of the other nanoparticle surface. Apart from metal, metal oxides like Al_2O_3, ZnO or TiO_2 have been employed to coat the surface of magnetic nanoparticles.

Surface Functionalization

Even after appropriate surface coating, surface functionalization is often necessary to serve specific purposes. For any biomedical application like biosensor development, cellular imaging or drug delivery, target specific delivery of nanoparticle is the first and foremost criteria. To identify the target and to deliver the nanoparticle (with cargo) to the targeted site, specific surface engineering is necessary. For specific biosensor development, it is necessary to identify the target specifically and precisely, which can be achieved by modifying the nanoparticle surface with a specific capturing ligand. For imaging applications, like MRI, it is important to trigger the desired nanoparticle in the targeted region for effective imaging. Our major focus here is drug delivery where we can specify our focus on tumor targeting. Solid tumors behave differently compared to normal cells. Normal tissues have tight blood capillary and efficient lymphatic drainage. Thus any nanoparticle that may enter the normal tissue is efficiently thrown out from the system by lymphatic drainage. But in solid tumors, there are leaky blood vessels and dysfunctional lymphatic drainage. Thus long circulating nanoparticles can enter the solid tumors and retain in the tumor tissue. Thus tumor targeting by the nanoparticles requires certain criteria [15] - (a) it must be long-circulating, that is, it must have the enhanced permeation and retention (EPR) effect. (b) It must be small enough in size to pass through the gaps of endothelial cells but large enough to avoid escape from tumor tissues. Typically size between 25-200 nm is ideal for tumor targeting, although nanoparticles with higher size ranges (<500 nm) have been employed in tumor targeting. (c) Colloidal stability and pH stability of the nanoparticles which avoid agglomeration of nanoparticles due to sudden change in pH in the tumor tissues. (d) surface charge which is preferred to be neutral to avoid nonspecific interactions. (e) biocompatibility, *etc.* Nonspecific interactions, which can lead the nanoparticles in either phagocytes or reticule endothelial system (RES), must be avoided by PEGylation. Due to different phenotypes of tumor tissues, different tumors have different receptors on their membrane [15, 110, 111]. These receptors are specific and vary depending upon the types and origin of the tumor tissue. These receptors are membrane proteins and overexpressed under certain conditions and distribute over the cell membrane. This distribution generally is not uniform and their density gradient depends upon several factors of specific tumor sites. They act as chemical channels for different species to enter into the tumor site as well as help in cell adhesion. These receptors are the main target for the drug delivery systems for active targeting. For specific receptors, specific ligands must have to be attached to the nanoparticle's surface. These ligands can be some protein, enzyme, carbohydrate, vitamins, *etc.* As for example, kupffer cell receptors can be actively targeted by mannose functionalized nanoparticles [112], biotin receptors are targeted by biotin functionalized nanoparticles [111], folate receptors are targeted by folic acid modified nanoparticles, *etc.* [113, 114]. Not only the choice of targeting ligand determines active targeting, ligand density on nanoparticle surface also plays an important role in this aspect. As for example, low density folic acid can target the drug-loaded nanoparticles to the nucleus of a tumor cell, whereas higher folic acid density leads to the accumulation of the nanoparticles in the endosomes

[102, 115]. This is due to different endocytosis mechanisms on the cell membrane. Low folic acid density triggers the caveolae-mediated endocytosis, whereas higher folic acid leads to clathrin-mediated endocytosis. Thus, it is extremely important to control the targeting ligand density on the nanoparticle surface [116].

Covalent conjugation chemistry is the most widely used and acceptable surface chemistry for nanoparticle surface engineering [117 - 120]. Various well-known covalent chemistries like 1-ethyl-3-(3-dimethylaminopropyl)carbodiimide (EDC)/ N,N'-dicyclohexylcarbodiimide (DCC) coupling, thiol-ene click chemistry, esterification, Cu mediated click chemistry, glutaraldehyde coupling, amine-NHS (N-hydroxysuccinamide) coupling, *etc.* are regularly employed for nanoparticle surface engineering (Table **2**). These conjugation chemistries do not require a harsh chemical environment and thus can protect the activity of the biologically active molecules. Water is used as a solvent in most of the conjugation chemistries. EDC coupling is efficient to conjugate a terminal amine group to an acid functional group. Although the conjugation chemistry is simple and performed in water, its yield and reproducibility are low. Alternatively, DCC coupling serves the same purpose but since it is performed in an organic medium, separation of by-product is easy and yield is better. Amine-NHS coupling is very efficient and can be performed in neutral conditions without much complication. For this purpose, the desired ligand or functional group is first modified with a terminal NHS group, which is then conjugated to amine-functionalized nanoparticles. In almost every case, the ligands or functional groups taken into consideration must have an active group on one end which attaches with the nanoparticle. In many instances, bifunctional small molecules act as a spacer which can tune the distance between nanoparticles and ligand as well as conjugate the nanoparticle with the ligand.

As for example, in glutaraldehyde coupling, glutaraldehyde having two terminal aldehyde groups acts as spacer which connects two amine groups *via* imine bond formation.

Non-covalent conjugation of targeting ligands to the nanoparticle surface is also abundant [121, 122]. Antibodies or fused proteins are often loaded on nanoparticle surface by non-covalent attachment which mostly depends on electrostatic attraction. In many cases, specific biological interactions play a vital role in non-covalent attachment. As for example, for biotin attachment, nanoparticle surfaces are first modified with Streptavidin or Avidin protein. Due to strong interaction between biotin-Streptavidin/Avidin, biotin is localized on the nanoparticle surface. Similar interaction is also useful for glucose functionalization with the help of glucose-Conconavalein A (glycoprotein) interactions.

DRUG LOADING ON MAGNETIC NANOPARTICLES

Drug loading on the nanoparticle surface is primarily carried out either by covalent attachment or by non-covalent interactions. Anticancer drugs can be hydrophilic (*e.g.*, doxorubicin (DOX), daunorubicin (DAUN), *etc.*) or hydrophobic (*e.g.*, curcumin, cisplatin, methoxarate, *etc.*). For covalent attachment of the drugs to the surface-functionalized magnetic nanoparticles, the drugs must possess active functional groups which do not hamper the drug activity upon conjugation. Most of the drugs contain surface hydroxyl groups which are not at all suitable for covalent drug loading. Thus, it is important to modify the drug molecules with active functional groups.

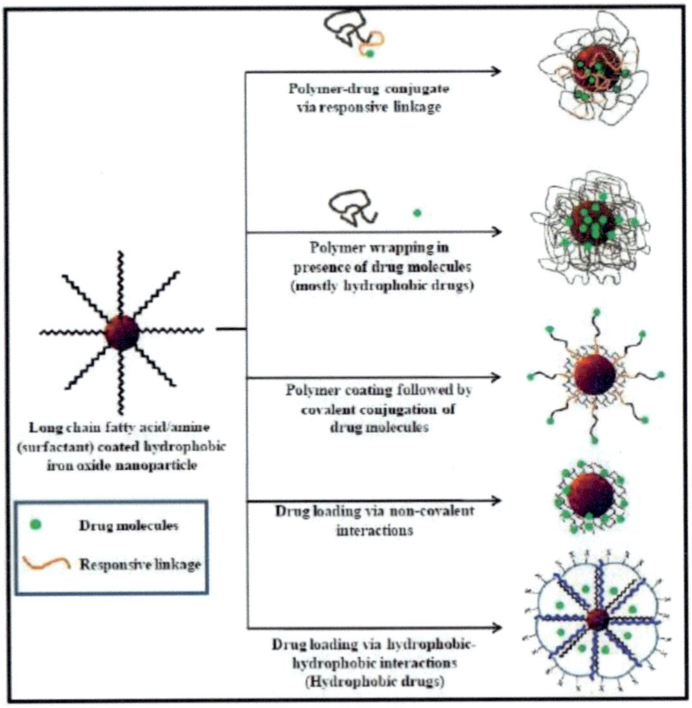

Fig. (3). Different strategies for drug loading.

Acryloyl chloride is often used to terminate the drug molecules with alkene groups which are susceptible to conjugate with thiol groups [102]. Glycidol or epichlorohydrin modification of drug molecules results in terminal epoxide groups, which are prone to attack by a primary amine or hydroxyl groups of other molecules. Drug modifications are done in such a way that does not affect the drug's effectiveness. In many cases, prodrugs are used rather than actual drugs.

The advantage is that prodrugs do not have any effectiveness until they are released in the targeted location. Covalent attachment of drugs is much more useful since it does not lead to drug release in undesired normal tissues, it can withstand biological and metabolical conditions inside the body and it preserves the drug's effectiveness for a long time.

Polymer-drug conjugates are the most extensively used drug loading system where a polymer of specific property is conjugated with the drug molecules *via* covalent linkage followed by coating on the nanoparticle surface [123 - 125]. The major advantages of these systems are that drugs can be loaded into the core of the system and in the shell or can be distributed over the entire region as per requirement, simultaneous coating and drug loading can be achieved and the covalent linkages between drug and polymer can be biodegradable as well as stimuli-responsive (Fig. **3**). PEG is the most common polymer and most extensively used polymer for this purpose [126 - 129]. Other than PEG, various biodegradable synthetic polymers have been explored, which can be dissolved in body after the drugs have been released. Poly-(amino acid) based drug carriers are well known for their biodegradable nature [130, 131]. Among them, poly-(aspartic acid) and poly-(glutamic acid) derivatives are well explored for drug delivery applications. These polymers have been conjugated with various anti-cancer drug molecules like - paclitaxel, DOX, *etc. via* enzyme responsive, pH responsive or redox responsive linkages. Various acrylic and vinyl polymers and copolymers have also been studied due to their improved hydrophilicity and active functional groups [132]. Poly-(acrylic acid), PVP, poly-(N-vinyl-pyrid-ne-N-oxide), poly-(styrene-co-maleic acid), HMPA, NIPAM, *etc.* are the common examples of this type of polymers. Other than this, PLA and PLGA and their copolymer with PEG have been extensively used for drug delivery due to its biocompatible nature and controlled micellization techniques. PCL, poly (acetal)s, poly (amido amine)s are other common examples of biodegradable polymers used for polymer-drug conjugate [133, 134].

Non-covalent drug loading has also received increased attention due to its simple methodology and applicability for both hydrophobic and hydrophilic drugs (Fig. **3**). The most common non-covalent interactions are electrostatic interaction, where nanoparticle surface is made highly cationic or anionic. Drugs of reverse polarity are adsorbed on the nanoparticle surface and can release by changing the polarity [135]. Hydrophobic drugs are often loaded *via* hydrophobic-hydrophobic interaction in the core of polymer-coated nanoparticles or inside the bilayer coating on the nanoparticles. As for example, in the case of interdigital bilayer coating over $ZnFe_2O_4$ nanoparticles or iron oxide nanoparticles, hydrophobic drug curcumin is encapsulated in the hydrophobic region of the bilayer coating (Fig. **3**) [59]. In many instances, drugs are loaded inside the hydrophobic micellar core during the polymerization reaction. In such cases, both the hydrophobic drug molecules and the hydrophobic nanoparticles are encapsulated in the core of the micelle. As for example, during polymerization of NIPAM on iron oxide nanoparticle surface, hydrophobic drug molecules are also loaded in the core of the nanosystem. This drug loading is particularly necessary for both hydrophilic and hydrophobic drug molecules. In the case of coating of nanoparticles with stimuli responsive polymers, this drug loading approach is most widely used.

STIMULI RESPONSIVE DRUG RELEASE

Once the drug is loaded in the system and administered to the desired location, it is the release that plays a pivotal role in the disease treatment. Normally tumor tissues have a different microenvironment compared to normal tissues. Tumor tissues have a lower pH (~4.5-5.0), higher temperature (45-50°C) and higher glutathione content due to higher rate of metabolism. Apart from this, there are several enzymes and proteins that are over-expressed in tumor tissues although it may depend on the type of tumor tissue and its origin. Hence, the drug should release depending on the external stimuli like temperature, pH, redox environment or specific enzyme. Stimuli responsive drug delivery has several advantages - (i) it avoids the release of drugs in normal tissues (ii) it can control the kinetics of the drug release that is sustained drug delivery (iii) it minimizes the loss of drug, that is it reduces the effective dose of the drugs (iv) it preserves the effectiveness of the drug molecules and many more. Responsive drug release entirely depends on the chemistry of drug loading and the design of nanoparticle surface. Responsive behaviour can be attained in both covalent and non-covalent drug loading. For covalent drug loading, responsive behaviour depends on the labile bond between nanoparticles and drugs. This labile bond can be cleavable by external stimuli. As for example, ester linkage or hydrazone linkages can be cleaved at lower pH (pH 5.0 or below), disulfide or selenosulfide bonds can be cleaved in reducing environment, Diels-Alder types of bonds or other ring opening-closing types of molecules can be light responsive or that can be controlled or tuned by changing the wavelength of the light. In the case of non-covalent drug loading, stimuli responsive polymers are generally used. Nanoparticles and drug molecules are often co-encapsulated inside the polymer *in situ* or *ex situ*. For thermoresponsive drug delivery, NIPAM is used as the monomer to coat the nanoparticle and drug molecules under suitable conditions. This coating generates p-NIPAM coated drug loaded magnetic nanoparticles [136, 137].

Table 2. List of few important small molecules and their role.

Name	Structure	Role
Tetraethyl orthosilicate (TEOS)	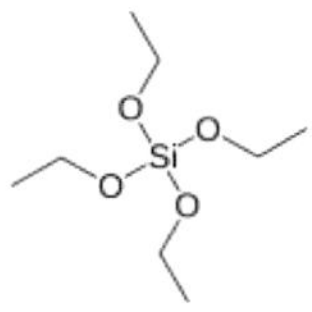	Silica coating
(3-Aminopropyl) triethoxysilane (APTES)	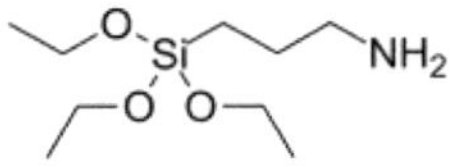	Silica coating with amine functionality
(3-dimethylaminopropyl)-N'-ethylcarbodiimide hydrochloride (EDC)	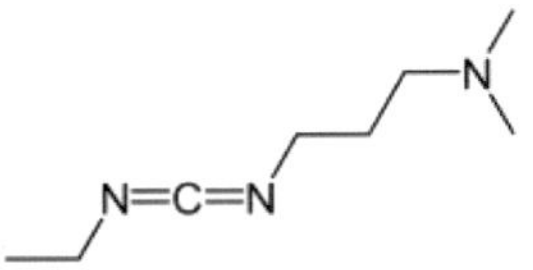	Reagent for coupling between primary amine and carboxylic acid groups.
N-hydroxysuccinamide	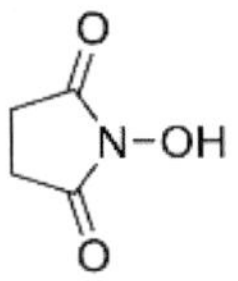	Activates carboxyl acid groups for coupling with primary amines.
Glutaraldehyde		Crosslinker to couple two primary amine groups.
succinimidyl 4-(N-maleimidomethyl) cyclohexane-1-carboxylate	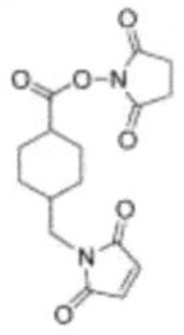	Crosslinker to couple between primary amine and sulfahydryl group.
Hydrazone	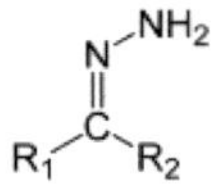	pH responsive bond
PEG-NHS disulfide	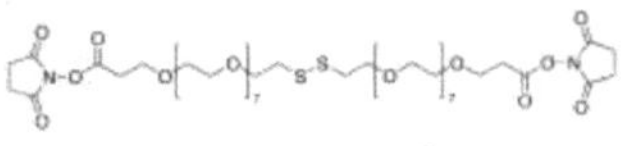	Redox responsive crosslinker to couple between two primary amine groups.

This polymer swells above lower critical solution temperature (LCST), which is around 32°C, and thus polymer opens up, facilitating the release of drug molecules (Fig. **4**). Again at lower temperature, the polymer contracts and thus inhibits the release of drugs. This temperature responsive switching depends on the reversal of hydrophilic-hydrophobic components on the polymer surface.

Below LCST, strong hydrogen bonding of the hydrophilic part collapses the polymer, whereas above LCST, hydrogen bond breaks down and hydrophobic part appears on the nanoparticle surface resulting in swelling of the polymer. This transition temperature can be controlled by varying the ratio of the hydrophilic and hydrophobic component during polymerization. Increased hydrophilic part enhances the LCST due to stronger hydrogen bonding whereas increased hydrophobic part lowers the LCST [138]. NIPAM coating can be controlled with respect to polymerization time and monomer to nanoparticle ratio which dictates the thickness of the NIPAM shell. Often, NIPAM is used with other polymers (*e.g.*, PEG) to enhance its biocompatibility. NIPAM can be used with other monomers in a simultaneous polymerization approach to provide other functionalities to the drug delivery system. As for example, NIPAM is mixed with amine monomer to provide primary amines on the nanoparticle surface. Other than NIPAM, poly (N, N′- diethylacrylamide) with an LCST in the range of 25−35°C and poly(dimethylaminoethyl methacrylate) with an LCST in the range of 14−50°C are also commonly employed. Various hydrophilic and hydrophobic polymers are copolymerized with NIPAM to form diblock copolymer which has been employed to keep NIPAM either in the core or to the shell depending on the requirements. Hydrophilic blocks like poly (butyl methacrylate) and hydrophobic blocks like poly-D,L-lactide (p-(DLLA)) are commonly used along with p-NIPAM for drug loading and thermoresponsive release properties [139].

PAA is commonly used for pH responsive coating. Acrylic acid monomer is often introduced to the nanoparticle drug micellar system followed by *in situ* polymerization under appropriate conditions. Researchers also used PAA directly to coat drug delivery systems. In this case, acid-responsive switching between swelling and collapsing takes place below pH 5.0, which is the suitable environment for tumor cells. PAA and p-NIPAM are often used simultaneously to generate pH and temperature dual responsive drug delivery system. Alternatively, acrylic acid and NIPAM are often polymerized with the drug molecules and the as-synthesized polymer-drug systems are used to coat the magnetic nanoparticles.

Other than PAA, poly-aspartates have also exhibited pH-responsive properties. Most pH responsive drug release has been realized by cleaving the pH responsive covalent bonds linking the coating layer of the magnetic nanoparticles and the drug molecules. Hydrazone, Schiff base (imines), acetals/ketals, esters, *etc.* are the most versatile pH-responsive linkages often employed (Fig. **4**, Table **2**) [140 - 142]. To keep the drug molecules safe inside the core of the polymer, drug-polymer conjugate with responsive linkages is often prepared followed by polymerization on the nanoparticle surface. Another very popular responsive drug delivery system involves enzyme responsive cleavage. This type of covalent bond enzymatically hydrolyses in the lysosome or endosome. The enzymes that are encountered commonly in the intercellular regions are phosphatases that cleaves the phosphate bonds, esterases that hydrolyse the ester linkages, glycosidases that cleave glycosidic bonds, proteases that cleave the oligopeptide linkages [143, 144]. The drug molecules are commonly conjugated with the polymer-coated nanoparticles *via* covalent bonds having oligopeptide, carbohydrate or phosphate linkers.

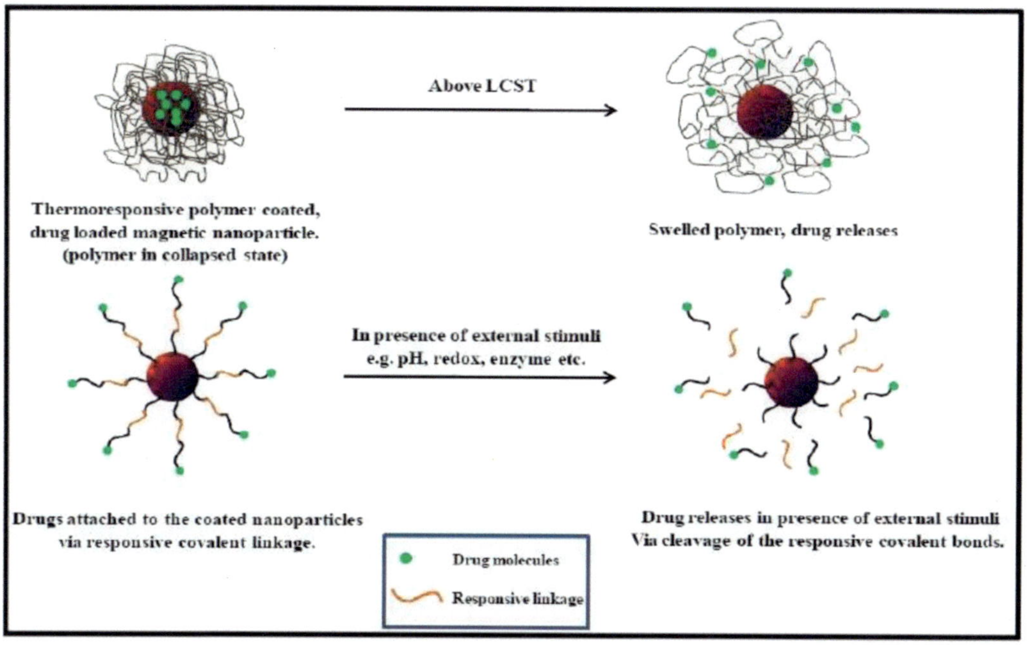

Fig. (4). Stimuli-responsive drug release strategies.

Another responsive drug delivery approach is redox responsive systems which regulate based on the redox environment of the target cells [145 - 147]. Cancer cells dramatically overexpress glutathione in the cellular compartment, which makes the environment reducing in nature and thus prevents activity of many drug molecules. Drug-nanoparticle conjugate *via* disulfide linkages or diselenide, seleno-sulfide linkages can be spontaneously cleaved in the reducing environment by reducing the -S-S-, -Se-Se- or -S-Se- bonds (Table **2**). In many instances, drugs are conjugated with the polymers *via* these specific linkages prior to the polymer coating on nanoparticles. On the other hand, inside inflammatory cells due to the presence of reactive oxygen species/H_2O_2, the environment is oxidative in nature. Arylboronic acid or thioketal linkages are responsive to the oxidative environment. Thus, this type of linkage helps in the delivery of drug molecules in oxidative environments. Multiresponsive systems are now well documented as part of advanced drug delivery systems where more than one stimuli-responsive linkages are assembled in the system. This allows more control and more specificity in drug delivery to the targeted sites.

The kinetics of drug release is another important factor which controls the drug release in the targeted site [148, 149]. Drug release can be instantaneous where most of the drugs release within few hours of administration or it can be sustained where drug release occurs slowly over the time period of 2 or 3 days or even more. Both the release kinetics is important depending on the purpose of drug delivery and the nature of the targeted site. In many instances, drug release has been found to be slow over one or two days and then burst release of drugs takes place within a few hours. This kinetics can be controlled by carefully designing the drug loading chemistry. Kinetics of drug release often depends on the length of the linker molecule between nanoparticle and drugs. The more is the length of the linker molecule, the slower is the release property. In many cases, drugs are entrapped inside the core of the drug delivery system. In this instance, drug release is slow over a few days' time periods. When drug-polymer conjugates are used to coat the nanoparticles, drug molecules are distributed over the polymeric shell as well as in the core. Sustained release of drug molecules is found initially in such systems with a burst release once the core is open up. Drug loading by simple adsorption has shown sustained-release property under a responsive environment. In many instances, dual responsive coatings or bonds are used to load drug molecules, which show no or little release initially. Although, after a few hours or days, drugs start to release in a sustained manner.

MULTIFUNCTIONAL APPLICATIONS OF MAGNETIC NANOPARTICLES

Other than drug delivery, magnetic nanoparticles have found several applications in the biomedical field, including imaging (MRI), therapy (hyperthermia), cell separation, protein enrichments, *etc.* (Fig. **5**).

Imaging

MRI depends on the proton relativity time of the water molecules inside the living tissues. MRI contrast agents are of two types, positive contrast agents (T_1 weighted contrast agents) that brighten the images and negative contrast agents (T_2 weighted contrast agents) that darken the images. On the application of a strong magnetic field, the spins of the hydrogen nuclei of water inside body tissues polarize from disorder to order state in the direction of the applied magnetic field. When a strong radiofrequency pulse is applied, the magnetization of the hydrogen nuclei interacts with the pulse and the spin polarization is observed. After the removal of the radiofrequency pulse, the net magnetization reverts back to its initial position. The relaxation time of the longitudinal magnetization corresponds to T_1 and the relaxation time of the transverse magnetization corresponds to T_2. Different amount of water in different living tissues generates different T_1/T_2 and that can be collected with MR images. Magnetic nanoparticles enhance the contrast of these images by interacting with the water protons as well as with the radiofrequency pulse. Currently, most of the MR contrast agents are based on Gd-Chelates. Various Gd_2O_3 nanoparticles have also emerged as positive (T_1) contrast agents [102]. At room temperature, Gd^{3+} exhibits paramagnetic property with seven unpaired electrons. Various coated and PEGylated Gd_2O_3 nanoparticles are currently employed as T_1 contrast agents due to their non-toxic nature and high contrasting power. However, recently Gd-based chelates have been found to cause nephrotoxicity, and thus, alternative contrast agents are in great demand.

Iron oxide nanoparticle-based negative (T_2) contrast agents have become popular [22, 150, 151]. Ferumoxide and Ferucarbotan are two popular commercialized T_2 contrast agents based on iron oxide formulation. Ferumoxide is dextran-coated iron oxide nanoparticle with hydrodynamic diameter ~120-180 nm and having transverse relaxivity (r_2) ~98 $mM^{-1}s^{-1}$. On the other hand, Ferucarbotan is a carboxydextran coated iron oxide nanoparticle with a much smaller hydrodynamic diameter (~45 to 60 nm) and high transverse relaxivity (r_2) ~150 $mM^{-1}s^{-1}$. The major advantages of iron oxides as negative contrast agents include high specificity, biocompatibility, and high clearance rate from the blood to the liver *via* RES. Although there are several advantages, negative contrast agents are often confused with the other pathogenic signals inside the body; these often destroy the surroundings due to their high magnetic moment and show slow clearance of the large particles. Thus, it is necessary to develop ultra-small iron oxide nanoparticles with positive contrast properties. Superparamagnetic iron oxide nanoparticles (SPIONs) have been found to suppress the longitudinal magnetization with enhanced r_1 relaxivity. High temperature decomposition of Fe-precursors is the most widely used method to generate ultra-small iron oxide nanoparticles (2-5 nm) with controlled shape and size. These nanoparticles exhibit superparamagnetic behaviour at room temperature. Due to their small size, they exhibit low magnetization values which can be attributed to the spin canting effect. These nanoparticles have been extensively used as positive contrast agents (T_1) in MRI with r_1 value between ~2-8 $mM^{-1}s^{-1}$ depending on their size. Generally, these ultra-small nanoparticles are coated with dextran or biocompatible polymer for colloidal stability and PEGylated for bioavailability.

Therapy

Magnetic nanoparticles have been documented to exhibit therapeutic effects *via* neutron therapy or hyperthermia. Gd_2O_3 nanoparticles and Gd-based chelates possess a high neutron capture cross-section and thus employed for neutron therapy. This therapy is especially important for inoperable radio-resistant tumors occurring anywhere in the body. Neutrons have high linear energy transfer (LET) radiation and thus do not depend on the oxygen species to damage a cancer cell as compared to normal radiation therapy. Thus, neutron therapy is very effective and has a high chance of damaging the affected cells. On the other hand, iron oxide nanoparticles are well known to generate localized hyperthermic effects to kill cancer cells. When the administered magnetic nanoparticles are placed in an alternating magnetic field, the external electromagnetic energy is transformed to internal thermal energy which damages the target tissue [152, 153]. This therapy is highly specific and effective with magnetic nanoparticles of ~50-100 nm size. Before administration, the magnetic nanoparticles are coated properly to enhance colloidal stability and bioavailability. Moreover, these are specifically conjugated with target ligand to target the desired tissue/cells/organs for treatment.

Cell/Protein Separation

Magnetic nanoparticles are also used for cell separation [154]. For this purpose, magnetic nanoparticles are specifically functionalized with targeting ligand, which has been used to capture the desired cell type. In vitro normal and cancer cell separation is well studied with folate functionalized iron oxide nanoparticles. For this purpose, mixtures of cells are incubated with functionalized iron oxide nanoparticles under optimized conditions and then separated using an external magnet. Nanoparticles of size ~25-50 nm are most suitable for this purpose. Similar strategy has been employed for protein enrichment. Here also, functionalizations of magnetic nanoparticles play a pivotal role. As for example, glucose functionalized iron oxide nanoparticles have been successfully employed for glycoprotein Con A enrichment in a mixture of proteins.

EFFECT OF SHAPE/SIZE OF MAGNETIC NANOPARTICLES ON DRUG LOADING/RELEASE

Drug loading and subsequently its release largely depends on the shape and size (hydrodynamic diameter) of the nanoparticles used [155, 156]. Again, this dependence varies with many factors like drug solubility, mode of preparation of the nanoparticles, nature of surface coating, *etc.* The general trend for spherical nanoparticles is particles with hydrodynamic size >100 nm possess high drug loading capacity compared to nanoparticles in the size range of ~25-100 nm due to their high surface area. On the other hand, nanoparticle size <25 nm is not as effective as others for drug loading purpose. There are many factors, as mentioned above, controlling the drug delivery and size is one of the crucial factors. Shape of nanoparticle is another important factor that governs drug loading. Different shapes of nanoparticles, sphere, cube, rod, plate, sheet, filament, *etc.* are currently being synthesized and have shown interesting property in terms of cellular internalization, sub cellular targeting and tissue distribution. Generally nanoparticles with anisotropic shape have less drug loading capacity due to their non-uniform surface energy, although surface coating plays a great role in drug loading capacity. This shape dependent drug loading is more prominent for hydrophobic drugs *via* hydrophobic-hydrophobic interactions. In one of our studies, we have observed that cubic Fe_3O_4 had less drug loading capacity (for both hydrophilic and hydrophobic drugs) compared to spherical one. This can be attributed primarily to their synthetic technique. Here, solvothermal technique has been used and the capping ligands are more prone to attach on the high-energy corners of the cubic nanoparticles whereas they are evenly distributed on the spherical nanoparticle surface. Thus, cubic nanoparticle has less number of capping ligands compared to spherical nanoparticle which eventually determines the surface coating *via* interdigited bilayer formation or ligand exchange method. On the other hand, higher loading does not mean higher release properties. Drug release depends much more on the surface property, loading nature and responsive linkage compared to its shape and size.

QUANTITATIVE ESTIMATION OF DRUG LOADING AND RELEASE

Quantitative estimation of the amount of loaded drug and released drug is equally important to understand the drug loading capability of magnetic nanoparticles. It is desirable to load a higher amount of drug molecules to overcome the toxic effect of nanoparticles as much as possible. Drug release profile is important to understand the effectiveness of drug delivery vehicle as well as stimuli-responsive release efficiency. Most of the drug molecules have specific absorption in the near UV region to visible region due to the conjugating electronic network, which is well documented in the literature [79, 102]. Many of the drug molecules possess specific emission properties along with their ability to absorb light. In Fig. (**5**), drug loading and release profile of two different drugs on magnetic Eu:Gd_2O_3 triangular nanoplates have been shown. Curcumin is a hydrophobic drug, whereas daunorubicin is a hydrophilic drug. Both the drug molecules have specific absorbance values (Fig. **5**). Curcumin strongly absorbs at ~420 nm.

Whereas Daunorubicin absorbs at ~480 nm. To calculate the quantitative loading efficiency, absorption spectra of bare nanoparticle and drug loaded nanoparticles are separately measured, keeping the nanoparticle concentration uniform. Next, the difference in absorbance value between these two spectra was calculated with respect to the respective wavelength of the two drugs.

From this value, concentration of loaded drug molecules has been calculated using Lambert-Beer's law. Alternatively, unloaded drug molecules are often collected after separating drug-loaded nanoparticles and their absorption spectra is measured. The concentration of unloaded drug molecules is subtracted from the concentration of added drug molecules which results in the concentration of loaded drug molecules. Loading efficiency can be calculated from the amount of drug loaded per milligram of nanoparticle taken and % of loading can be calculated from the amount of drug loaded with respect to the total drug added.

To estimate the amount of released drug in a specific time interval, emission spectra is more useful. Drug loaded nanoparticles are packed inside a dialysis tube in small quantity and placed inside the desired buffer solution. After a specific time interval, the samples are taken out from the dialysis tube and emission spectra are recorded. The emission spectra is compared with the initial emission spectra of the drug loaded nanoparticles, which results in % release of the drug at that specific time interval. To obtain the release profile, % release of drug molecules at different time intervals is plotted against time which shows how the drugs are released with time and when the release gets saturated with time (Fig. **5**). Alternatively, absorbance spectra have also been used frequently to obtain the release profile.

ADVANTAGES AND DISADVANTAGES OF MAGNETIC NANOPARTICLES

There are many advantages of magnetic nanoparticles in drug delivery and hence tremendously growing interest is noticed in the field of magnetic nanoparticle-based drug delivery. The major advantages are (i) Magnetically targeted drug delivery: Once the magnetic drug delivery system is administered inside the body, it can be targeted to the affected tissue by using an external magnetic field through blood circulation. This process is completely non-invasive and requires a low magnetic field to operate. This helps in dual targeting - surface attached targeting ligand and magnetic targeting, and thus minimizes the effect on normal tissues (Fig. **6**). For magnetic targeting, nanoparticle should have high colloidal stability and high blood circulation time to avoid aggregation in undesired locations.

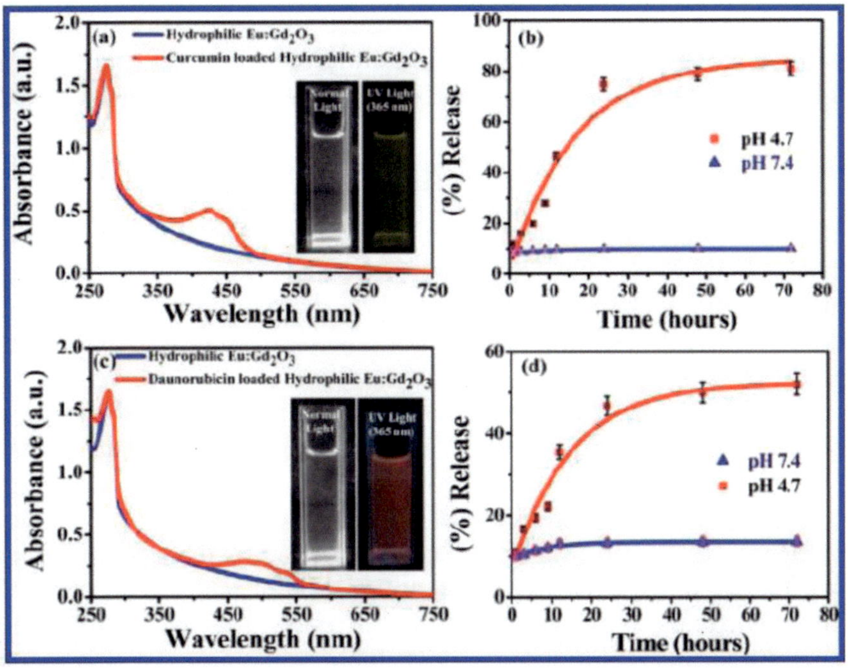

Fig. (5). UV–visible spectra showing curcumin loading on functionalized Eu:Gd2O3 triangular nanoplates. (**a**) Enhanced absorbance at ~420 nm, confirming the curcumin loading on the nanoplates. Inset showing digital images of curcumin-loaded samples under normal light and 365 nm excitation. The percent of curcumin release with time at pH 4.7 and 7.4 was calculated from photoluminescence spectroscopy. (**b**) Higher release was observed at lower pH compared to physiological pH. Daunorubicin loading on functionalized Eu:Gd2O3 triangular nanoplates was also confirmed from the enhanced absorbance at ~480 nm for drug-loaded samples. (**c**) Inset showing the digital images of drug-loaded samples under normal light and 365 nm excitation. Daunorubicin release profile at pH 4.7 and 7.4. (**d**) More drug release was encountered at lower pH compared to physiological pH. Error bars indicate the standard deviation obtained after collecting the data for five different experiments. (Reprinted with permission from ACS Appl. Mater. Interfaces 2017, 9, 4126−4141. Copyright 2017).

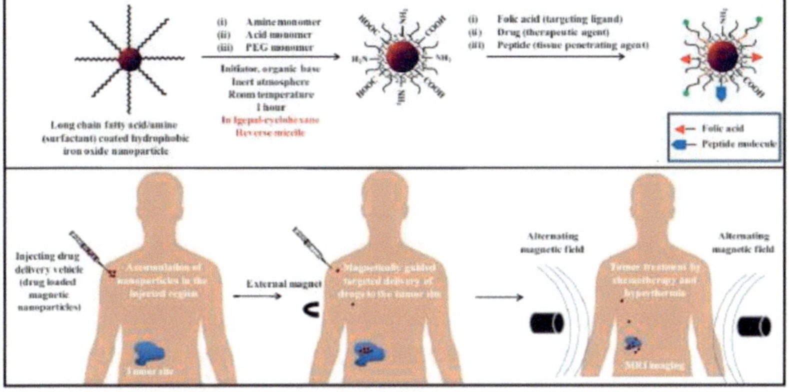

Fig. (6). A typical example of polyacrylate coating and functionalization strategies of magnetic nanoparticles and its use in therapy and imaging on the living body.

i. Tracking of drug delivery system: Magnetic nanoparticles can act as contrast agents in MRI which is a potential imaging technique to track the magnetic drug delivery system *in-vivo*. Polymeric nanoparticles, micelles and other drug delivery systems lack this integrated imaging technique and thus difficult to track these systems from outside. Magnetic nanoparticles, on the other hand, can be easily monitored through a non-invasive MRI based imaging system.
ii. Hyperthermia: Magnetic nanoparticles are particularly useful for hyperthermic therapy of the targeted cells. Thus, magnetic drug delivery system can achieve dual therapy, one is chemotherapy based on the drug payload and the other is hyperthermic therapy where heat is generated by external magnetic field to destroy the affected tissue.
iii. Biocompatibility: Iron is biocompatible since our body possesses iron in the form of ferritin, haemoglobin, *etc.* Thus iron nanoparticles can accumulate in liver, spleen, *etc.* in small amounts without any toxic effect, although a large amount of accumulation is always toxic to the body.
iv. Controlled synthesis and functionalization techniques: There are several well-documented methods available to synthesize size and shape controlled iron oxide nanoparticles with good size distribution. Apart from that, several surface engineering techniques are available to design the nanoparticles for efficient drug delivery.
v. Cost-effective, convenient and simple: Iron oxide nanoparticle based drug delivery systems are comparatively inexpensive and easy to handle. Their synthetic and functionalization techniques are simple and require less time for preparation. Moreover, these nanoparticles are comparatively stable in harsh conditions of surface engineering and drug loading.

Although there are several advantages, magnetic nanoparticles still possess a few challenges for their clinical applications. The major issue is biodegradability. Although these are biocompatible, they accumulate inside our body as oxide nanoparticles and cause toxicity in the long run. Normally, oxide nanoparticles dissolve in acidic pH inside our body and generate Fe^{3+} salts, which are less toxic and can be cleared *via* RES. But, if the surface coating is not properly controlled, nanoparticles remains as it is and accumulate slowly in liver, spleen, RES, *etc.* Moreover, depending on the surface coating, renal clearance of the dissolved nanoparticles can take a few months to few years for complete clearance particularly if the nanoparticle size is large. Another problem is remnant magnetization which can cause particle aggregation inside blood vessels and can be detrimental. This factor also makes renal clearance difficult due to aggregation. The potential solution is to use superparamagnetic iron oxide nanoparticles, which possess zero remnant magnetization at room temperature.

CONCLUSION AND FUTURE CHALLENGES

With the development of new synthetic and coating methodology, controlling nanoparticles surface architecture and their conjugation with drug molecules have become effortless. Manipulating shape, size and surface of nanoparticles provide suitable access to attach drug molecules and their active and/or passive targeting to living tissues. The magnetic nanoparticles have been demonstrated to exhibit several advantages over conventional drug delivery systems like micellar system, polymeric system, liposomes, *etc.* Many therapeutics based on magnetic nanoparticle-drug conjugates are already available in the market and many are under development. Various novel magnetic systems are under study for drug delivery approaches such as magnetic nanoclusters, magnetic liposomes, nanorobots, sophisticated stimuli responsive magnetic systems, *etc.* Moreover, many new systems are entering in clinical development and evolution stage which includes polymer-drug conjugates with magnetic nanoparticles, PEGylated drug delivery systems, responsive magnetic drug delivery systems, *etc.* Magnetic nanoparticles are often combined with other systems for multimodal therapy and imaging approaches. Composite magnetic nanoparticles are becoming more and more popular for *in vivo* applications. Magnetic nanoparticles not only carrying drug molecules, but other payloads like gene, SiRNA, *etc.* have been successfully delivered through magnetic nanoparticle-based systems. Still a lot of focus is required by the researchers and regulatory authorities to develop combined theranostic approaches using this nanosystem and to apply them in the process of preclinical development and clinical tests.

There are few challenges that still remain. One major challenge is upscaling of the product from lab scale to industrial scale. Many of the lab-scale drug delivery systems have been tested *in vitro* and *in vivo*, and even in preclinical studies with excellent results; although during large scale synthesis, the properties of drug delivery systems, like shape, size, colloidal stability, drug loading capability, the effectiveness of the drug molecules and therapeutic capabilities, are compromised. So this challenge must be considered seriously to overcome the pitfall of magnetic drug delivery systems. Another major challenge is biodegradability and clearance of the drug delivery systems/vehicles from the body. This issue has been currently addressed by many scientists to develop more biocompatible drug delivery systems with improved renal clearance. Detailed toxicity studies (genotoxicity, neurotoxicity, *etc.*) and more clinical study data are required to introduce this system for long-term practical use. Deep tissue penetration of the drug delivery system is another vital challenge to the scientific community. Targeting ligands can target cellular systems or superficial tumor tissues. But, in tumor systems, tissues are closely packed and penetration of these systems for drug delivery vehicles will still be a challenge. Many recent works have been focused on deep tissue penetration, although their clinical study is yet to be performed.

CONSENT FOR PUBLICATION

Not applicable.

CONFLICT OF INTEREST

The author/editor has no conflicts of interest, financial or otherwise, to declare.

ACKNOWLEDGEMENTS

AS acknowledges DST for the INSPIRE Fellowship program.

REFERENCES

[1] Moriarty P. Nanostructured materials. Rep Prog Phys 2001; 64: 297-381. [http://dx.doi.org/10.1088/0034-4885/64/3/201]

[2] Drexler E. There's plenty of room at the bottom. Metamodern: The Trajectory of Technology 2009; 12: 29.

[3] Pan H, Feng YP. Semiconductor nanowires and nanotubes: effects of size and surface-to-volume ratio. ACS Nano 2008; 2(11): 2410-4. [http://dx.doi.org/10.1021/nn8004872] [PMID: 19206409]

[4] Takagahara T, Takeda K. Theory of the quantum confinement effect on excitons in quantum dots of indirect-gap materials. Phys Rev B Condens Matter 1992; 46(23): 15578-81. [http://dx.doi.org/10.1103/PhysRevB.46.15578] [PMID: 10003694]

[5] Dabbousi BO, Rodriguez-Viejo J, Mikulec FV, *et al.* (CdSe)ZnS Core–Shell quantum dots: Synthesis and characterization of a size series of highly luminescent nanocrystallites. J Phys Chem B 1997; 101(46): 9463-75. [http://dx.doi.org/10.1021/jp971091y]

[6] Shen C, Li K, Hou Q, Feng H, Dong X. White LED based on YAG: Ce, Gd phosphor and CdSe–ZnS core/shell quantum dots. IEEE Photonics Technol Lett 2010; 22(12): 884-5. [http://dx.doi.org/10.1109/LPT.2010.2046724]

[7] Shipway AN, Katz E, Willner I. Nanoparticle arrays on surfaces for electronic, optical, and sensor applications. ChemPhysChem 2000; 1(1): 18-52. [http://dx.doi.org/10.1002/1439-7641(20000804)1:1<18::AID-CPHC18>3.0.CO;2-L] [PMID: 23696260]

[8] Yohan D, Chithrani BD. Applications of nanoparticles in nanomedicine. J Biomed Nanotechnol 2014; 10(9): 2371-92. [http://dx.doi.org/10.1166/jbn.2014.2015] [PMID: 25992462]

[9] Shenhar R, Rotello VM. Nanoparticles: scaffolds and building blocks. Acc Chem Res 2003; 36(7): 549-61. [http://dx.doi.org/10.1021/ar020083j] [PMID: 12859216]

[10] Wang Y, Xia Y. Bottom-up and top-down approaches to the synthesis of monodispersed spherical colloids of low melting-point metals. Nano Lett 2004; 4(10): 2047-50. [http://dx.doi.org/10.1021/nl048689j]

[11] Burda C, Chen X, Narayanan R, El-Sayed MA. Chemistry and properties of nanocrystals of different

shapes. Chem Rev 2005; 105(4): 1025-102.
[http://dx.doi.org/10.1021/cr030063a] [PMID: 15826010]

[12] Tao AR, Habas S, Yang P. Shape control of colloidal metal nanocrystals. Small 2008; 4(3): 310-25.
[http://dx.doi.org/10.1002/smll.200701295]

[13] Brannon-Peppas L, Blanchette JO. Nanoparticle and targeted systems for cancer therapy. Adv Drug Deliv Rev 2004; 56(11): 1649-59.
[http://dx.doi.org/10.1016/j.addr.2004.02.014] [PMID: 15350294]

[14] Davis ME, Chen ZG, Shin DM. Nanoparticle therapeutics: an emerging treatment modality for cancer. Nat Rev Drug Discov 2008; 7(9): 771-82.
[http://dx.doi.org/10.1038/nrd2614] [PMID: 18758474]

[15] Ulbrich K, Holá K, Šubr V, Bakandritsos A, Tuček J, Zbořil R. Targeted drug delivery with polymers and magnetic nanoparticles: covalent and noncovalent approaches, release control, and clinical studies. Chem Rev 2016; 116(9): 5338-431.
[http://dx.doi.org/10.1021/acs.chemrev.5b00589] [PMID: 27109701]

[16] Mornet S, Vasseur S, Grasset F, Duguet E. Magnetic nanoparticle design for medical diagnosis and therapy. J Mater Chem 2004; 14: 2161-75.
[http://dx.doi.org/10.1039/b402025a]

[17] Arruebo M, Fernández-Pacheco R, Ibarra MR, Santamaría J. Magnetic nanoparticles for drug delivery. Nano Today 2007; 2(3): 22-32.
[http://dx.doi.org/10.1016/S1748-0132(07)70084-1]

[18] Issa B, Obaidat IM, Albiss BA, Haik Y. Magnetic nanoparticles: surface effects and properties related to biomedicine applications. Int J Mol Sci 2013; 14(11): 21266-305.
[http://dx.doi.org/10.3390/ijms141121266] [PMID: 24232575]

[19] Wu K, Wang JP. Magnetic hyperthermia performance of magnetite nanoparticle assemblies under different driving fields. AIP Adv 2017; 7: 056327.
[http://dx.doi.org/10.1063/1.4978458]

[20] Lee JY, Crake C, Teo B, *et al.* Ultrasound-enhanced siRNA delivery using magnetic nanoparticle-loaded chitosan-deoxycholic acid nanodroplets. Adv Healthc Mater 2017; 6(8): 1601246.
[http://dx.doi.org/10.1002/adhm.201601246] [PMID: 28195673]

[21] Yang CL, Chen JP, Wei KC, Chen JY, Huang CW, Liao ZX. Release of doxorubicin by a folate-grafted, chitosan-coated magnetic nanoparticle. Nanomaterials (Basel) 2017; 7(4): 85.
[http://dx.doi.org/10.3390/nano7040085] [PMID: 28406429]

[22] Shen Z, Wu A, Chen X. Iron oxide nanoparticle based contrast agents for magnetic resonance imaging. Mol Pharm 2017; 14(5): 1352-64.
[http://dx.doi.org/10.1021/acs.molpharmaceut.6b00839] [PMID: 27776215]

[23] Lu HD, Yang SS, Wilson BK, McManus SA, Chen CVHH, Prud'homme RK. Nanoparticle targeting of Gram-positive and Gram-negative bacteria for magnetic-based separations of bacterial pathogens. Appl Nanosci 2017; 7: 83-93.
[http://dx.doi.org/10.1007/s13204-017-0548-0]

[24] Patra S, Roy E, Das R, *et al.* RETRACTED: Bimetallic magnetic nanoparticle as a new platform for fabrication of pyridoxine and pyridoxal-5′-phosphate imprinted polymer modified high throughput electrochemical sensor. Biosens Bioelectron 2015; 73: 234-44.
[http://dx.doi.org/10.1016/j.bios.2015.06.005] [PMID: 26086443]

[25] Jemal A, Siegel R, Ward E, Murray T, Xu J, Thun MJ. Cancer statistics. CA Cancer J Clin 2007; 57(1): 43-66.
[http://dx.doi.org/10.3322/canjclin.57.1.43] [PMID: 17237035]

[26] Khan N, Afaq F, Mukhtar H. Lifestyle as risk factor for cancer: Evidence from human studies. Cancer Lett 2010; 293(2): 133-43.

[http://dx.doi.org/10.1016/j.canlet.2009.12.013] [PMID: 20080335]

[27] Anselmo AC, Mitragotri S. An overview of clinical and commercial impact of drug delivery systems. J Control Rel 2014; 190: 15-28.
[http://dx.doi.org/10.1016/j.jconrel.2014.03.053] [PMID: 24747160]

[28] Anselmo AC, Mitragotri S. A review of clinical translation of inorganic nanoparticles. AAPS J 2015; 17(5): 1041-54.
[http://dx.doi.org/10.1208/s12248-015-9780-2] [PMID: 25956384]

[29] Li P, Zhang Z, Su Z, Wei G. Thermosensitive polymeric micelles based on the triblock copolymer poly(D,L-lactide)-b-poly(N-isopropyl acrylamide)-b-poly(D,L-lactide) for controllable drug delivery. J Appl Polym Sci 2017; 134(37): 45304.
[http://dx.doi.org/10.1002/app.45304]

[30] Yang W, Yu C, Wu C, Yao SQ, Wu S. Cell-penetrating poly(disulfide)-based star polymers for simultaneous intracellular delivery of miRNAs and small molecule drugs. Polym Chem 2017; 8: 4043-51.
[http://dx.doi.org/10.1039/C7PY00666G]

[31] Zununi Vahed S, Salehi R, Davaran S, Sharifi S. Liposome-based drug co-delivery systems in cancer cells. Mater Sci Eng C 2017; 71: 1327-41.
[http://dx.doi.org/10.1016/j.msec.2016.11.073] [PMID: 27987688]

[32] Mustafa R, Luo Y, Wu Y, Guo R, Shi X. Dendrimer-functionalized laponite nanodisks as a platform for anticancer drug delivery. Nanomaterials (Basel) 2015; 5(4): 1716-31.
[http://dx.doi.org/10.3390/nano5041716] [PMID: 28347091]

[33] Heidari-Kharaji M, Taheri T, Doroud D, Habibzadeh S, Rafati S. Solid lipid nanoparticle loaded with paromomycin: *in vivo* efficacy against Leishmania tropica infection in BALB/c mice model. Appl Microbiol Biotechnol 2016; 100(16): 7051-60.
[http://dx.doi.org/10.1007/s00253-016-7422-y] [PMID: 26960322]

[34] Devajaran PV, Jain S, Eds. Targeted drug delivery: Concepts and design. Heidelberg: Springer 2015.
[http://dx.doi.org/10.1007/978-3-319-11355-5]

[35] Bogdanov A Jr, Wright SC, Marecos EM, *et al.* A long-circulating co-polymer in "passive targeting" to solid tumors. J Drug Target 1997; 4(5): 321-30.
[http://dx.doi.org/10.3109/10611869708995848] [PMID: 9169989]

[36] Hobbs SK, Monsky WL, Yuan F, *et al.* Regulation of transport pathways in tumor vessels: role of tumor type and microenvironment. Proc Natl Acad Sci USA 1998; 95(8): 4607-12.
[http://dx.doi.org/10.1073/pnas.95.8.4607] [PMID: 9539785]

[37] Guo L, Huang QJ, Li XY, Yang SH. Iron nanoparticles: Synthesis and applications in surface enhanced Raman scattering and electro catalysis. Phys Chem Chem Phys 2001; 3: 1661-5.
[http://dx.doi.org/10.1039/b009951l]

[38] Huber DL. Synthesis, properties, and applications of iron nanoparticles. Small 2005; 1(5): 482-501.
[http://dx.doi.org/10.1002/smll.200500006] [PMID: 17193474]

[39] Su Y, Adeleye AS, Huang Y, Zhou X, Keller AA, Zhang Y. Direct synthesis of novel and reactive sulfide-modified nano iron through nanoparticle seeding for improved cadmium contaminated water treatment. Sci Rep 2016; 6: 24358.
[http://dx.doi.org/10.1038/srep24358] [PMID: 27095387]

[40] Majima T, Miyahara T, Haneda K, Takami M. Preparation of γ-iron ultrafine particles using a transversely excited atmospheric CO_2 laser. Jpn J Appl Phys 1994; 33(2): L223.
[http://dx.doi.org/10.1143/JJAP.33.L223]

[41] Caro DD, Bradley JS. Surface spectroscopic study of carbon monoxide adsorption on nanoscale nickel colloids prepared from a zerovalent organometallic precursor. Langmuir 1997; 13(12): 3067-9.
[http://dx.doi.org/10.1021/la9620824]

[42] Pileni MP. Self-assemblies of nanocrystals: fabrication and collective properties. Appl Surf Sci 2001; 171: 1-14.
[http://dx.doi.org/10.1016/S0169-4332(00)00553-5]

[43] Doppiu S, Langlais V, Sort J, *et al.* Controlled reduction of NiO using reactive ball milling under hydrogen atmosphere leading to Ni–NiO nanocomposites. Chem Mater 2004; 16(26): 5664-9.
[http://dx.doi.org/10.1021/cm048810n]

[44] Gubin SP, Ed. Magnetic nanoparticles. Wiley-VCH Verlag GmbH & Co. 2009.
[http://dx.doi.org/10.1002/9783527627561]

[45] Li XG, Chiba A, Takahashi S. Preparation and magnetic properties of ultrafine particles of Fe-Ni alloys. J Magn Magn Mater 1997; 170: 339-45.
[http://dx.doi.org/10.1016/S0304-8853(97)00039-5]

[46] Chen M, Liu JP, Sun S. One-step synthesis of FePt nanoparticles with tunable size. J Am Chem Soc 2004; 126(27): 8394-5.
[http://dx.doi.org/10.1021/ja047648m] [PMID: 15237993]

[47] Sourmail T. Near equiatomicFeCo alloys: Constitution, mechanical and magnetic properties. Prog Mater Sci 2005; 50(7): 816-80.
[http://dx.doi.org/10.1016/j.pmatsci.2005.04.001]

[48] Qiu JM, Wang JP. Monodispersed and highly ordered L10 FePt nanoparticles prepared in the gas phase. Appl Phys Lett 2006; 88: 192505.
[http://dx.doi.org/10.1063/1.2202130]

[49] Hansen MF, Koch CB, Morup S. Magnetic dynamics of weakly and strongly interacting hematite nanoparticles. Phys Rev B Condens Matter Mater Phys 2000; 62(2): 1124-35.
[http://dx.doi.org/10.1103/PhysRevB.62.1124]

[50] Nguyen MT, Seriani N, Gebauer R. Defective α-Fe_2O_3(0001): an ab initio study. ChemPhysChem 2014; 15(14): 2930-5.
[http://dx.doi.org/10.1002/cphc.201402153] [PMID: 25045043]

[51] Kubickova S, Niznansky D, Herrero MPM, Salas G, Vejpravova J. Structural disorder *versus* spin canting in monodisperse maghemite nanocrystals. Appl Phys Lett 2014; 104(22): 223105.
[http://dx.doi.org/10.1063/1.4881331]

[52] Sun S, Zeng H. Size-controlled synthesis of magnetite nanoparticles. J Am Chem Soc 2002; 124(28): 8204-5.
[http://dx.doi.org/10.1021/ja026501x] [PMID: 12105897]

[53] Ding J, Miao WF, Pirault E, Street R, McCormick PG. J Alloys Compd 1998; 161: 199.
[http://dx.doi.org/10.1016/S0925-8388(97)00502-1]

[54] Maiti D, Manju U, Velaga S, Devi PS. Phase evolution and growth of iron oxide nanoparticles: effect of hydrazine addition during sonication. Cryst Growth Des 2013; 13(8): 3637-44.
[http://dx.doi.org/10.1021/cg400627c]

[55] Feng J, Zeng HC. Size-controlled growth of Co_3O_4 nanocubes. Chem Mater 2003; 15(14): 2829-35.
[http://dx.doi.org/10.1021/cm020940d]

[56] Biju V, Khadar AM. Dielectric properties of nanostructured nickel oxide. J Mater Sci 2003; 38(19): 4055-63.
[http://dx.doi.org/10.1023/A:1026131103898]

[57] Sharma HB, Singh NB, Devi KN, Lee JH, Singh SB. Structural and optical properties of manganese substituted nanocrystalline bismuth ferrite thin films by sol–gel process. J Alloys Compd 2014; 583: 106-10.
[http://dx.doi.org/10.1016/j.jallcom.2013.08.037]

[58] Mosivand S, Kazeminezhad I. Synthesis of electrocrystallized cobalt ferrite nanopowders by tuning

the cobalt salt concentration. RSC Adv 2015; 5: 14796-803.
[http://dx.doi.org/10.1039/C4RA17162D]

[59] Maiti D, Saha A, Devi PS. Surface modified multifunctional $ZnFe_2O_4$ nanoparticles for hydrophobic and hydrophilic anti-cancer drug molecule loading. Phys Chem Chem Phys 2016; 18(3): 1439-50.
[http://dx.doi.org/10.1039/C5CP05840F] [PMID: 26524183]

[60] Lee CS, Lee H, Westervelt RM. Microelectromagnets for the control of magnetic nanoparticles. Appl Phys Lett 2001; 79: 3308.
[http://dx.doi.org/10.1063/1.1419049]

[61] Ketteler G, Weiss W, Ranke W, Schloegel R. Bulk and surface phases of iron oxides in an oxygen and water atmosphere at low pressure. Phys Chem Chem Phys 2001; 3: 1114-22.
[http://dx.doi.org/10.1039/b009288f]

[62] Ding J, Tao K, Li J, Song S, Sun K. Cell-specific cytotoxicity of dextran-stabilized magnetite nanoparticles. Colloids Surf B Biointerfaces 2010; 79(1): 184-90.
[http://dx.doi.org/10.1016/j.colsurfb.2010.03.053] [PMID: 20427159]

[63] Cheng C, Xu F, Gu H. Facile synthesis and morphology evolution of magnetic iron oxide nanoparticles in different polyol processes. New J Chem 2011; 35: 1072-9.
[http://dx.doi.org/10.1039/c0nj00986e]

[64] Wu JH, Ko SP, Liu HL, Kim S, Ju JS, Kim YK. Sub 5 nm magnetite nanoparticles: Synthesis, microstructure, and magnetic properties. Mater Lett 2007; 61: 3124-9.
[http://dx.doi.org/10.1016/j.matlet.2006.11.032]

[65] Breulmann M, Colfen H, Hentze HP, Antonietti M, Walsh D, Mann S. Elastic magnets: Template-controlled mineralization of iron oxide colloids in a sponge-like gel matrix. Adv Mater 1998; 10(3): 237-41.
[http://dx.doi.org/10.1002/(SICI)1521-4095(199802)10:3<237::AID-ADMA237>3.0.CO;2-6]

[66] Jana NR, Chen Y, Peng X. Size- and shape-controlled magnetic (Cr, Mn, Fe, Co, Ni) oxide nanocrystals *via* a simple and general approach. Chem Mater 2004; 16(20): 3931-5.
[http://dx.doi.org/10.1021/cm049221k]

[67] Chen F, Gao Q, Hong G, Ni JJ. Synthesis and characterization of magnetite dodecahedron nanostructure by hydrothermal method. MagnMagn Mater 2008; 320(11): 1775-80.
[http://dx.doi.org/10.1016/j.jmmm.2008.02.117]

[68] Hemery G, Keyes AC Jr, Garaio E, *et al.* Tuning sizes, morphologies, and magnetic properties of monocore *versus* multicore iron oxide nanoparticles through the controlled addition of water in the polyol synthesis. Inorg Chem 2017; 56(14): 8232-43.
[http://dx.doi.org/10.1021/acs.inorgchem.7b00956] [PMID: 28671822]

[69] Bang JH, Suslick KS. Applications of ultrasound to the synthesis of nanostructured materials. Adv Mater 2010; 22(10): 1039-59.
[http://dx.doi.org/10.1002/adma.200904093] [PMID: 20401929]

[70] Strobel R, Pratsinis SE. Direct synthesis of maghemite, magnetite and wustite nanoparticles by flame spray pyrolysis. Adv Powder Technol 2009; 20(2): 190-4.
[http://dx.doi.org/10.1016/j.apt.2008.08.002]

[71] Cabrera L, Gutierrez S, Menendes N, Morales MP, Herrast P. Magnetite nanoparticles: Electrochemical synthesis and characterization. Electrochim Acta 2008; 53(8): 3436-41.
[http://dx.doi.org/10.1016/j.electacta.2007.12.006]

[72] Salazar-Alvarez G, Muhammed M, Zagorodni AA. Novel flow injection synthesis of iron oxide nanoparticles with narrow size distribution. Chem Eng Sci 2006; 61(14): 4625-33.
[http://dx.doi.org/10.1016/j.ces.2006.02.032]

[73] Eckert CA, Knutson BL, Debenedetti PG. Supercritical fluids as solvents for chemical and materials processing. Nature 1996; 383: 313-8.

[http://dx.doi.org/10.1038/383313a0]

[74] Vereda F, Rodríguez-González B, de Vicente J, Hidalgo-Álvarez R. Evidence of direct crystal growth and presence of hollow microspheres in magnetite particles prepared by oxidation of $Fe(OH)_2$. J Colloid Interface Sci 2008; 318(2): 520-4.
[http://dx.doi.org/10.1016/j.jcis.2007.11.003] [PMID: 18061607]

[75] Moon JW, Rawn CJ, Rondinone AJ, *et al.* Large-scale production of magnetic nanoparticles using bacterial fermentation. J Ind Microbiol Biotechnol 2010; 37(10): 1023-31.
[http://dx.doi.org/10.1007/s10295-010-0749-y] [PMID: 20544257]

[76] Gupta AK, Curtis ASG. Surface modified superparamagnetic nanoparticles for drug delivery: interaction studies with human fibroblasts in culture. J Mater Sci Mater Med 2004; 15(4): 493-6.
[http://dx.doi.org/10.1023/B:JMSM.0000021126.32934.20] [PMID: 15332623]

[77] Ma DL, Guan JW, Normandin F, *et al.* Multifunctional nano-architecture for biomedical applications. Chem Mater 2006; 18(7): 1920-7.
[http://dx.doi.org/10.1021/cm052067x]

[78] Chen S, Li Y, Guo C, *et al.* Temperature-responsive magnetite/PEO-PPO-PEO block copolymer nanoparticles for controlled drug targeting delivery. Langmuir 2007; 23(25): 12669-76.
[http://dx.doi.org/10.1021/la702049d] [PMID: 17988160]

[79] Basiruddin SK, Saha A, Pradhan N, Jana NR. Advances in coating chemistry in deriving soluble functional nanoparticle. J Phys Chem C 2010; 114(25): 11009-17.
[http://dx.doi.org/10.1021/jp100844d]

[80] Bourlinos AB, Simopoulos A, Petridis D. Synthesis of capped ultrafine γ-Fe_2O_3 particles from iron(III) hydroxide caprylate: A novel starting material for readily attainable organosols. Chem Mater 2002; 14(2): 899-903.
[http://dx.doi.org/10.1021/cm010794w]

[81] Sahoo Y, Goodarzi A, Swihart MT, *et al.* Aqueous ferrofluid of magnetite nanoparticles: Fluorescence labeling and magnetophoretic control. J Phys Chem B 2005; 109(9): 3879-85.
[http://dx.doi.org/10.1021/jp045402y] [PMID: 16851439]

[82] Sousa MH, Rubim JC, Sobrinho PG, Tourinho FA. Biocompatible magnetic fluid precursors based on aspartic and glutamic acid modified maghemite nanostructures. J Magn Magn Mater 2001; 225: 67-72.
[http://dx.doi.org/10.1016/S0304-8853(00)01229-4]

[83] Morais PC, Oliveira AC, Tronconi AL, Goetze T, Buske N. Photoacoustic spectroscopy: a promising technique to investigate magnetic fluids. IEEE Trans Magn 2003; 39(5): 2654-6.
[http://dx.doi.org/10.1109/TMAG.2003.815544]

[84] Liao N, Wu M, Pan F, *et al.* Poly (dopamine) coated superparamagnetic iron oxide nanocluster for noninvasive labeling, tracking, and targeted delivery of adipose tissue-derived stem cells. Sci Rep 2016; 6: 18746.
[http://dx.doi.org/10.1038/srep18746] [PMID: 26728448]

[85] Chakraborty A, Jana NR. Vitamin C-conjugated nanoparticle protects cells from oxidative stress at low doses but induces oxidative stress and cell death at high doses. ACS Appl Mater Interfaces 2017; 9(48): 41807-17.
[http://dx.doi.org/10.1021/acsami.7b16055] [PMID: 29135217]

[86] Banerjee SS, Chen D. Magnetic nanoparticles grafted with cyclodextrin for hydrophobic drug delivery. Chem Mater 2007; 19(25): 6345-9.
[http://dx.doi.org/10.1021/cm702278u]

[87] Shao C, Li X, Pei Z, *et al.* Facile fabrication of glycopolymer-based iron oxide nanoparticles and their applications in the carbohydrate–lectin interaction and targeted cell imaging. Polym Chem 2016; 7: 1337-44.
[http://dx.doi.org/10.1039/C5PY01954K]

[88] Foster EL, Xue Z, Roach CM, Larsen ES, Bielawski CW, Johnston KP. Iron oxide nanoparticles grafted with sulfonated and zwitterionic polymers: High stability and low adsorption in extreme aqueous environments. ACS Macro Lett 2014; 3(9): 867-71. [http://dx.doi.org/10.1021/mz5004213]

[89] Mikhaylova M, Kim DK, Bobrysheva N, *et al.* Superparamagnetism of magnetite nanoparticles: dependence on surface modification. Langmuir 2004; 20(6): 2472-7. [http://dx.doi.org/10.1021/la035648e] [PMID: 15835712]

[90] Kim DK, Mikhaylova M, Wang FH, *et al.* Starch-coated superparamagnetic nanoparticles as MR contrast agents. Chem Mater 2003; 15(23): 4343-51. [http://dx.doi.org/10.1021/cm031104m]

[91] Olsen D, Yang C, Bodo M, *et al.* Recombinant collagen and gelatin for drug delivery. Adv Drug Deliv Rev 2003; 55(12): 1547-67. [http://dx.doi.org/10.1016/j.addr.2003.08.008] [PMID: 14623401]

[92] Li G, Huang K, Jiang Y, Ding P, Yang D. Preparation and characterization of carboxyl functionalization of chitosan derivative magnetic nanoparticles. Biochem Eng J 2008; 40(3): 408-14. [http://dx.doi.org/10.1016/j.bej.2008.01.018]

[93] Paul KG, Frigo TB, Groman JY, Groman EV. Synthesis of ultrasmall superparamagnetic iron oxides using reduced polysaccharides. Bioconjug Chem 2004; 15(2): 394-401. [http://dx.doi.org/10.1021/bc034194u] [PMID: 15025537]

[94] D'Souza AJM, Schowen RL, Topp EM. Polyvinylpyrrolidone-drug conjugate: synthesis and release mechanism. J Control Rel 2004; 94(1): 91-100. [http://dx.doi.org/10.1016/j.jconrel.2003.09.014] [PMID: 14684274]

[95] Gómez-Lopera SA, Arias JL, Gallardo V, Delgado AV. Colloidal stability of magnetite/poly(lactic acid) core/shell nanoparticles. Langmuir 2006; 22(6): 2816-21. [http://dx.doi.org/10.1021/la0530079] [PMID: 16519488]

[96] Ma HL, Xu YF, Qi XR, Maitani Y, Nagai T. Superparamagnetic iron oxide nanoparticles stabilized by alginate: pharmacokinetics, tissue distribution, and applications in detecting liver cancers. Int J Pharm 2008; 354(1-2): 217-26. [http://dx.doi.org/10.1016/j.ijpharm.2007.11.036] [PMID: 18191350]

[97] Singh H, Laibinis PE, Hatton TA. Rigid, superparamagnetic chains of permanently linked beads coated with magnetic nanoparticles. Synthesis and rotational dynamics under applied magnetic fields. Langmuir 2005; 21(24): 11500-9. [http://dx.doi.org/10.1021/la0517843] [PMID: 16285833]

[98] Shan GB, Xing JM, Luo MF, Liu HZ, Chen JY. Immobilization of *Pseudomonas delafieldii* with magnetic polyvinyl alcohol beads and its application in biodesulfurization. Biotechnol Lett 2003; 25(23): 1977-81. [http://dx.doi.org/10.1023/B:BILE.0000004388.15751.8c] [PMID: 14719809]

[99] Zhou D, Sun Y, Zheng Y, *et al.* Superparamagnetic PLGA–iron oxide microspheres as contrast agents for dual-imaging and the enhancement of the effects of high-intensity focused ultrasound ablation on liver tissue. RSC Advances 2015; 5: 35693-703. [http://dx.doi.org/10.1039/C5RA00880H]

[100] Kurzhals S, Zirbs R, Reimhult E. Synthesis and magneto-thermal actuation of iron oxide core–PNIPAM shell nanoparticles. ACS Appl Mater Interfaces 2015; 7(34): 19342-52. [http://dx.doi.org/10.1021/acsami.5b05459] [PMID: 26270412]

[101] Ahmed N, Ahmad NM, Fessi H, Elaissari A. *In vitro* MRI of biodegradable hybrid (iron oxide/polycaprolactone) magnetic nanoparticles prepared *via* modified double emulsion evaporation mechanism. Colloids Surf B Biointerfaces 2015; 130: 264-71. [http://dx.doi.org/10.1016/j.colsurfb.2015.04.022] [PMID: 25960142]

[102] Saha A, Mohanta SC, Deka K, Deb P, Devi PS. Surface-engineered multifunctional Eu:Gd_2O_3 nanoplates for targeted and pH-responsive drug delivery and imaging applications. ACS Appl Mater Interfaces 2017; 9(4): 4126-41.
[http://dx.doi.org/10.1021/acsami.6b12804] [PMID: 28098453]

[103] Jana NR, Patra PK, Saha A. Imidazole based biocompatible polymer coating in deriving< 25 nm functional nanoparticle probe for cellular imaging and detection. J Phys Chem C 2009; 113(52): 21484-92.
[http://dx.doi.org/10.1021/jp905685a]

[104] Sinha A, Jana NR. Nanoparticle-incorporated functional mesoporous silica colloid for diverse applications. Eur J Inorg Chem 2012; 2012(28): 4470-8.
[http://dx.doi.org/10.1002/ejic.201200496]

[105] Sinha A, Jana NR. Functional, mesoporous, superparamagnetic colloidal sorbents for efficient removal of toxic metals. Chem Commun (Camb) 2012; 48(74): 9272-4.
[http://dx.doi.org/10.1039/c2cc33893a] [PMID: 22872025]

[106] Rossi LM, Shi L, Quina FH, Rosenzweig Z. Stöber synthesis of monodispersed luminescent silica nanoparticles for bioanalytical assays. Langmuir 2005; 21(10): 4277-80.
[http://dx.doi.org/10.1021/la0504098] [PMID: 16032835]

[107] Bae H, Ahmad T, Rhee I, Chang Y, Jin SU, Hong S. Carbon-coated iron oxide nanoparticles as contrast agents in magnetic resonance imaging. Nanoscale Res Lett 2012; 7: 44.
[http://dx.doi.org/10.1186/1556-276X-7-44] [PMID: 22221912]

[108] Mandal M, Kundu S, Ghosh SK, *et al.* Magnetite nanoparticles with tunable gold or silver shell. J Colloid Interface Sci 2005; 286(1): 187-94.
[http://dx.doi.org/10.1016/j.jcis.2005.01.013] [PMID: 15848416]

[109] Teng X, Black D, Watkins NJ, Gao Y, Yang H. Platinum-maghemite core–shell nanoparticles using a sequential synthesis. Nano Lett 2003; 3(2): 261-4.
[http://dx.doi.org/10.1021/nl025918y]

[110] Le Droumaguet B, Nicolas J, Brambilla D, *et al.* Versatile and efficient targeting using a single nanoparticulate platform: application to cancer and Alzheimer's disease. ACS Nano 2012; 6(7): 5866-79.
[http://dx.doi.org/10.1021/nn3004372] [PMID: 22725248]

[111] D'Souza AA, Devarajan PV. Asialoglycoprotein receptor mediated hepatocyte targeting - strategies and applications. J Control Rel 2015; 203: 126-39.
[http://dx.doi.org/10.1016/j.jconrel.2015.02.022] [PMID: 25701309]

[112] Dahms NM, Lobel P, Kornfeld S. Mannose 6-phosphate receptors and lysosomal enzyme targeting. J Biol Chem 1989; 264(21): 12115-8.
[http://dx.doi.org/10.1016/S0021-9258(18)63825-6] [PMID: 2545698]

[113] Sudimack J, Lee RJ. Targeted drug delivery *via* the folate receptor. Adv Drug Deliv Rev 2000; 41(2): 147-62.
[http://dx.doi.org/10.1016/S0169-409X(99)00062-9] [PMID: 10699311]

[114] Maity AR, Saha A, Roy A, Jana NR. Folic acid functionalized nanoprobes for fluorescence-, dark-field-, and dual-imaging-based selective detection of cancer cells and tissue. ChemPlusChem 2013; 78(3): 259-67.
[http://dx.doi.org/10.1002/cplu.201200296]

[115] Dalal C, Saha A, Jana NR. Nanoparticle multivalency directed shifting of cellular uptake mechanism. J Phys Chem C 2016; 120(12): 6778-86.
[http://dx.doi.org/10.1021/acs.jpcc.5b11059]

[116] Saha A, Basiruddin SK, Maity AR, Jana NR. Synthesis of nanobioconjugates with a controlled average number of biomolecules between 1 and 100 per nanoparticle and observation of multivalency

dependent interaction with proteins and cells. Langmuir 2013; 29(45): 13917-24. [http://dx.doi.org/10.1021/la402699a] [PMID: 24117157]

[117] Dubertret B, Skourides P, Norris DJ, Noireaux V, Brivanlou AH, Libchaber A. *In vivo* imaging of quantum dots encapsulated in phospholipid micelles. Science 2002; 298(5599): 1759-62. [http://dx.doi.org/10.1126/science.1077194] [PMID: 12459582]

[118] De M, Ghosh PS, Rotello VM. Applications of nanoparticles in biology. Adv Mater 2008; 20: 4225-41. [http://dx.doi.org/10.1002/adma.200703183]

[119] Erathodiyil N, Ying JY. Functionalization of inorganic nanoparticles for bioimaging applications. Acc Chem Res 2011; 44(10): 925-35. [http://dx.doi.org/10.1021/ar2000327] [PMID: 21648430]

[120] Sapsford KE, Algar WR, Berti L, *et al.* Functionalizing nanoparticles with biological molecules: developing chemistries that facilitate nanotechnology. Chem Rev 2013; 113(3): 1904-2074. [http://dx.doi.org/10.1021/cr300143v] [PMID: 23432378]

[121] Pechar M, Pola R, Ulbrich K, Carlisle RC, Willemsen RA, Seymour LW. Non-covalent attachment of recombinant proteins to polymer therapeutics. 2nd World Conference on Nanomedicine and Drug Delivery. Kottayam, India. 2011.

[122] Pola R, Laga R, Ulbrich K, *et al.* Polymer therapeutics with a coiled coil motif targeted against murine bcl1 leukemia. Biomacromolecules 2013; 14(3): 881-9. [http://dx.doi.org/10.1021/bm3019592] [PMID: 23373696]

[123] Canal F, Sanchis J, Vicent MJ. Polymer--drug conjugates as nano-sized medicines. Curr Opin Biotechnol 2011; 22(6): 894-900. [http://dx.doi.org/10.1016/j.copbio.2011.06.003] [PMID: 21724381]

[124] Larson N, Ghandehari H. Polymeric conjugates for drug delivery. Chem Mater 2012; 24(5): 840-53. [http://dx.doi.org/10.1021/cm2031569] [PMID: 22707853]

[125] Kopeček J. Polymer-drug conjugates: origins, progress to date and future directions. Adv Drug Deliv Rev 2013; 65(1): 49-59. [http://dx.doi.org/10.1016/j.addr.2012.10.014] [PMID: 23123294]

[126] Pechar M, Ulbrich K, Šubr V, Seymour LW, Schacht EH. Poly(ethylene glycol) multiblock copolymer as a carrier of anti-cancer drug doxorubicin. Bioconjug Chem 2000; 11(2): 131-9. [http://dx.doi.org/10.1021/bc990092l] [PMID: 10725088]

[127] Berna M, Dalzoppo D, Pasut G, *et al.* Novel monodisperse PEG-dendrons as new tools for targeted drug delivery: synthesis, characterization and cellular uptake. Biomacromolecules 2006; 7(1): 146-53. [http://dx.doi.org/10.1021/bm050480s] [PMID: 16398509]

[128] Schluep T, Hwang J, Cheng J, *et al.* Preclinical efficacy of the camptothecin-polymer conjugate IT-101 in multiple cancer models. Clin Cancer Res 2006; 12(5): 1606-14. [http://dx.doi.org/10.1158/1078-0432.CCR-05-1566] [PMID: 16533788]

[129] Sedlák M, Drabina P, Bílková E, Simůnek P, Buchta V. New targeting system for antimycotic drugs: beta-glucosidase sensitive amphotericin B-star poly(ethylene glycol) conjugate. Bioorg Med Chem Lett 2008; 18(9): 2952-6. [http://dx.doi.org/10.1016/j.bmcl.2008.03.065] [PMID: 18396401]

[130] Soyez H, Seymour LW, Schacht E. Macromolecular derivatives of N,N-di-(2-chloroethyl)-4-phenylene diamine mustard. 2. *In vitro* cytotoxicity and *in vivo* anticancer efficacy. J Control Rel 1999; 57(2): 187-96. [http://dx.doi.org/10.1016/S0168-3659(98)00117-5] [PMID: 9971901]

[131] Kricheldorf HR. Polypeptides and 100 years of chemistry of α-amino acid N-carboxyanhydrides. Angew Chem Int Ed Engl 2006; 45(35): 5752-84. [http://dx.doi.org/10.1002/anie.200600693] [PMID: 16948174]

[132] Maeda H, Sawa T, Konno T. Mechanism of tumor-targeted delivery of macromolecular drugs, including the EPR effect in solid tumor and clinical overview of the prototype polymeric drug SMANCS. J Control Rel 2001; 74(1-3): 47-61.
[http://dx.doi.org/10.1016/S0168-3659(01)00309-1] [PMID: 11489482]

[133] Ferruti P, Marchisio MA, Duncan R. Poly(amido-amine)s: Biomedical applications. Macromol Rapid Commun 2002; 23: 332-55.
[http://dx.doi.org/10.1002/1521-3927(20020401)23:5/6<332::AID-MARC332>3.0.CO;2-I]

[134] Tomlinson R, Heller J, Brocchini S, Duncan R. Polyacetal-doxorubicin conjugates designed for pH-dependent degradation. Bioconjug Chem 2003; 14(6): 1096-106.
[http://dx.doi.org/10.1021/bc030028a] [PMID: 14624622]

[135] Maiti D, Mukhopadhyay S, Mohanta SC, Saha A, Devi PS. A multifunctional nanocomposite of magnetic γ-Fe_2O_3 and mesoporous fluorescent ZnO. J Alloys Compd 2015; 653: 187-94.
[http://dx.doi.org/10.1016/j.jallcom.2015.08.230]

[136] Talelli M, Hennink WE. Thermosensitive polymeric micelles for targeted drug delivery. Nanomedicine (Lond) 2011; 6(7): 1245-55.
[http://dx.doi.org/10.2217/nnm.11.91] [PMID: 21929459]

[137] Hrubý M, Filippov SK, Štěpánek P. Smart polymers in drug delivery systems on crossroads: Which way deserves following? Eur Polym J 2015; 65: 82-97.
[http://dx.doi.org/10.1016/j.eurpolymj.2015.01.016]

[138] Huber S, Jordan R. Modulation of the lower critical solution temperature of 2-Alkyl-2-oxazoline copolymers. Colloid Polym Sci 2008; 286: 395-402.
[http://dx.doi.org/10.1007/s00396-007-1781-y]

[139] Roy D, Brooks WL, Sumerlin BS. New directions in thermoresponsive polymers. Chem Soc Rev 2013; 42(17): 7214-43.
[http://dx.doi.org/10.1039/c3cs35499g] [PMID: 23450220]

[140] El-Dakdouki MH, Zhu DC, El-Boubbou K, *et al.* Development of multifunctional hyaluronan-coated nanoparticles for imaging and drug delivery to cancer cells. Biomacromolecules 2012; 13(4): 1144-51.
[http://dx.doi.org/10.1021/bm300046h] [PMID: 22372739]

[141] Zhu L, Wang D, Wei X, *et al.* Multifunctional pH-sensitive superparamagnetic iron-oxide nanocomposites for targeted drug delivery and MR imaging. J Control Rel 2013; 169(3): 228-38.
[http://dx.doi.org/10.1016/j.jconrel.2013.02.015] [PMID: 23485450]

[142] Zolata H, Abbasi Davani F, Afarideh H. Synthesis, characterization and theranostic evaluation of Indium-111 labeled multifunctional superparamagnetic iron oxide nanoparticles. Nucl Med Biol 2015; 42(2): 164-70.
[http://dx.doi.org/10.1016/j.nucmedbio.2014.09.007] [PMID: 25311750]

[143] Lee GY, Qian WP, Wang L, *et al.* Theranostic nanoparticles with controlled release of gemcitabine for targeted therapy and MRI of pancreatic cancer. ACS Nano 2013; 7(3): 2078-89.
[http://dx.doi.org/10.1021/nn3043463] [PMID: 23402593]

[144] Ansari C, Tikhomirov GA, Hong SH, *et al.* Development of novel tumor-targeted theranostic nanoparticles activated by membrane-type matrix metalloproteinases for combined cancer magnetic resonance imaging and therapy. Small 2014; 10(3): 566-575, 417.
[http://dx.doi.org/10.1002/smll.201301456] [PMID: 24038954]

[145] Wilson DS, Dalmasso G, Wang L, Sitaraman SV, Merlin D, Murthy N. Orally delivered thioketal nanoparticles loaded with TNF-α-siRNA target inflammation and inhibit gene expression in the intestines. Nat Mater 2010; 9(11): 923-8.
[http://dx.doi.org/10.1038/nmat2859] [PMID: 20935658]

[146] Stephen ZR, Kievit FM, Veiseh O, *et al.* Redox-responsive magnetic nanoparticle for targeted convection-enhanced delivery of O6-benzylguanine to brain tumors. ACS Nano 2014; 8(10): 10383-

95.
[http://dx.doi.org/10.1021/nn503735w] [PMID: 25247850]

[147] Lee N, Yoo D, Ling D, Cho MH, Hyeon T, Cheon J. Iron oxide based nanoparticles for multimodal imaging and magnetoresponsive therapy. Chem Rev 2015; 115(19): 10637-89.
[http://dx.doi.org/10.1021/acs.chemrev.5b00112] [PMID: 26250431]

[148] Fu Y, Kao WJ. Drug release kinetics and transport mechanisms of non-degradable and degradable polymeric delivery systems. Expert Opin Drug Deliv 2010; 7(4): 429-44.
[http://dx.doi.org/10.1517/17425241003602259] [PMID: 20331353]

[149] Modi S, Anderson BD. Determination of drug release kinetics from nanoparticles: overcoming pitfalls of the dynamic dialysis method. Mol Pharm 2013; 10(8): 3076-89.
[http://dx.doi.org/10.1021/mp400154a] [PMID: 23758289]

[150] Ploussi AG, Gazouli M, Stathis G, Kelekis NL, Efstathopoulos EP. Iron oxide nanoparticles as contrast agents in molecular magnetic resonance imaging: Do they open new perspectives in cardiovascular imaging? Cardiol Rev 2015; 23(5): 229-35.
[http://dx.doi.org/10.1097/CRD.0000000000000055] [PMID: 25688663]

[151] Meloni MM, Barton S, Xu L, Kaski JC, Song W, He T. Contrast agents for cardiovascular magnetic resonance imaging: an overview. J Mater Chem B Mater Biol Med 2017; 5(29): 5714-25.
[http://dx.doi.org/10.1039/C7TB01241A] [PMID: 32264204]

[152] Thorat ND, Bohara RA, Noor MR, Dhamecha D, Soulimane T, Tofail SAM. Effective cancer theranostics with polymer encapsulated superparamagnetic nanoparticles: Combined effects of magnetic hyperthermia and controlled drug release. ACS Biomater Sci Eng 2017; 3(7): 1332-40.
[http://dx.doi.org/10.1021/acsbiomaterials.6b00420] [PMID: 33429691]

[153] Liwinska W, Stanislawska I, Lyp M, Mackiewicz M, Stojek Z, Zabost E. A degradable nanogel drug carrier crosslinked with three-oligonucleotide hybrids for two-way drug release in mild and high hyperthermia treatment. J Mater Chem B Mater Biol Med 2017; 5(24): 4713-24.
[http://dx.doi.org/10.1039/C7TB00092H] [PMID: 32264314]

[154] Saha A, Basiruddin SK, Pradhan N, Jana NR. Ligand exchange approach in deriving magnetic-fluorescent and magnetic-plasmonic hybrid nanoparticle. Langmuir 2010; 26(6): 4351-6.
[http://dx.doi.org/10.1021/la903428r] [PMID: 20214397]

[155] Champion JA, Katare YK, Mitragotri S. Particle shape: a new design parameter for micro- and nanoscale drug delivery carriers. J Control Rel 2007; 121(1-2): 3-9.
[http://dx.doi.org/10.1016/j.jconrel.2007.03.022] [PMID: 17544538]

[156] Uskoković V, Batarni SS, Schweicher J, King A, Desai TA. Effect of calcium phosphate particle shape and size on their antibacterial and osteogenic activity in the delivery of antibiotics *in vitro.* ACS Appl Mater Interfaces 2013; 5(7): 2422-31.
[http://dx.doi.org/10.1021/am4000694] [PMID: 23484624]

SUBJECT INDEX

A

B

Surendra Nimesh, Nidhi Gupta and Ramesh Chandra (Eds.)

C

D

E

F

G

H

I

L

M

N

S

T

U

V

W

X

Y

Made in the USA
Las Vegas, NV
03 November 2026

57423145R00188